Underwater Intervention (UI 2023)

New Orleans, Louisiana, USA
29 November - 1 December 2023

ISBN: 978-1-7138-8680-8

Printed from e-media with permission by:

Curran Associates, Inc.
57 Morehouse Lane
Red Hook, NY 12571

Some format issues inherent in the e-media version may also appear in this print version.

Underwater Intervention
5206 Cypress Creek Parkway
Suite 202
Houston, Texas 77069
USA

Phone: 281-893-8539

www.underwaterintervention.com

Additional copies of this publication are available from:

Curran Associates, Inc.
57 Morehouse Lane
Red Hook, NY 12571 USA
Phone: 845-758-0400
Fax: 845-758-2633
Email: curran@proceedings.com
Web: www.proceedings.com

TABLE OF CONTENTS

NEW TECHNOLOGY ENABLING LOW COST, LIGHT-WEIGHT HYBRID UNDERWATER WETMATE CONNECTORS

James Cairns and Andrew Clark
Pontus Subsea Connectors (PSC)
www.pontus-subea.com

Underwater Intervention 2023

INVENTORS HALL OF FAME INDUCTEE

James L. Cairns, Ph.D.

CEO, Abyssal Systems, Inc.

62 U.S. Patents

James L. Cairns is CEO of Abyssal Systems, Inc., and the Founder and Executive Director of the Cairns Foundation. Dr. Cairns' pioneering underwater electrical, fiber-optic connector technologies have revolutionized subsea surveillance, anti-submarine warfare, and ocean energy production.

Dr. Cairns' electrical/electronic and fiber-optic connector technology allowed circuits for the first time to be mated and de-mated reliably within the extremely harsh environment of the deep ocean. The need for deep-ocean electrical connectors goes back to the World War II development of SONAR for antisubmarine warfare. Prior to the mid 1950's, the "dry mate" connectors existing then had to be assembled and disassembled in a dry environment before and after each submersion. In Dr. Cairns' seminal 1968 patent, the electric sockets were housed in sealed fluid-filled chambers where the pressures are balanced to the exterior environment by a movable portion of the chamber wall. His principle of oil-filled and pressure-balanced construction has become the universal standard for subsea connectors.

His innovations have revolutionized anti-submarine warfare and subsea surveillance by providing sophisticated subsea systems for homeland security and military surveillance; expanded the global area available for conventional deepwater oil and gas production in addition to facilitating the emerging and burgeoning industry of sustainable offshore windmill farms; and made possible real-time oceanographic observatories by providing internet access on the seafloor thereby enabling tsunami, volcanic and earthquake early warning systems for coastal communities around the world.

OCTOBER 2023

TOWER CABLE REPAIR (1965)

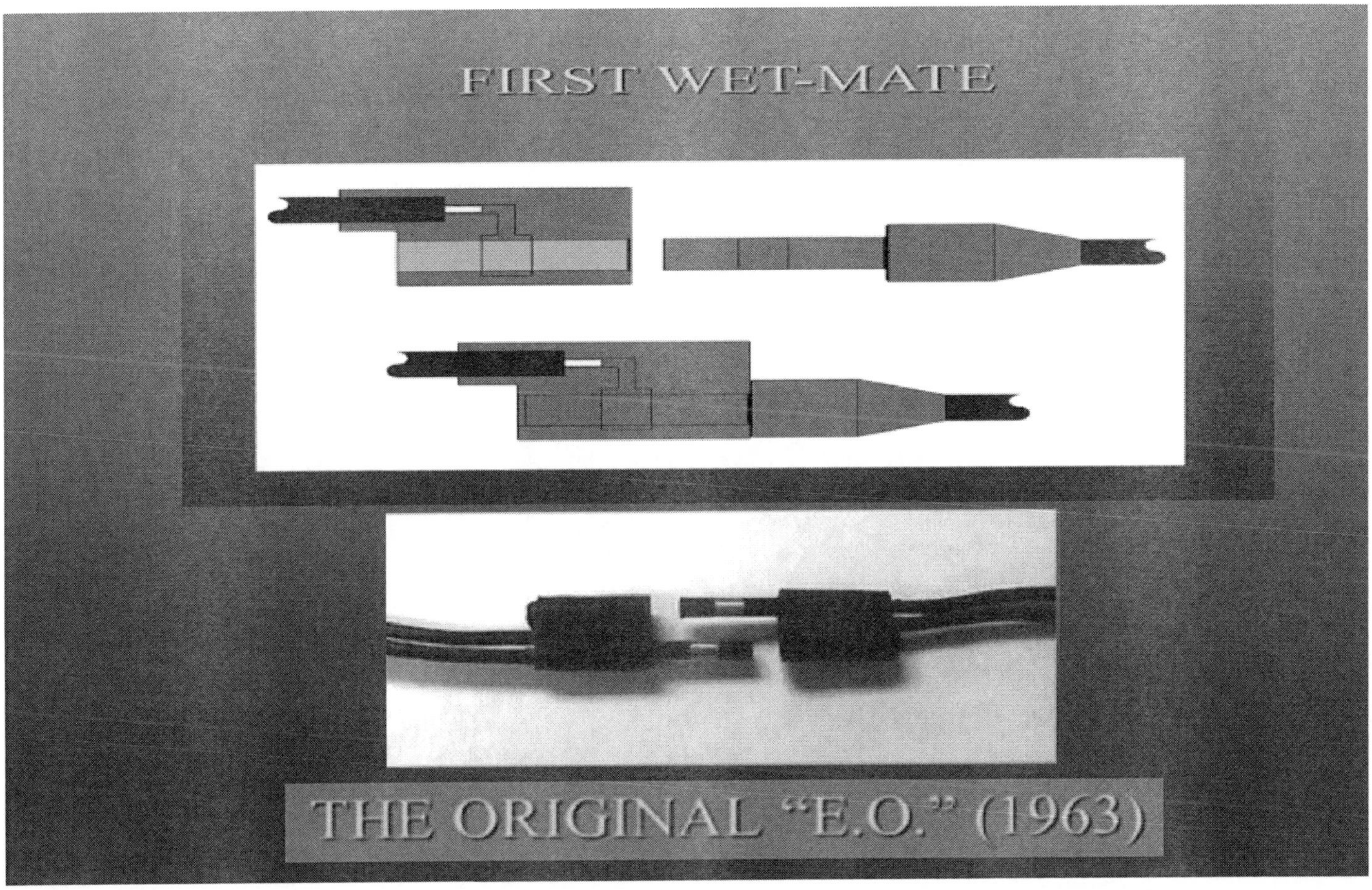
FIRST WET-MATE
THE ORIGINAL "E.O." (1963)

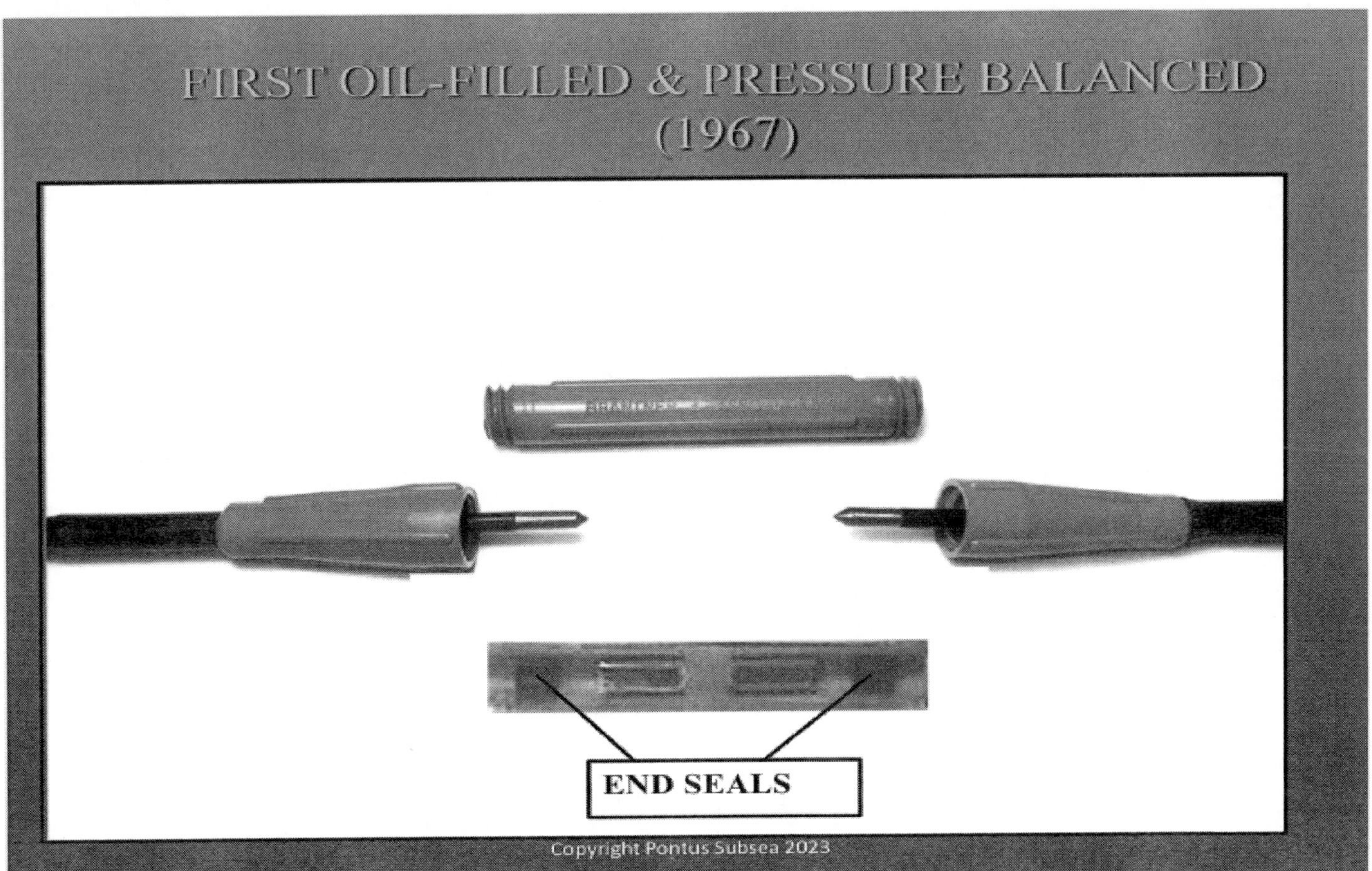
FIRST OIL-FILLED & PRESSURE BALANCED (1967)
END SEALS
Copyright Pontus Subsea 2023

CHALLENGER MARINE COAX (1981)
AFTER 15 AGONIZING YEARS OF FAILURE TO BE ACCEPTED BY THE MARKETPLACE
A GREAT SUCCESS!

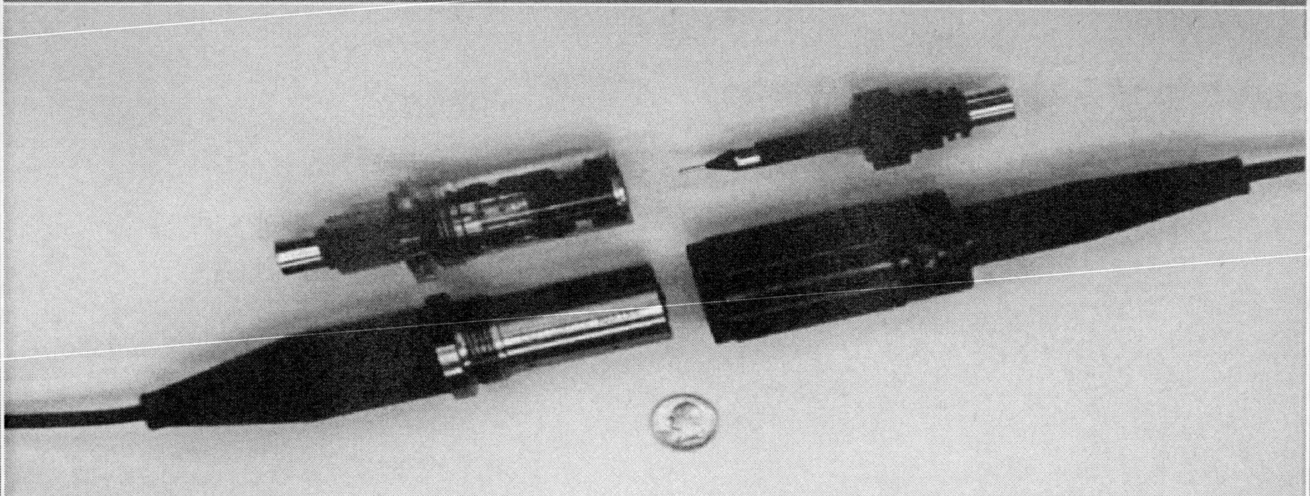

OIL-FILLED TECHNOLOGY FINALLY ACCEPTED!

THE CURRENT STATE OF THE ART OF FLUID-FILLED WET-MATEABLE CONNECTORS...

HAS REMAINED SUBSTANTIALLY *UNCHANGED* FOR THE LAST 30 YEARS

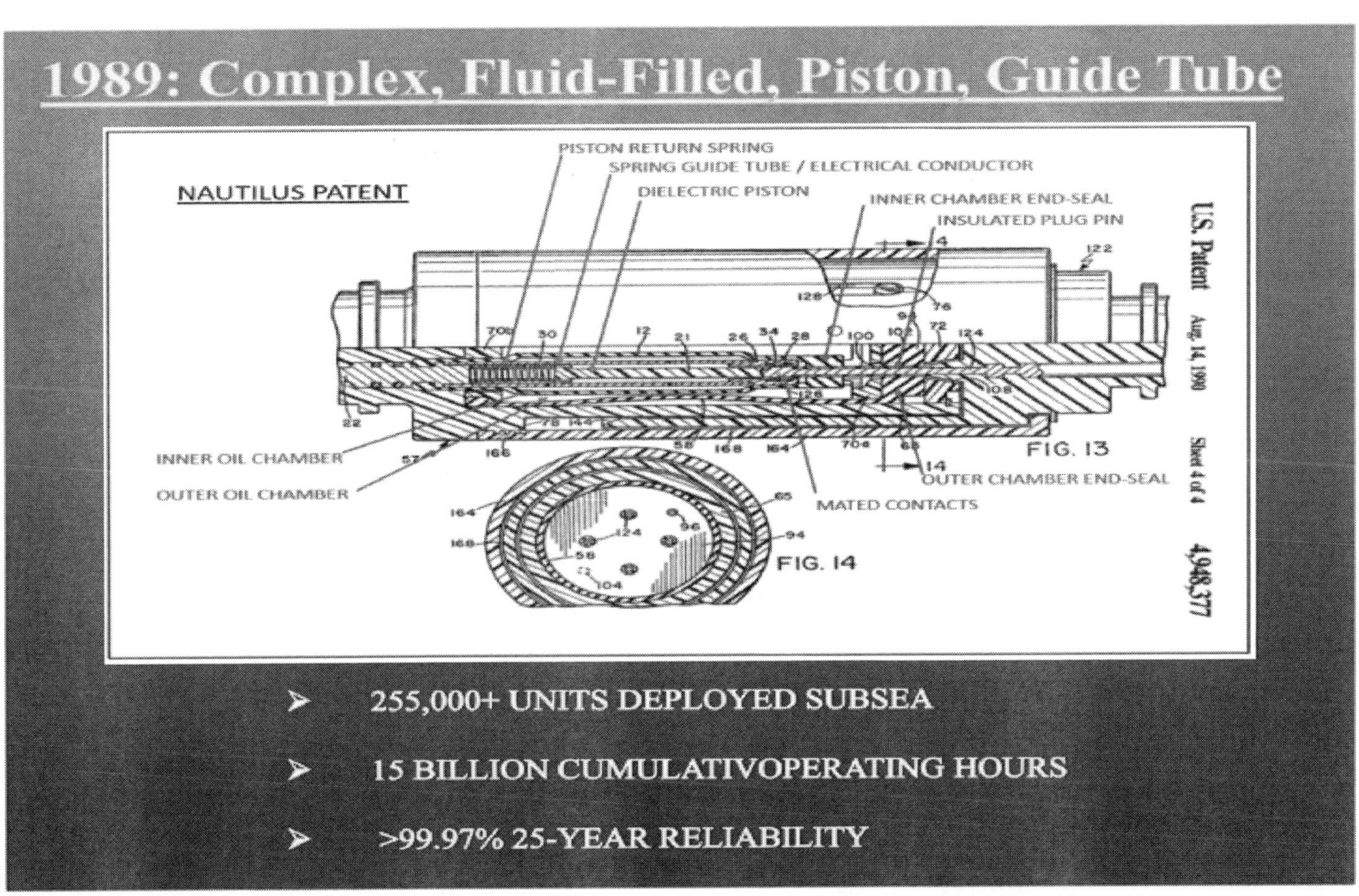
1989: Complex, Fluid-Filled, Piston, Guide Tube
NAUTILUS PATENT
PISTON RETURN SPRING
SPRING GUIDE TUBE / ELECTRICAL CONDUCTOR
DIELECTRIC PISTON
INNER CHAMBER END-SEAL
INSULATED PLUG PIN
INNER OIL CHAMBER
OUTER OIL CHAMBER
FIG. 13
OUTER CHAMBER END-SEAL
MATED CONTACTS
FIG. 14
U.S. Patent Aug. 14, 1990 Sheet 4 of 4 4,948,377
➢ 255,000+ UNITS DEPLOYED SUBSEA
➢ 15 BILLION CUMULATIVOPERATING HOURS
➢ >99.97% 25-YEAR RELIABILITY

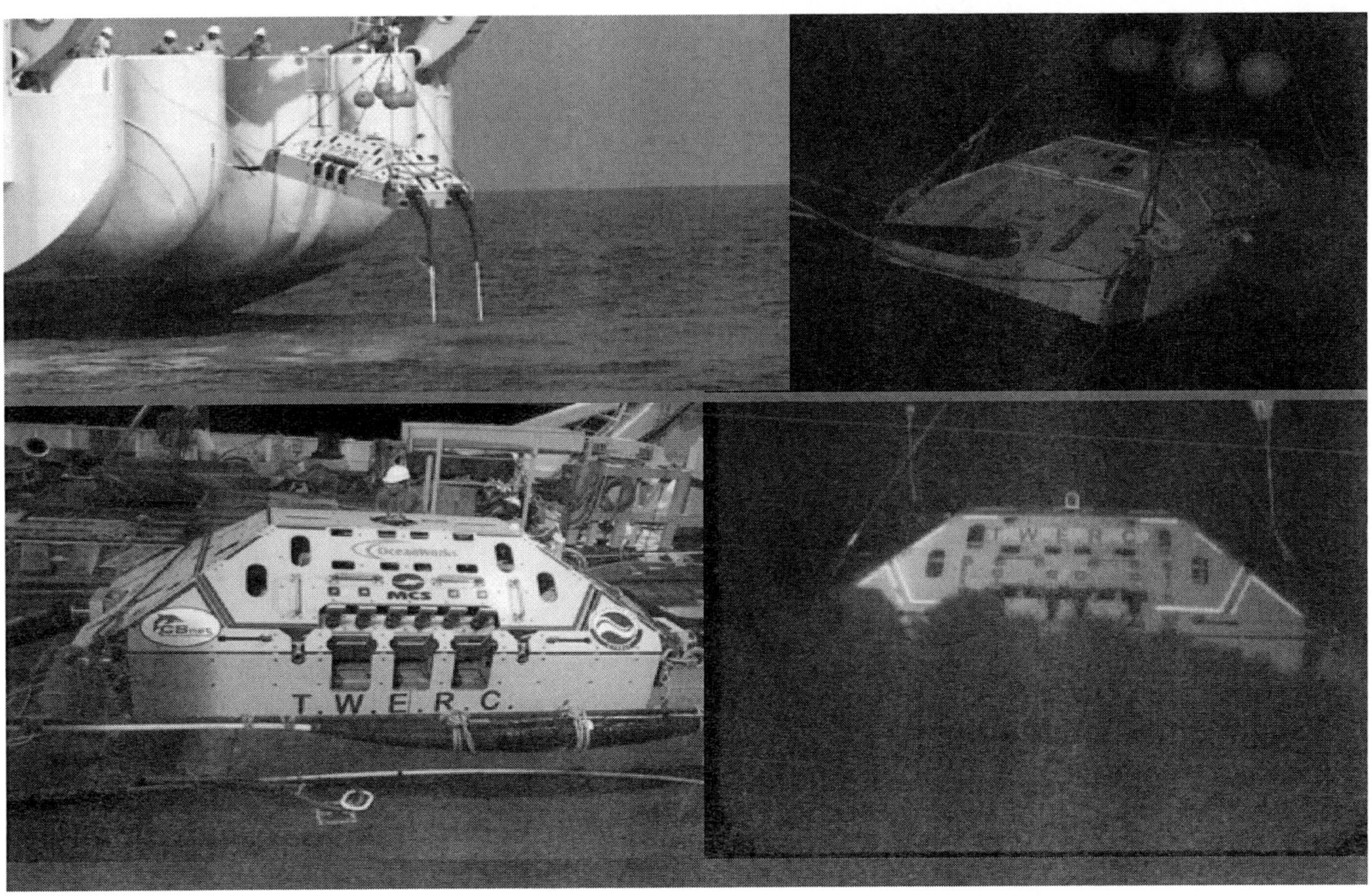
T.W.E.R.C.

IF THE CURRENT STATE-OF-THE-ART PRODUCTS ARE SO RELIABLE, <u>*WHY DO WE NEED NEW ONES?*</u>

- COST
- SIZE
- WEIGHT
- COMPLEXITY
- CIRCUIT COUNT
- DESIGN FLEXIBILITY

THERE HAS NEVER BEEN A BETTER OPPORTUNITY THAN THIS TO MAKE A MAJOR IMPACT ON THE UNDERWATER CONNECTOR INDUSTRY

GLOBAL ANNUAL WET-MATE CONNECTR SALES ESTIMATED TO BE WELL IN EXCESS OF $500 MILLION AND GROWING*

- Pontus Subsea – Advancing State of the Art

THE FOLLOWING SLIDES SHOW SOME OF THE NEW REVOLUTIONARY PONTUS TECHNOLOGY

PONTUS "TAB SEALS" : NEW ENABLING TECHNOLOGY

"TAB SEALS" ARE AN EXTREMELY SIMPLE MEANS TO PROVIDE SEALED, PENETRABLE INTERFACES FOR CONNECTOR FLUID CHAMBERS. THEY ELIMINATE THE COMPLEXITY OF PRIOR ART SPRINGS, PISTONS AND TUBES, OR ROLLERS, AXLES, AND ACTUATING RODS EXISTING IN CURRENTLY AVAILABLE WET- MATE CONNECTORS.

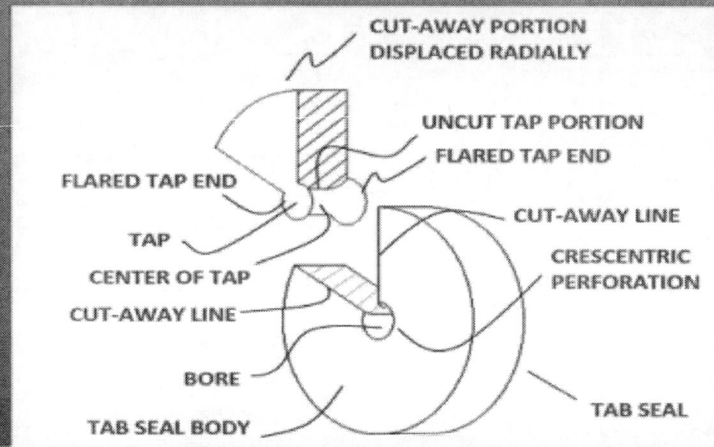

THERE ARE NO ANCILLARY PARTS.... SIMPLY A DEFORMABLE, PENETRABLE, ELASTOMERIC INTERFACE.
WHAT COULD BE SIMPLER?

FOXHOLE

- Designed as a Fiber optic connector
- Both plug and receptacle fluid-filled and pressure-compensated
- Shown here as a hand-mateable connector, but there are many packaging options including ROV stab-plate operations

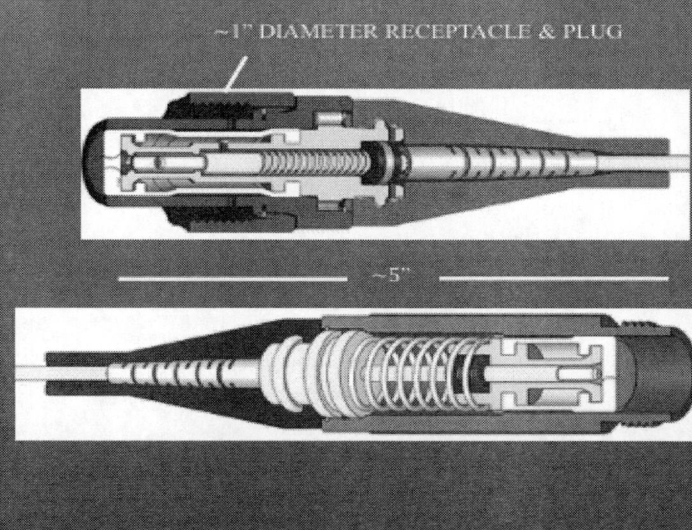

STINGER

- Designed as a hybrid electro-optical connector
- Clean, simple design (compare to Rolling Seal!)
- Both plug and receptacle are fluid-filled and pressure-compensated
- Shown here without a latching mechanism
- There are many packaging options including ROV stab-plate operations

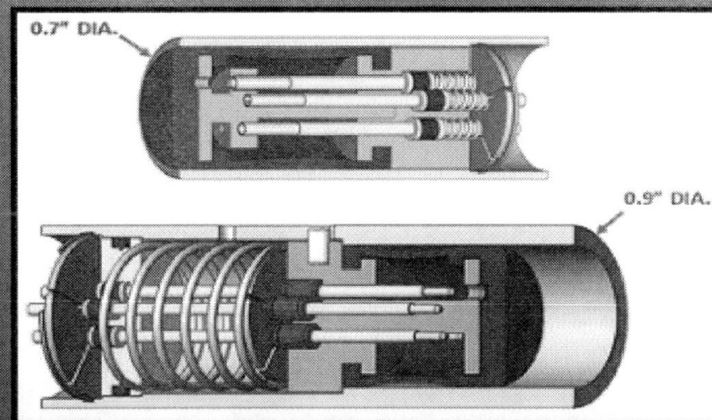

WORKHORSE

- ROBUST
- THE SAME LEVEL OF ELECTRICAL ISOLATION AS NAUTILUS
- WILL COMPETE WELL IN EXISTING NAUTILUS MARKETS
- CAN BE MINIATURIZED
- NO PUSH-APART FORCES AND SMALLER MATING FORCES
- NO LATCHING MECHANISM FOR MANY APPLICATIONS
- GOOD FOR DROPPABLE BALLAST APPLICATIONS (WILL SEPARATE)
- LOWER PART COUNT= FEWER QC INSPECTION POINTS AND SCRAP PARTS
- LOW-COST MOLDABLE SHELLS FOR MANY APPLICATIONS

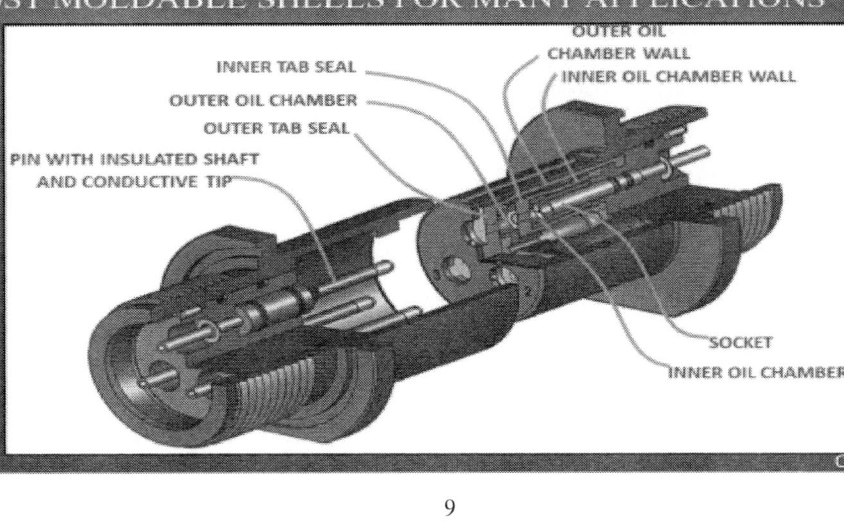

TAB-SEAL CONNECTOR SUMMARY

- SIMPLE TAB-SEAL CONNECTORS CAN DO THINGS NEVER BEFORE POSSIBLE
- THEY ELIMINATE THE MECHANICAL COMPLEXITY OF CURRENT PRODUCTS
- THEY ENABLE A BROAD SPECTRUM OF POTENTIAL NEW, FLEXIBLE DESIGNS
- THE NEW DESIGNS WILL OPEN NEW MARKETS
- LOWER-COST VERSATILE PRODUCTS SHOULD FACILITATE MARKET ENTRY
- SHOULD DISRUPT THE CURRENT WET-MATE CONNECTOR MARKETPLACE
- TECHNOLOGY SHOULD COMMAND A SECURE MARKET NICHE FOR MANY YEARS

CABLE TERMINATIONS

- ADAPTABLE FOR ANY CONNECTOR WITH A SUITABLE BACKSHELL
- FIELD INSTALLABLE AND TESTABLE
- FEWER COMPONENT PARTS
- SMALLER AND LIGHTER WEIGHT THAN CURRENT ART
- SHOULD COMPETE WITH OVERMOLDED TERMINATIONS

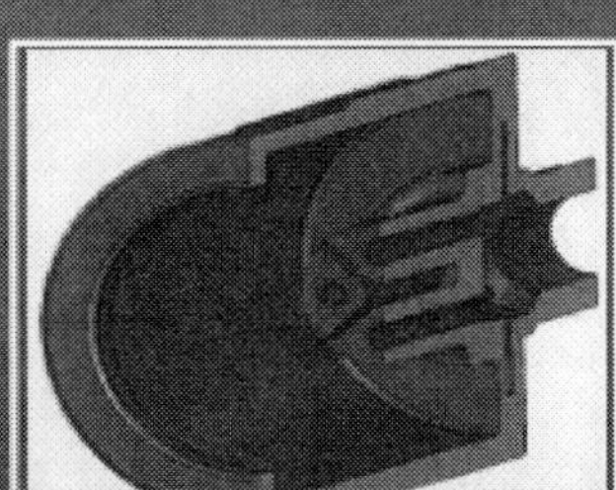

CUTAWAY OF ONE-PIECE SEAL

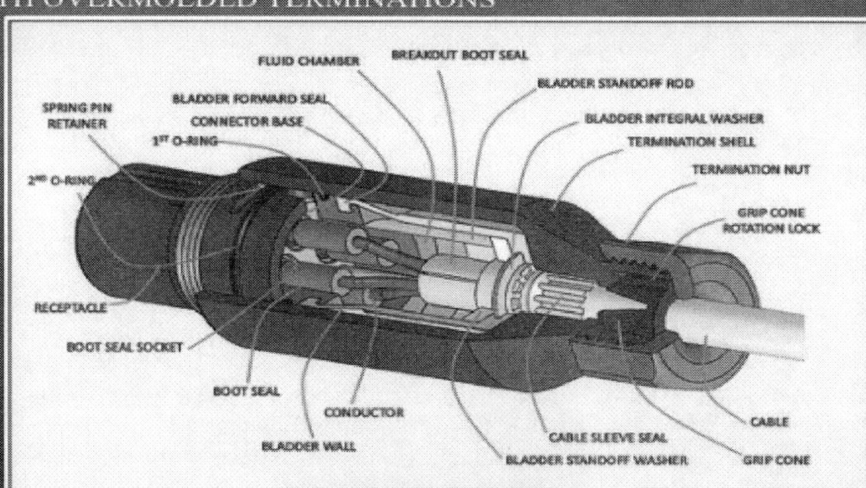

HERMES RECEPTACLE TERMINATED

CABLE SPLICES

- SUITABLE FOR SPLICING CABLES AND HOUSING IN-LINE FIBER SPLICES
- CONSISTS OF BACK-TO-BACK CABLE TERMINATIONS WITH CENTRAL COUPLER
- FIELD INSTALLABLE AND TESTABLE
- NO COMPARABLE CURRENT ART

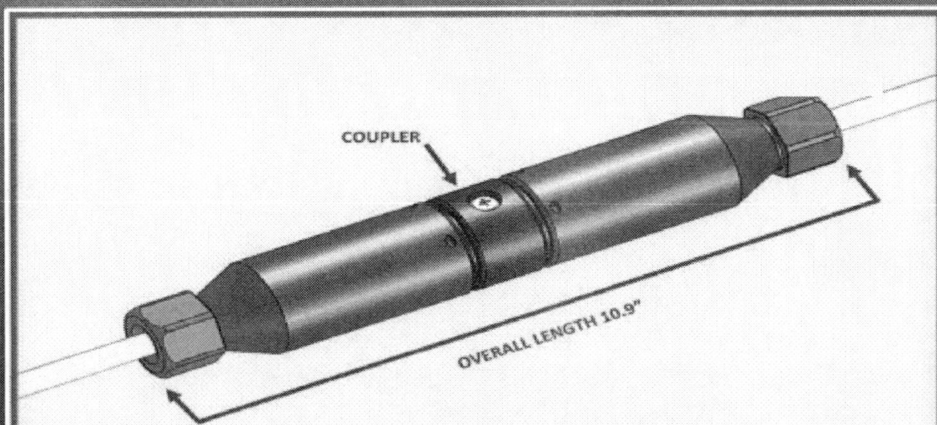

PSC CABLE SPLICE

THERE HAS NEVER BEEN A BETTER OPPORTUNITY THAN THIS TO MAKE A MAJOR IMPACT ON THE UNDERWATER CONNECTOR INDUSTRY

GLOBAL ANNUAL WET-MATE CONNECTR SALES ESTIMATED TO BE WELL IN EXCESS OF $500 MILLION AND GROWING

- Pontus Subsea – Advancing State of the Art

OCEAN
TRACKING NETWORK

Moored Asset Recovery Using an Observation-Class ROV

2023 International Workboat Show

Iago Gradin, Field Technician
Fiona Lynch, ROV Technician

ABOUT OTN

OTN is a global aquatic research, data management and partnership platform headquartered at Dalhousie University in Nova Scotia, Canada.

Since 2008, OTN has been deploying state-of-the-art ocean monitoring equipment and marine autonomous vehicles (gliders) in key aquatic locations.

A global community of researchers is using OTN's infrastructure and analytical tools to track the movements and survival of keystone, commercially important and endangered species.

Knowledge generated by OTN is used by scientists, managers, policy-makers, industry, Indigenous and coastal communities and public stakeholders.

TRACK, CONNECT, TRANSFORM

Together, OTN and its collaborators are tracking oceans, connecting people and transforming knowledge.

TRACK

Using electronic tags to track more than 365 keystone, commercially important and endangered species.

CONNECT

Connecting infrastructure, communities, resources, researchers, policy-makers, industry, and other stakeholders at local and global scales

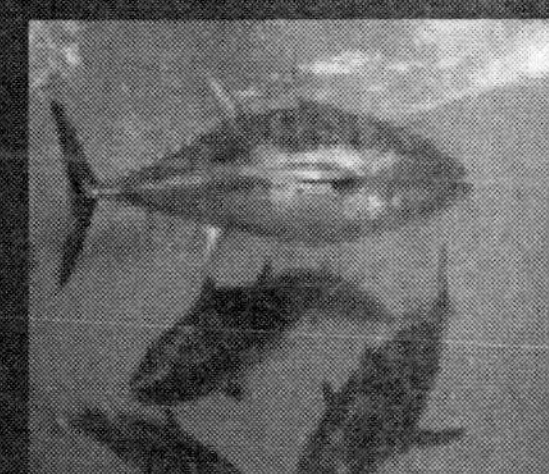

TRANSFORM

Transforming our understanding of the oceans by developing global data standards and a paradigm of collaborative research and management.

3 OCEAN TRACKING NETWORK

WHY TRACK?

Filling knowledge gaps in the animal telemetry index

- Fisheries management (equitable quotas; monitoring)
- Design of Marine Protected Areas
- Impact assessment for coastal/offshore (Maritime Link; tidal power)
- Endangered species research (white shark, Atlantic salmon)
- Wild fish interactions with farmed fish
- Environmental conditions and the prediction of future animal distributions
- Assess habitat repairs and fate of hatchery fish
- Engagement of the public

4 OCEAN TRACKING NETWORK

HOW TRACKING WORKS

1. Acoustic tags are small transmitters that allow researchers to track aquatic animals underwater.
2. Larger animals can carry Vemco Mobile Transceivers (VMTs), which act as both transmitters and receivers.
3. Wave Gliders float on the ocean's surface. They are emission free, propelled by wave energy and solar powered.
4. Slocum gliders dive up and down collecting information underneath the waves.
5. Acoustic receivers capture animal movement data.

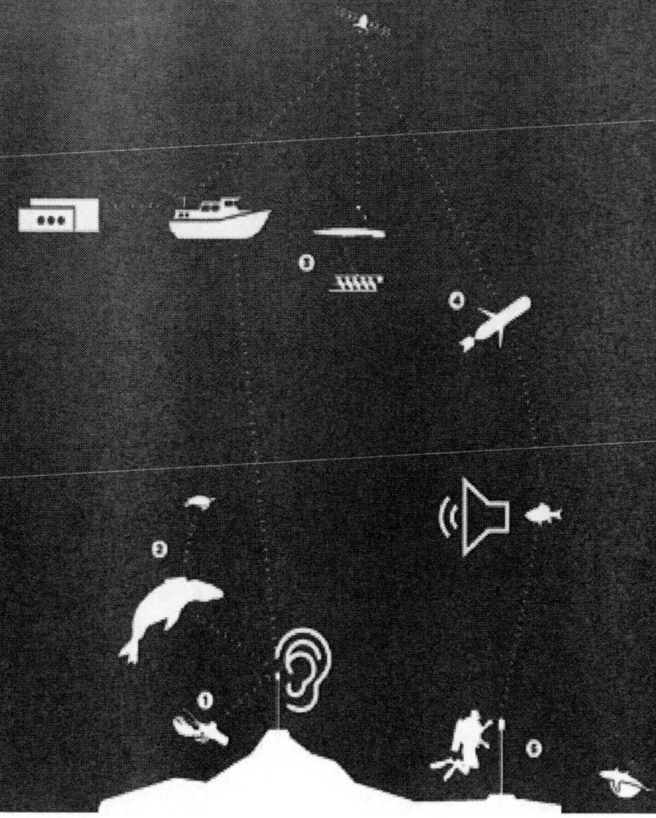

5 OCEAN TRACKING NETWORK

Number of Animals Tagged

The Ocean Tracking Network manages data for 344 different species of ecological and economic importance. Telemetry technology allows researchers to track a broad range of taxa and describe their interspecies interactions, define their habitat usage, and evaluate any potential migration pathways and timings. The colour coding used by this chart denotes the species' threat level as reported by the IUCN Red List of Threatened Species™.

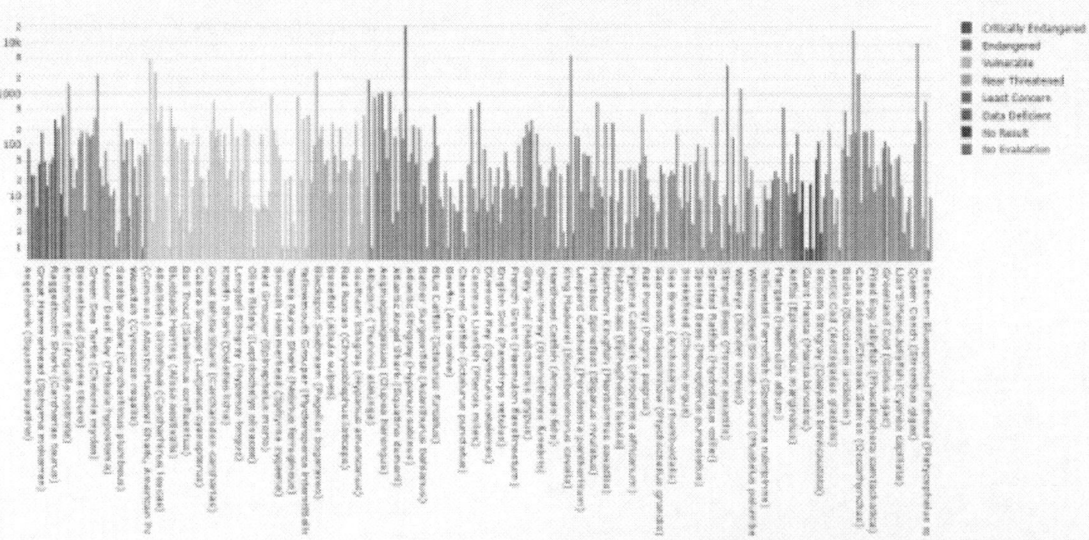

6 OCEAN TRACKING NETWORK

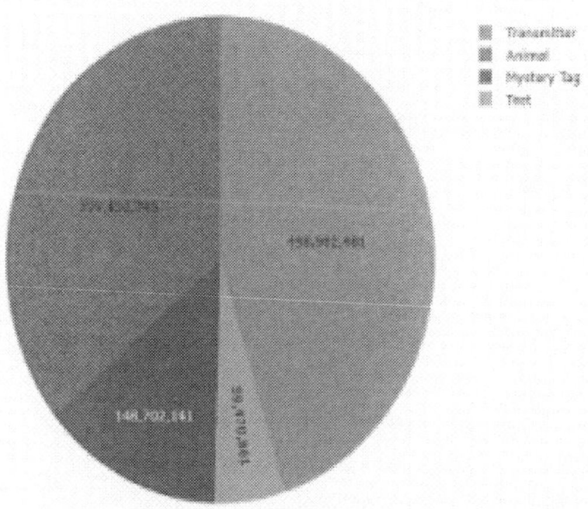

Total Detections by Type

OTN has a total of **1,106,606,229** detections divided up into the following:

Animal: These detections are confirmed as belonging to a tagged animal deployed by an OTN-affiliated researcher or data partner.

Transmitter: These detections are from tags that were deployed in fixed locations (i.e. sentinel tags), or tag codes used by transceiver units, i.e. for inter-receiver synchronization.

Mystery Tag: These detections are not yet associated with any deployed tag. The mystery tag lookup tool can be used to discover matches to your tags in the OTN database.

Test: These detections are of tags that have been identified by manufacturers as reserved for testing purposes only, but for which we have no associated information (i.e in the case of shipboard testing or sentinel transmitters).

ACOUSTIC RECEIVER MOORINGS

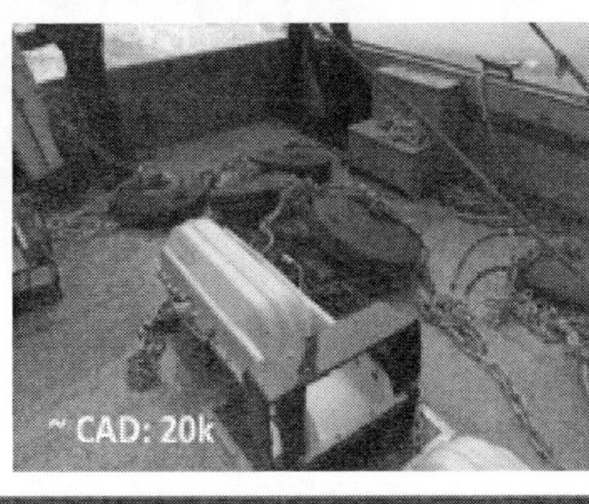

Currently Active Moored Receivers

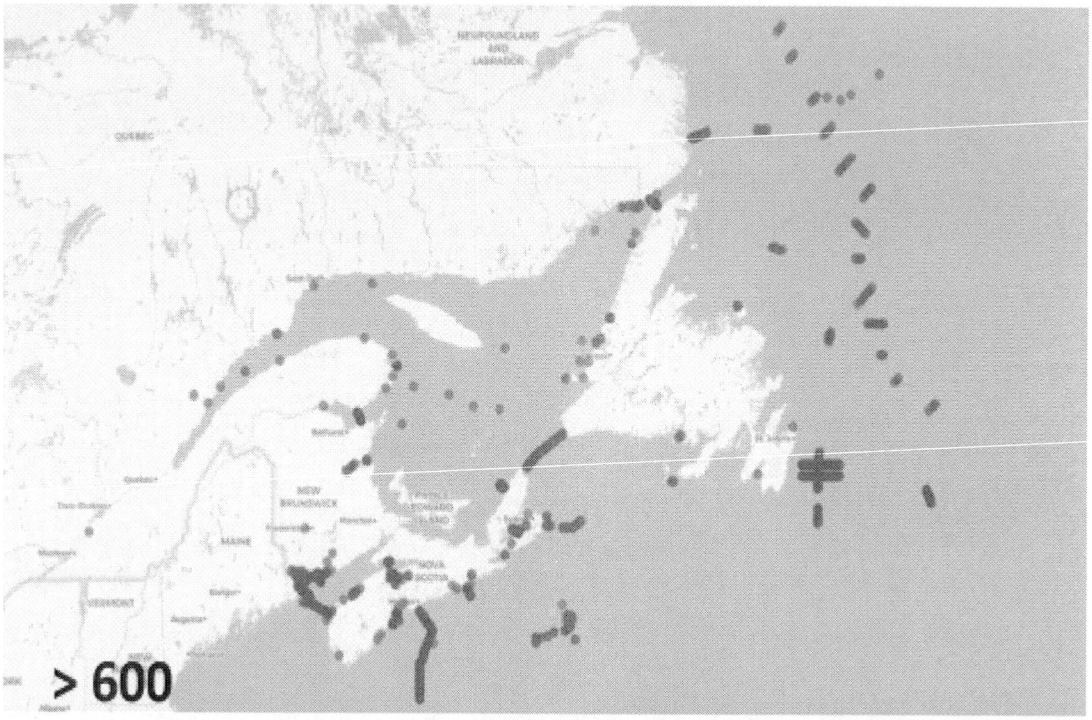

Stuck stations

ROV TO THE RESCUE

SAAB Falcon DR 1000

- 300m and 1000m tether setups
- Hydraulic skid (5-point manipulator or rope cutter attachment)
- Sonar
- HD cameras

ROV TO THE RESCUE

First attempt to recover a stuck station

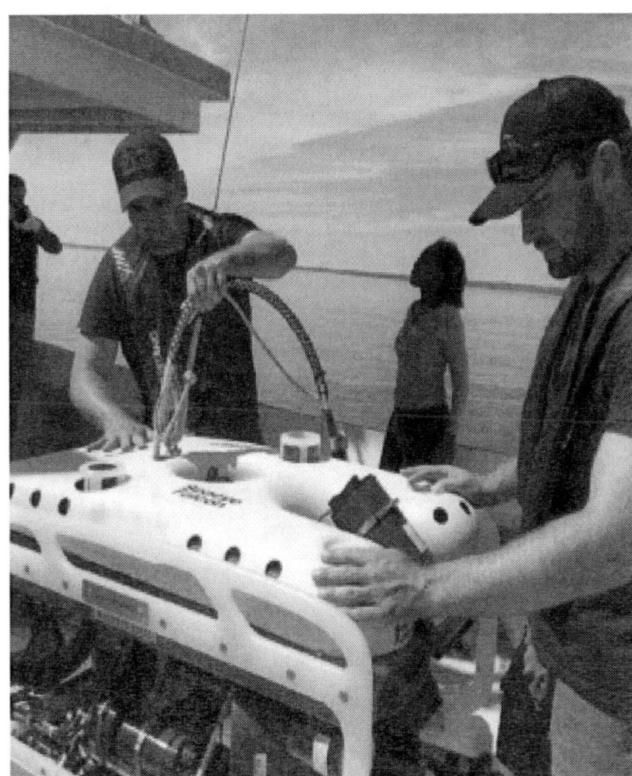

ROV TO THE RESCUE

Improving the operations to be more successful

Added an USBL system to help navigating

Purchased an Edge Tech 4205 Side Scan Sonar to survey the area prior to ROV dive

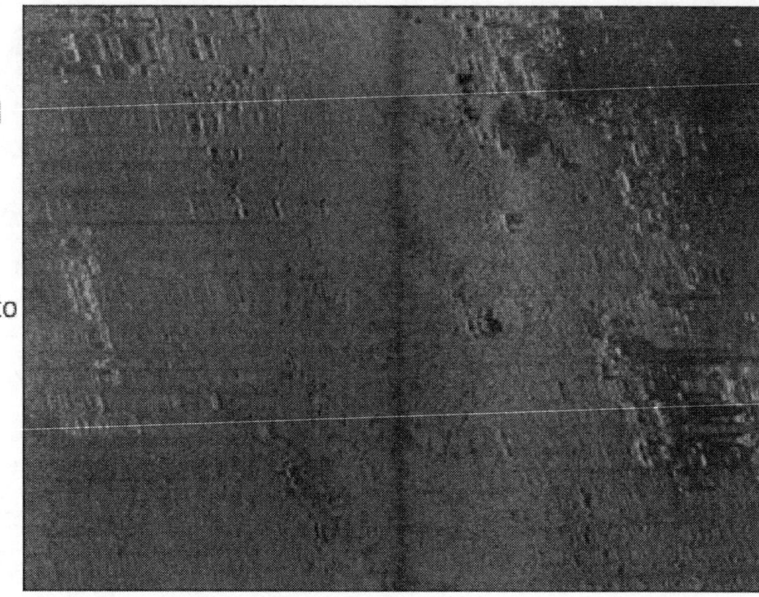

ROV TO THE RESCUE

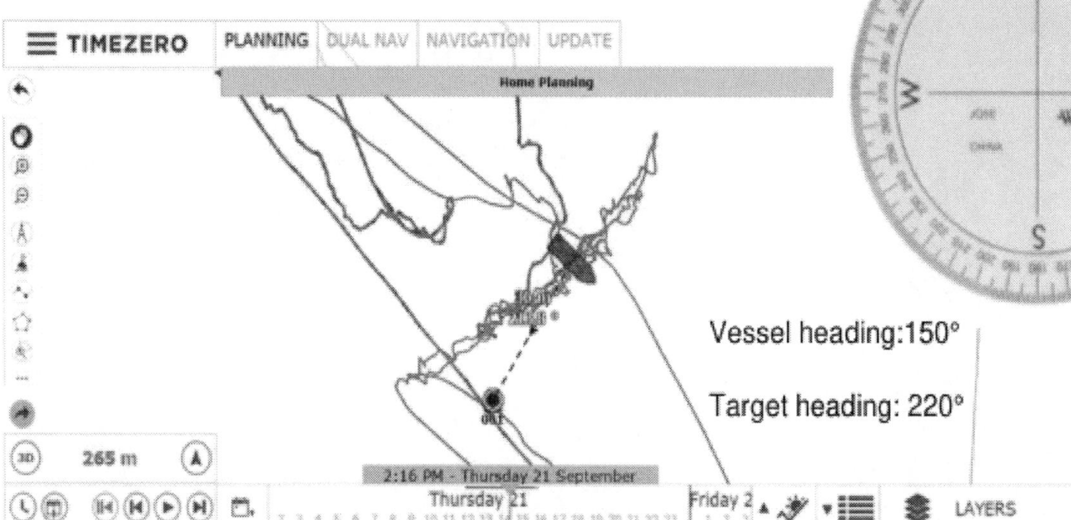

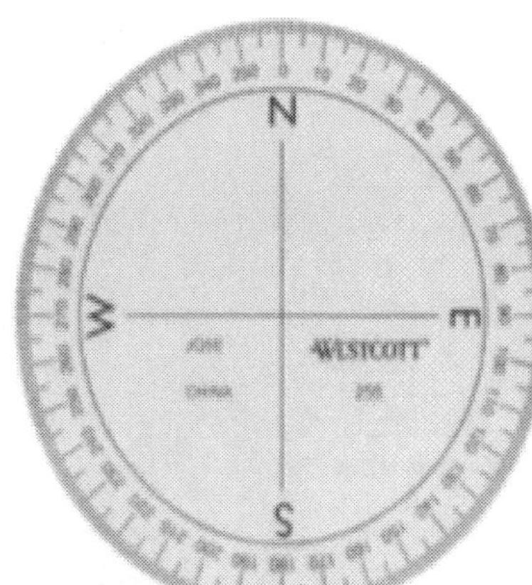

Vessel heading:150°

Target heading: 220°

ROV TO THE RESCUE

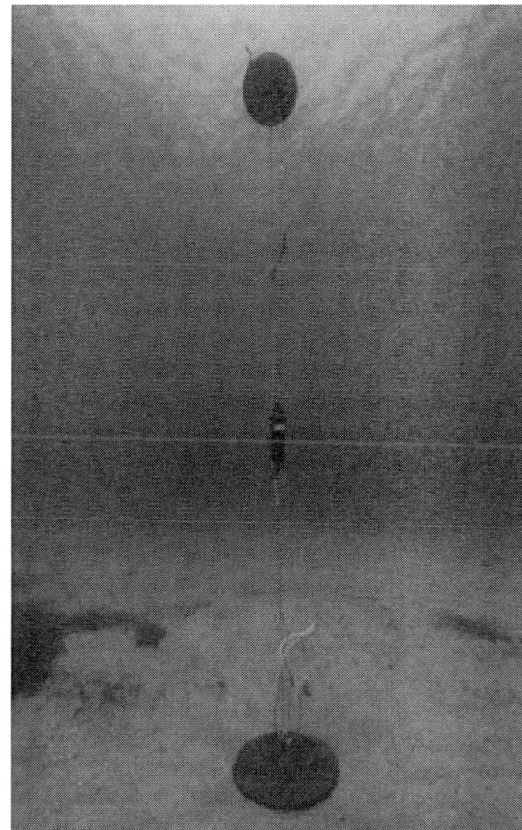

Rope cutter issue

Not enough tension in the rope to be able to cut it

5/16" Dyneema rope can hold roughly 14,000 lbs and make a knife dull in no time

15 OCEAN TRACKING NETWORK

ROV TO THE RESCUE

Using a secondary rope attached to a carabiner and clip the hook to the mooring's rope.

16 OCEAN TRACKING NETWORK

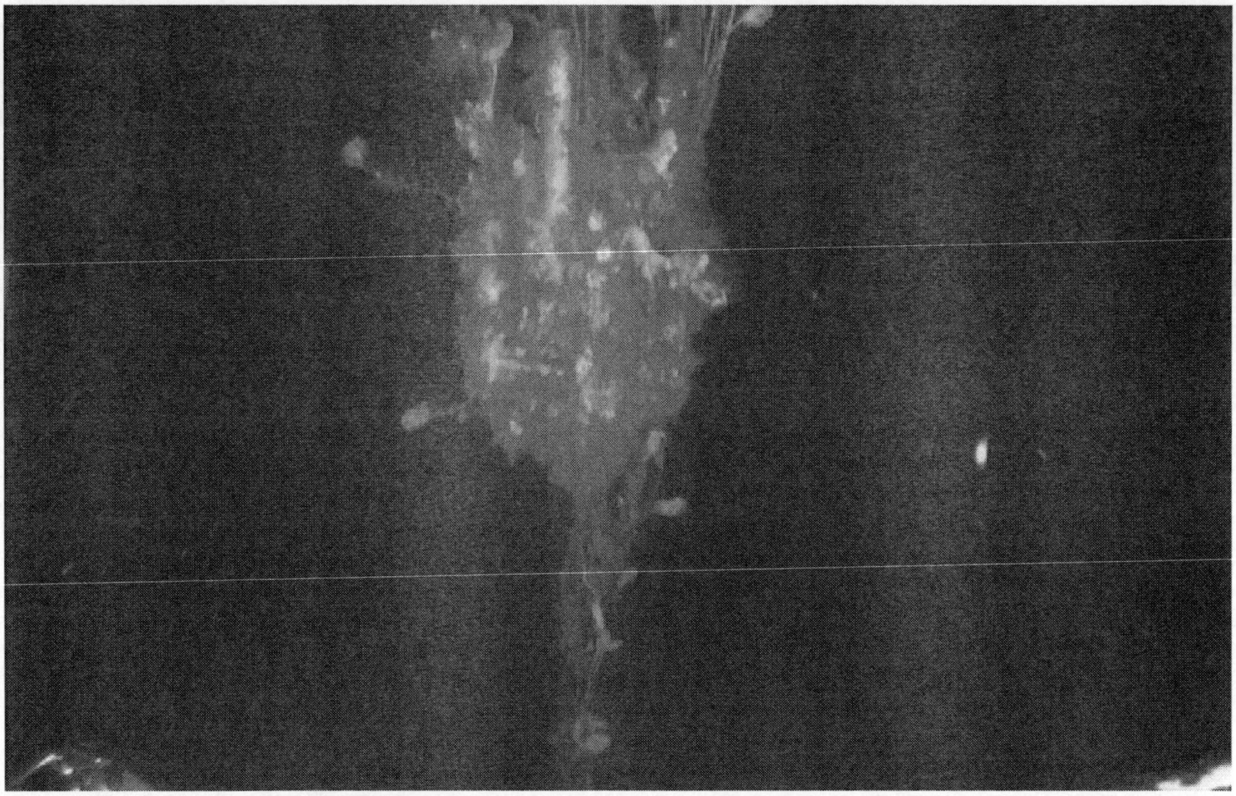

ROV TO THE RESCUE

To date the ROV has recovered:

- 18 Receiver stations (14 VR2AR's and 6 VR4)
- 3 Emera Ice Sonar stations
- 1 BRUV

OTN IN NUMBERS

90,000+ TAGGED ANIMALS TRACKED

1,600+ CONTRIBUTORS

340+ SPECIES TRACKED GLOBALLY

138,000+ KM COVERED BY GLIDERS

OCEAN TRACKING NETWORK

INNOVATION Canada Foundation for Innovation Fondation canadienne pour l'innovation

DALHOUSIE UNIVERSITY

NSERC CRSNG

researchNS

Fisheries and Oceans Canada Pêches et Océans Canada

Transport Canada Transports Canada

OBIS OCEAN BIODIVERSITY INFORMATION SYSTEM

Marine Life

@oceantrackingnetwork @oceantracking oceantrackingnetwork.org

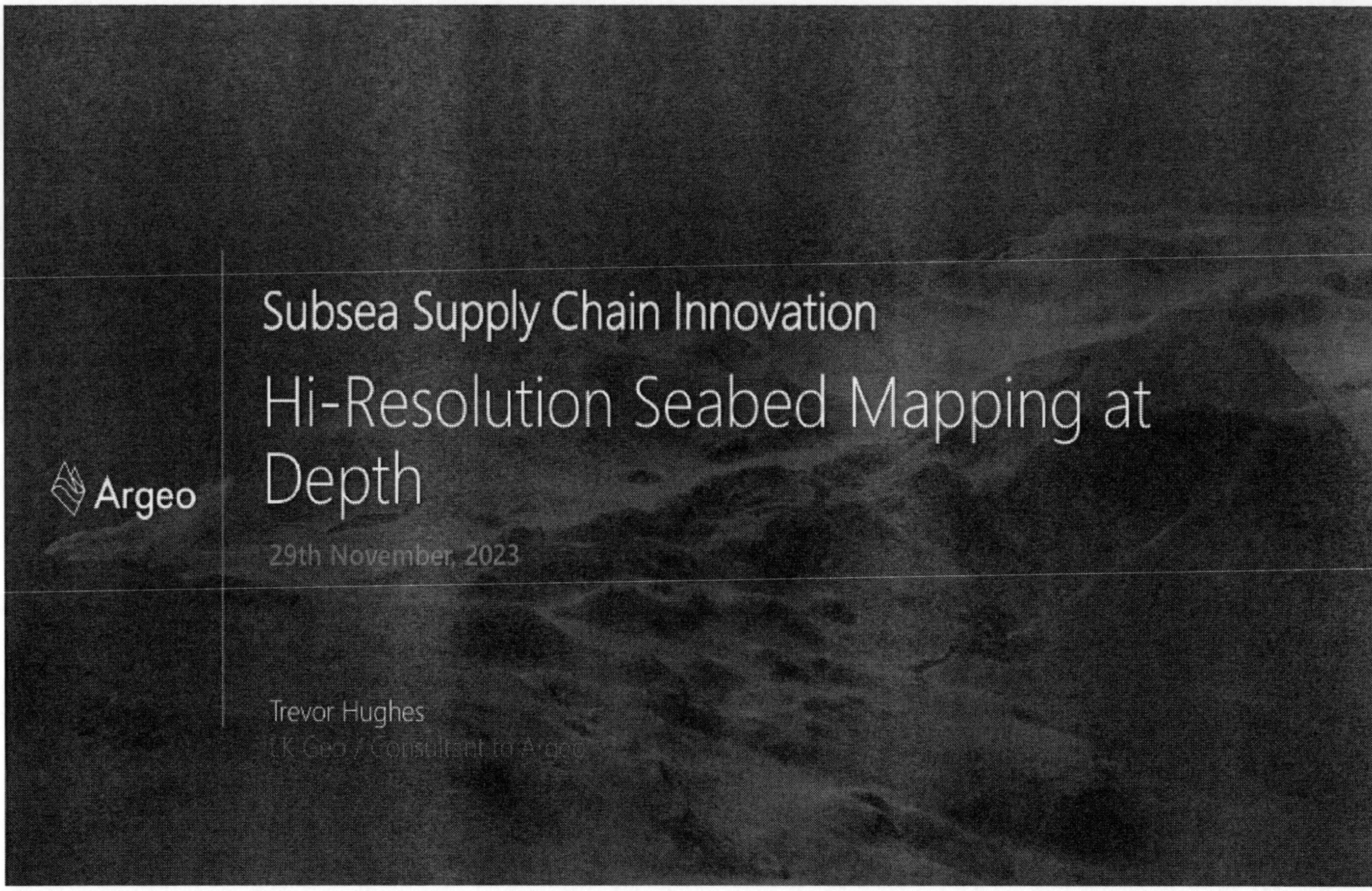
Subsea Supply Chain Innovation
Hi-Resolution Seabed Mapping at Depth
Argeo
29th November, 2023
Trevor Hughes
LK Geo / Consultant to Argeo

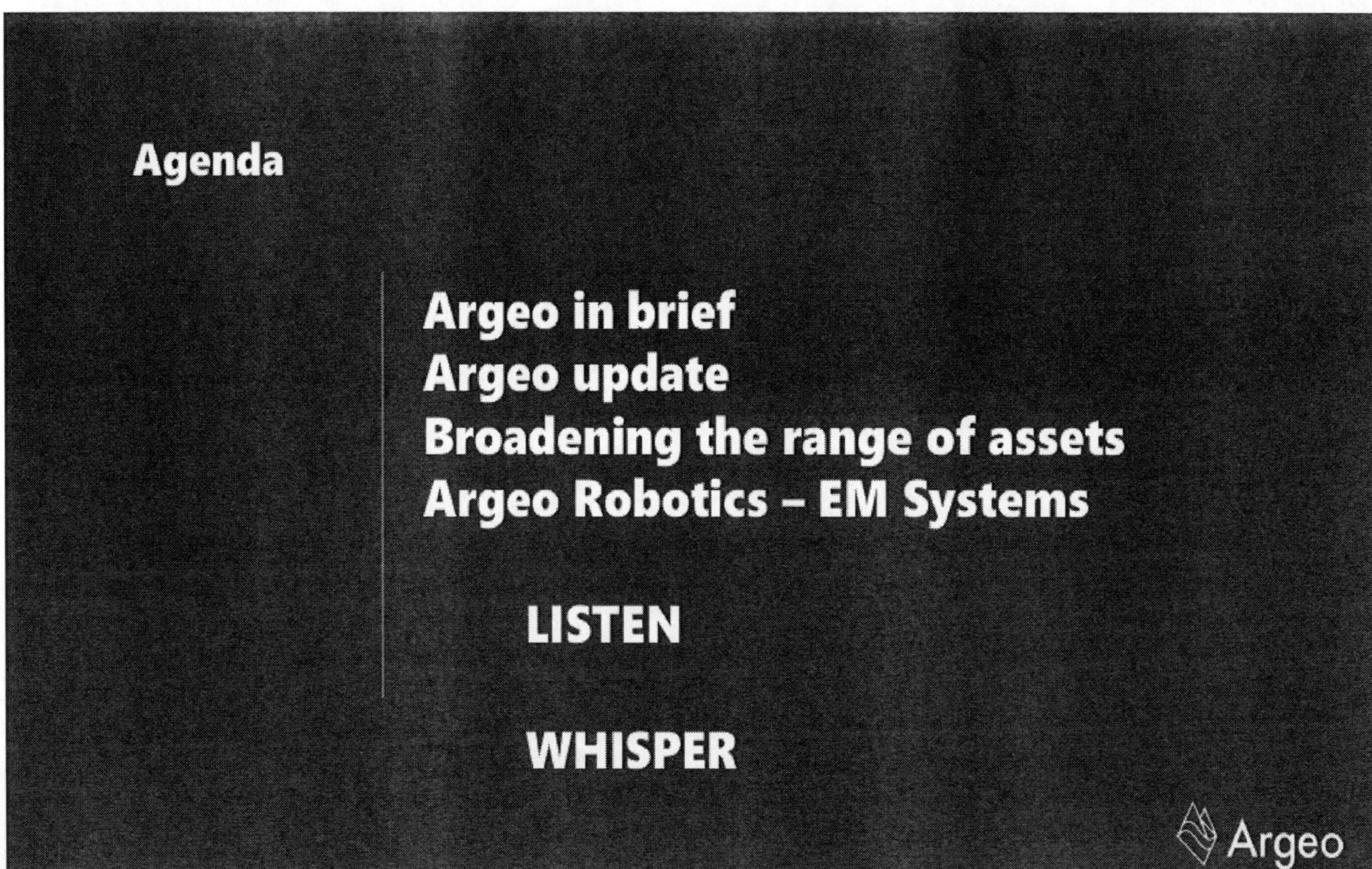
Agenda
Argeo in brief
Argeo update
Broadening the range of assets
Argeo Robotics – EM Systems
LISTEN
WHISPER
Argeo

Complete subsea service provider

From acquisition to actionable data

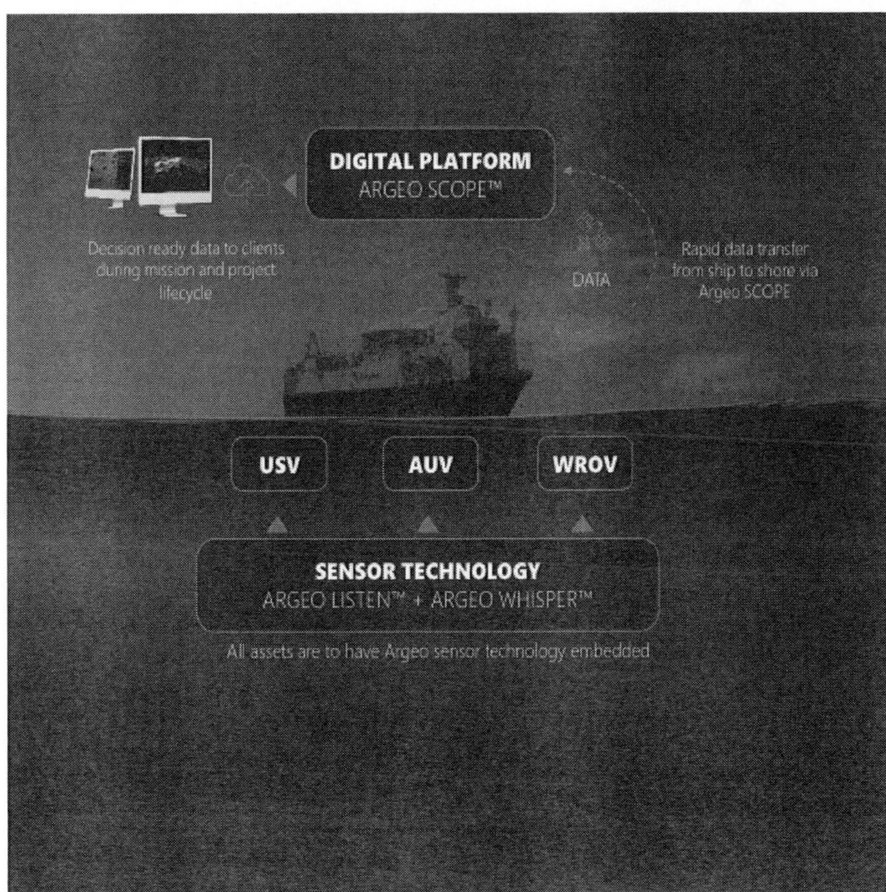

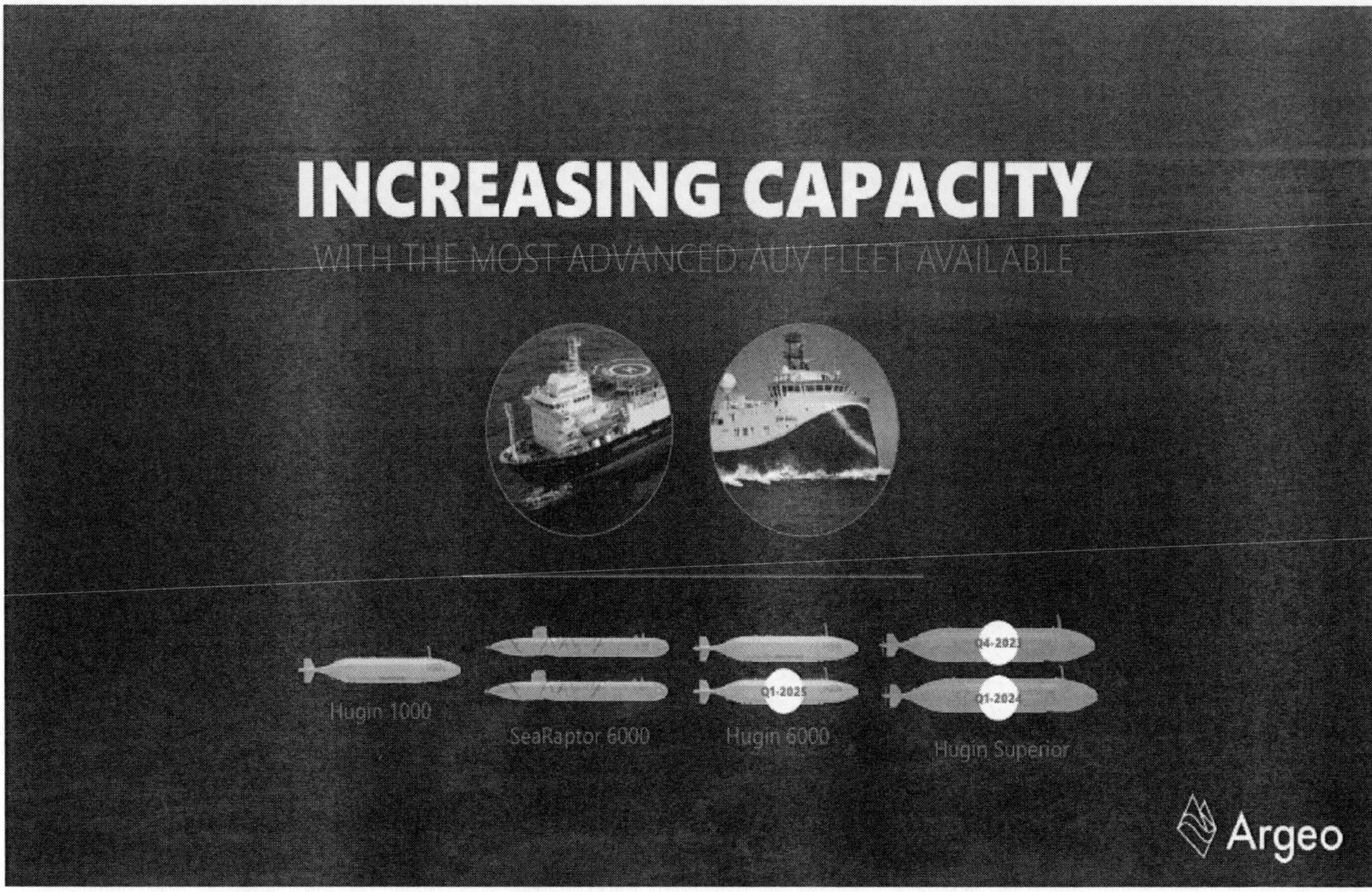
INCREASING CAPACITY
WITH THE MOST ADVANCED AUV FLEET AVAILABLE
Q4-202
Q1-2025
Q1-2024
Hugin 1000
SeaRaptor 6000
Hugin 6000
Hugin Superior
Argeo

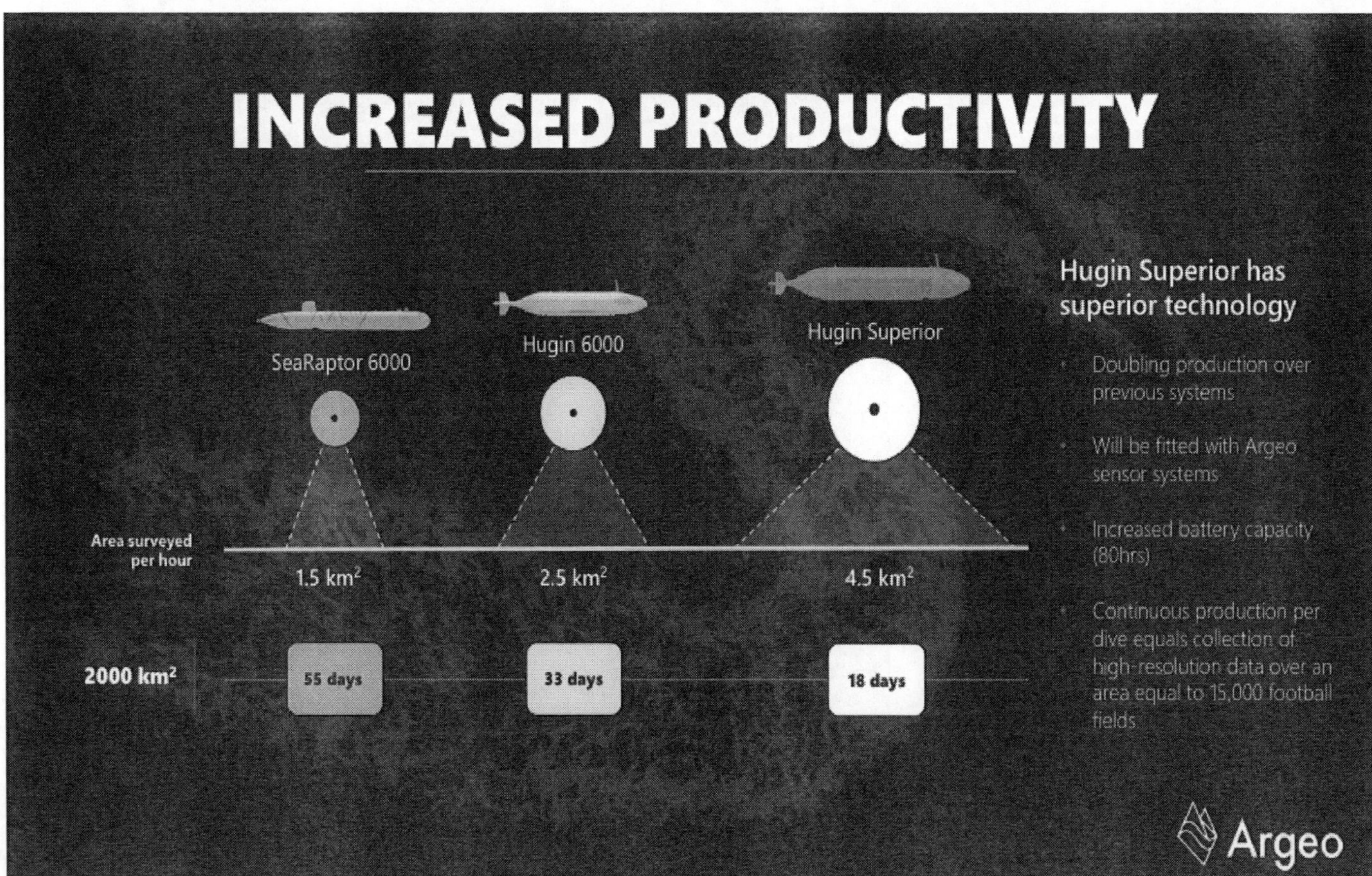
INCREASED PRODUCTIVITY
SeaRaptor 6000
Hugin 6000
Hugin Superior
Area surveyed per hour
1.5 km^2
2.5 km^2
4.5 km^2
2000 km^2
55 days
33 days
18 days
Hugin Superior has superior technology
Doubling production over previous systems
Will be fitted with Argeo sensor systems
Increased battery capacity (80hrs)
Continuous production per dive equals collection of high-resolution data over an area equal to 15,000 football fields
Argeo

Argeo Robotics – Electromagnetic Sensor Systems

Argeo

ADVANCED ROBOTICS

ARGEO ELECTROMAGNETIC SENSOR SYSTEM

ARGEO LISTEN

- ✓ Positioning and burial depth of ACTIVE power cables
- ✓ Inspection of subsea cathodic protection systems
- ✓ Marine Mineral exploration
- ✓ General site survey

ARGEO WHISPER

- ✓ Tracking/burial depth of PASSIVE power cables
- ✓ Tracking buried pipelines
- ✓ Detection of Unexploded Ordnance (UXO)
- ✓ Marine Minerals exploration

ARGEO DISCOVER

- ✓ Marine Minerals exploration

Argeo

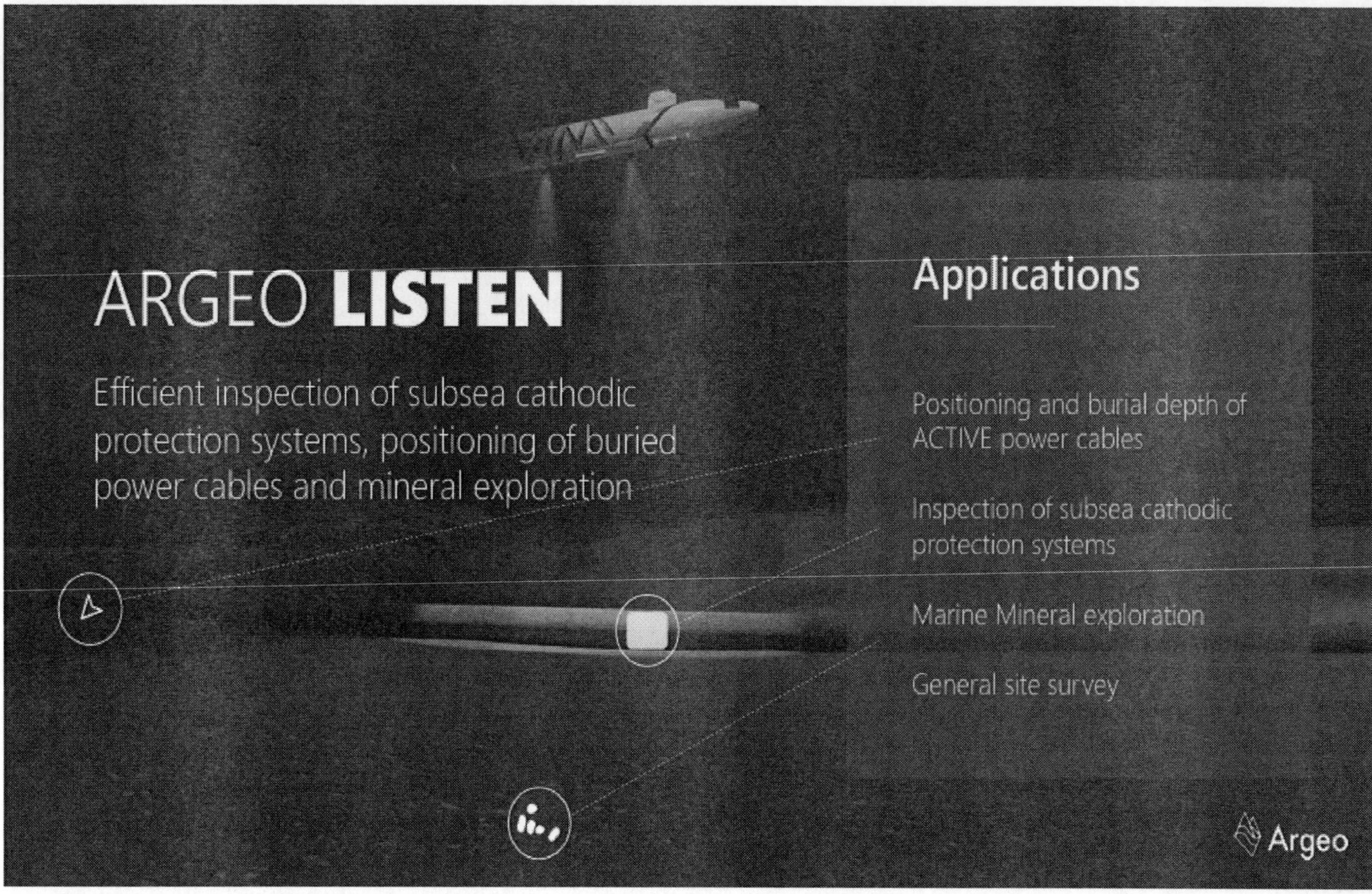
ARGEO LISTEN
Efficient inspection of subsea cathodic protection systems, positioning of buried power cables and mineral exploration
Applications
Positioning and burial depth of ACTIVE power cables
Inspection of subsea cathodic protection systems
Marine Mineral exploration
General site survey
Argeo

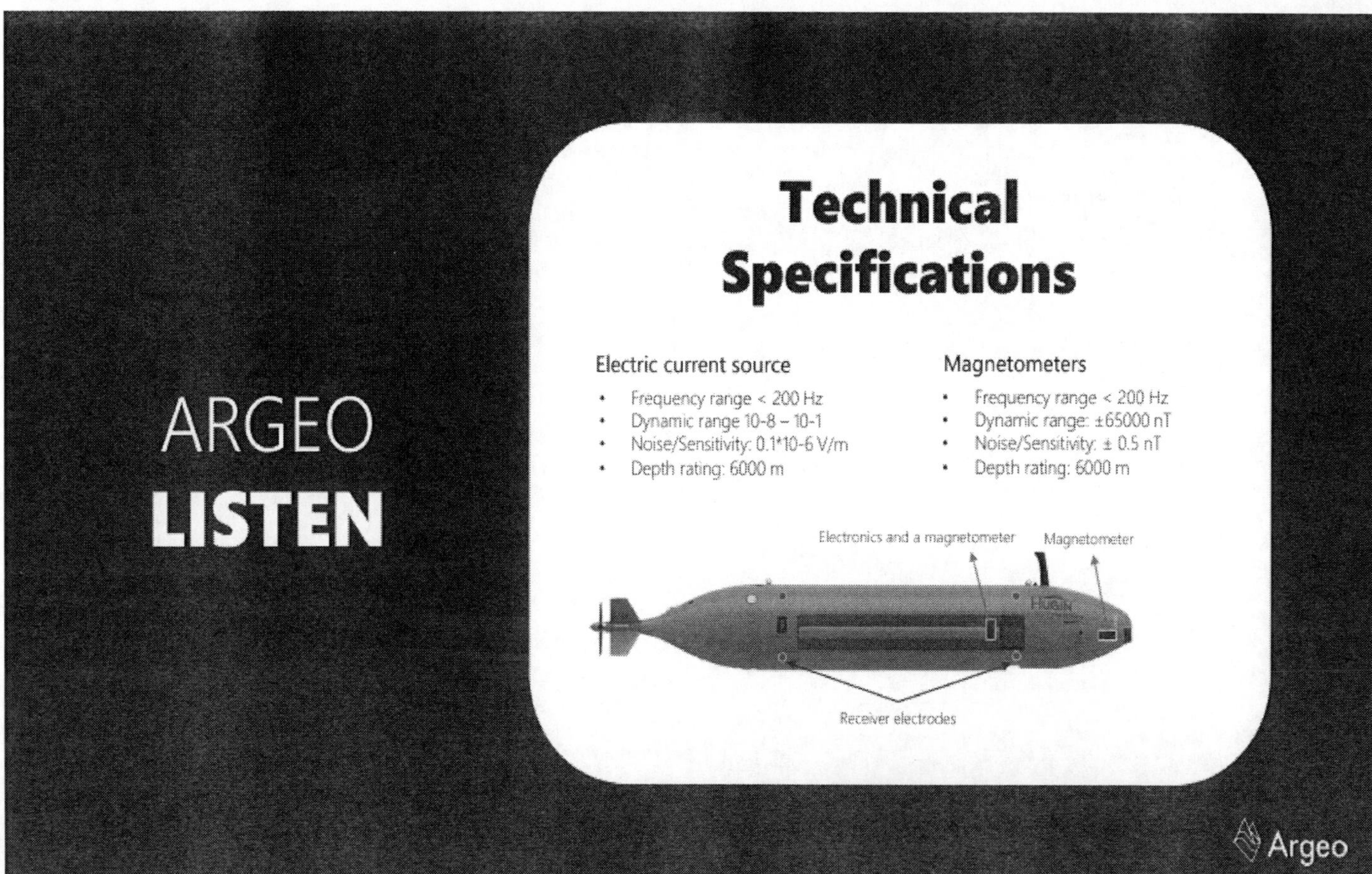
ARGEO
LISTEN
Technical Specifications
Electric current source
• Frequency range < 200 Hz
• Dynamic range 10-8 – 10-1
• Noise/Sensitivity: 0.1*10-6 V/m
• Depth rating: 6000 m
Magnetometers
• Frequency range < 200 Hz
• Dynamic range: ±65000 nT
• Noise/Sensitivity: ± 0.5 nT
• Depth rating: 6000 m
Electronics and a magnetometer
Magnetometer
Receiver electrodes
Argeo

Shorter time to high quality results

UP TO 150 km
Of high productivity inspection

3-3,5 knots
Survey speed

NEAR **REAL-TIME**
Data analysis during field operations

- A passive electrode- and magnetometer system for electric- and magnetic field measurements in sea water.
- Modular system which can be implemented into various underwater vehicles.
- Currently the system is fully integrated on Argeo's AUV platforms which can operate down to 6000m water depth.

Passive electrode- and magnetometer system for electric- and magnetic field measurements in sea water

- ✓ **Efficient inspection** of cables and pipelines, both buried and on seafloor
- ✓ **Rapid estimation** of cathodic protection status and remaining lifetime
- ✓ **Simultaneous data acquisition** with other sensors
- ✓ **Complete characterisation** through integration of all sensor data
- ✓ **Faster project turnaround** providing safer operations with lower HSE risk
- ✓ **Accurate** and high-quality data with low noise levels

Argeo

ARGEO **LISTEN**

EXAMPLE

Cathodic Protection System - AUV Survey

AUQUISITION METHOD

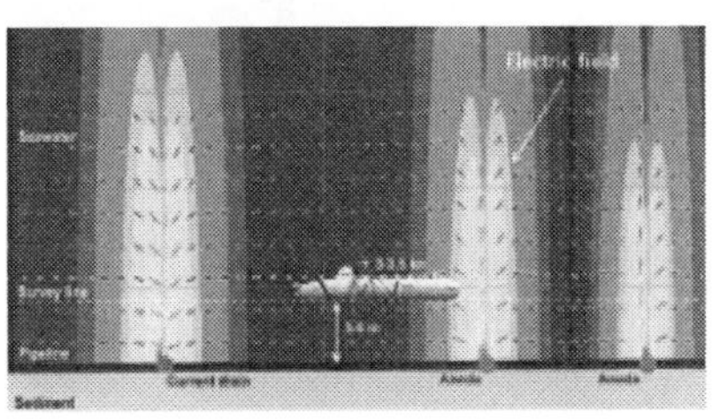

MEASURED DATA

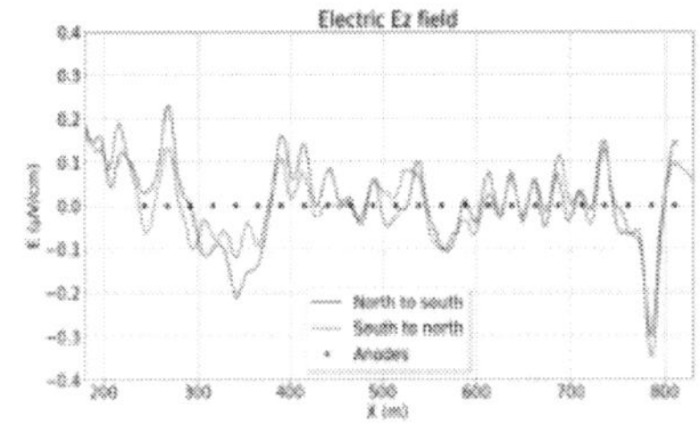

PROCESSED DATA

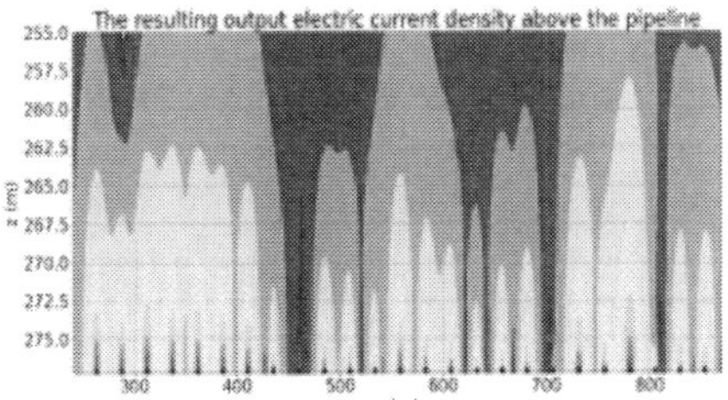

INTEGRATED DATA

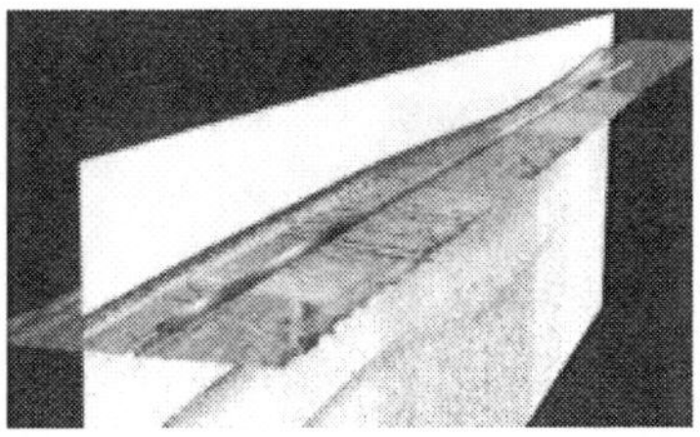

Visualization in the **Argeo SCOPE** platform.

Argeo

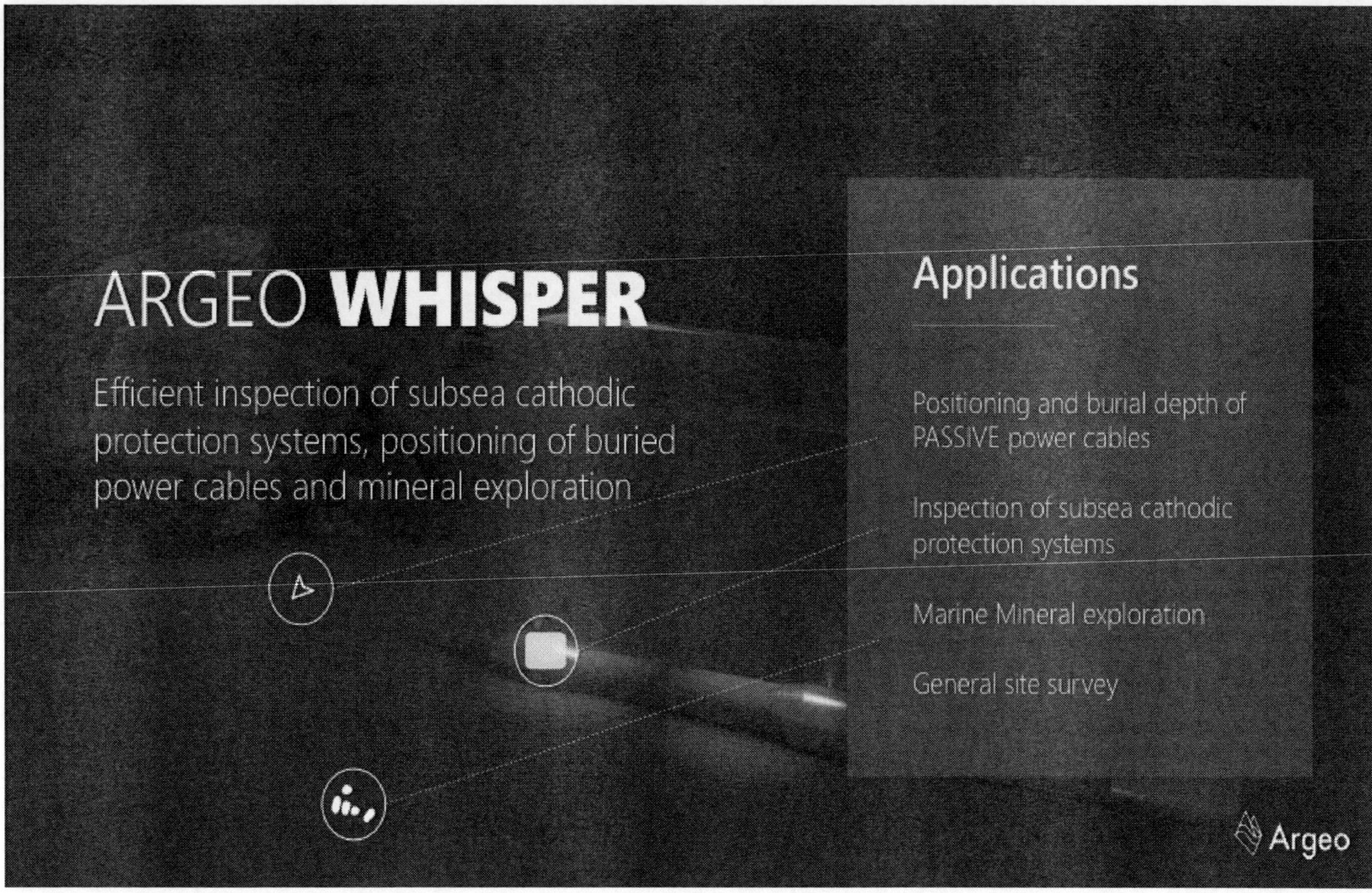
ARGEO WHISPER
Efficient inspection of subsea cathodic protection systems, positioning of buried power cables and mineral exploration
Applications
Positioning and burial depth of PASSIVE power cables
Inspection of subsea cathodic protection systems
Marine Mineral exploration
General site survey
Argeo

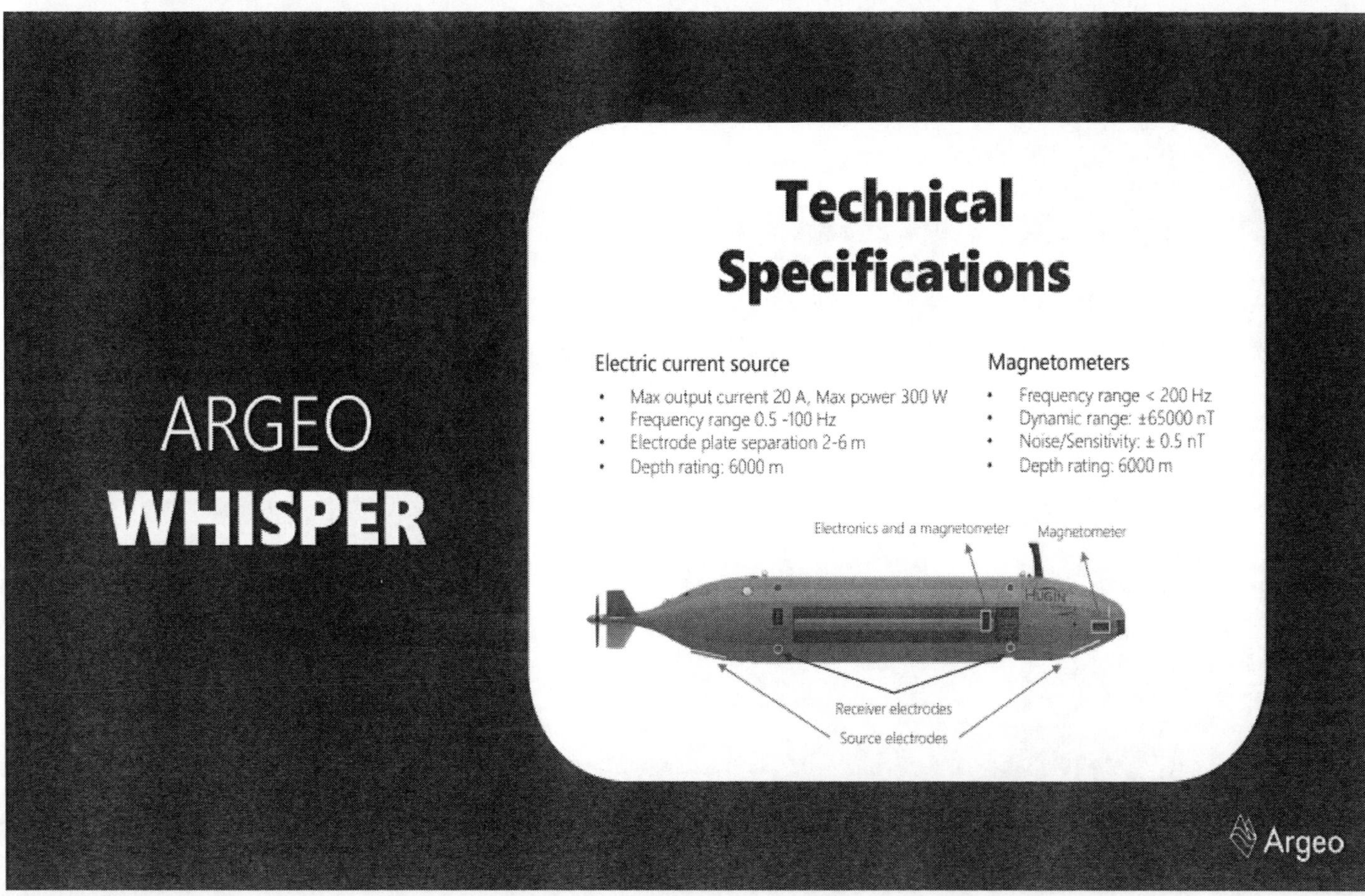
ARGEO
WHISPER
Technical Specifications
Electric current source
• Max output current 20 A, Max power 300 W
• Frequency range 0.5 -100 Hz
• Electrode plate separation 2-6 m
• Depth rating: 6000 m
Magnetometers
• Frequency range < 200 Hz
• Dynamic range: ±65000 nT
• Noise/Sensitivity: ± 0.5 nT
• Depth rating: 6000 m
Electronics and a magnetometer
Magnetometer
Receiver electrodes
Source electrodes
Argeo

Save time on inspection

UP TO 150 km	3-3,5 knots	NEAR REAL-TIME
Of high productivity inspection	Survey speed	Data analysis during field operations

- An active electro-magnetic system for detection of buried metal objects and cables in marine environments utilizing underwater vehicles
- Integrated on Argeo's AUV platforms which can operate down to 6000m water depth
- Emits electrical signals from the underwater vehicle and measures the response, detecting objects with conductive properties that stand out from their surroundings.

Fast electromagnetic tracking of buried cables and efficient detection of buried metal objects

- ✓ Detection of **buried objects**
- ✓ **Positioning** of cables and burial depth, as well as "dead" power line cables
- ✓ **Simultaneous** data acquisition with other sensors
- ✓ Faster project turnaround providing **safer operations** with lower HSE risk
- ✓ High signal to noise ratio
- ✓ The system is modular and can be installed on a variety of underwater platforms

Argeo

ARGEO **WHISPER** - TEST EXAMPLE

Detection of buried conductive objects

Buried object detection after processing using the active EM-system

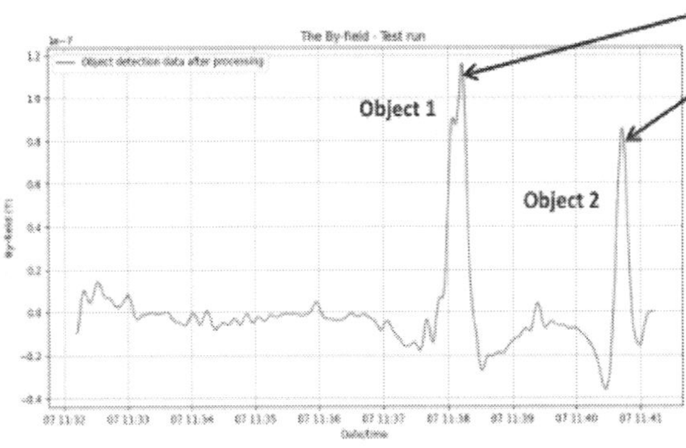

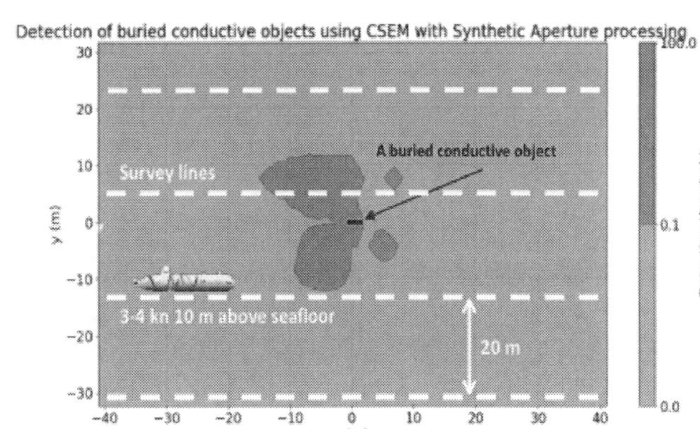

- Good signal to noise ratio
- About 10-12 m detection range on each side of the system → 15-20 m survey line separation
- AUV speed 3-4 kn 5-10 m above seafloor

Argeo

How inappropriate to call this planet Earth when it is clearly Ocean.

ARTHUR C. CLARKE

Remote Work for USV's

CHANCE
MARITIME TECHNOLOGIES

- Line-of-Sight Uncrewed Surface Vehicles (USVs) are a proven cost-effective means of increasing data collection volume.
  - However, they require personnel and subject matter experts to go to sea, driving additional operational costs, in addition to launch and recovery equipment that can take up vital deck space on a mothership
- Driving towards bringing the "remote work" paradigm to offshore operations

27-Sep-23 2

Customer Need

CHANCE
MARITIME TECHNOLOGIES

- University of Louisiana at Lafayette has an ongoing program, LO-SPAT to understand and develop oyster populations
- Sought spatial data to supplement their permanent station in Southwest Pass, connecting Vermilion Bay to the Gulf of Mexico

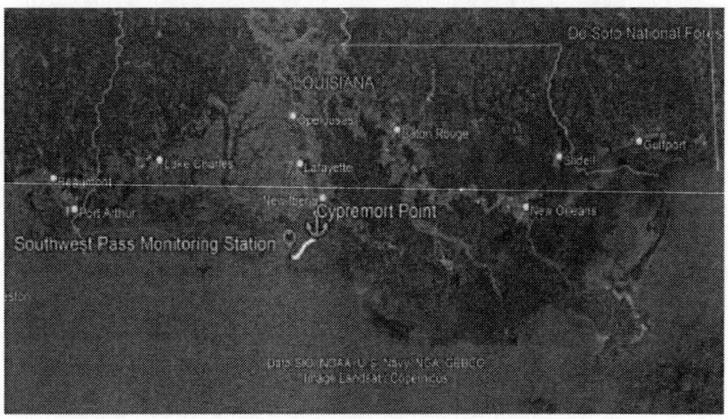

LO-SPAT

http://www.lospat.org/

27-Sep-23 3

USV Configuration

CHANCE
MARITIME TECHNOLOGIES

- Line of sight USV modified for over-the-horizon operation [1]
- ~15 ft long vessel
- 80 hp marine diesel engine
- ~7kt top speed, 4kt cruise speed

27-Sep-23 4

Payload Sensors

CHANCE
MARITIME TECHNOLOGIES

- YSI Sonde configured with:
  - Salinity / Conductivity
  - Temperature
  - Chlorophyll Flourometer
  - Turbidity
- Airmar Single Beam Depth Sounder
- Sensor package was hull mounted
- Data logged onboard sensor

Safety Considerations

CHANCE
MARITIME TECHNOLOGIES

- ✓ Follow NAVSAC 16-01 Unmanned Maritime Systems (UMS) Best Practices
- ✓ Follow IMO MSC.1-Circ.1604 Interim Guidelines For MASS Trials
- ✓ Staff qualified with:
  - ✓ Standard seamanship training
  - ✓ Practical uncrewed training in controlled environment (private test lake)
  - ✓ Supervision from USCG Certified Master

Copyright © 2023 Chance Technologies, LLC
All Rights Reserved

Risk Management

CHANCE
MARITIME TECHNOLOGIES

Risks	Mitigations
High traffic near public launch	Escort vessel with ability to take control using dedicated radio
Loss of control	Remote and local ESTOP, Automatic stop on loss of communications
Navigation hazards & situational awareness	Onboard Radar, Visual Cameras, Thermal Camera, AIS, Depth Sounder, Depth Charts
Loss of propulsion	Broadcast AIS location, remote engine restart, rope cutters on shaft, bow tow point recovery system
Cybersecurity	Layered security approach including encrypted communications, CMMC 2.0 IT system
Fire	Onboard automatic suppression system with engine and ventilation shutdown

27-Sep-23

7

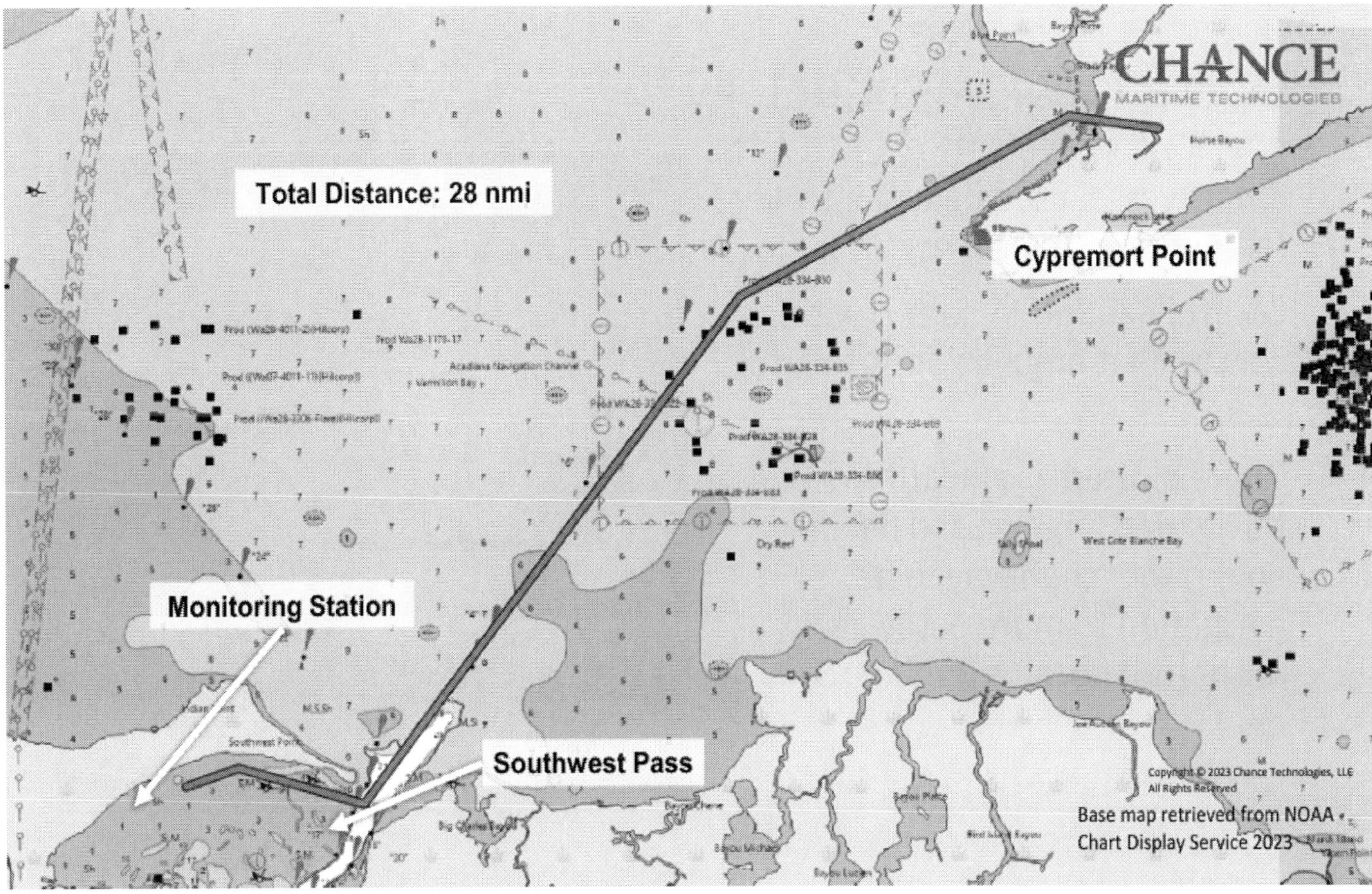

Remote Operation Center

CHANCE
MARITIME TECHNOLOGIES

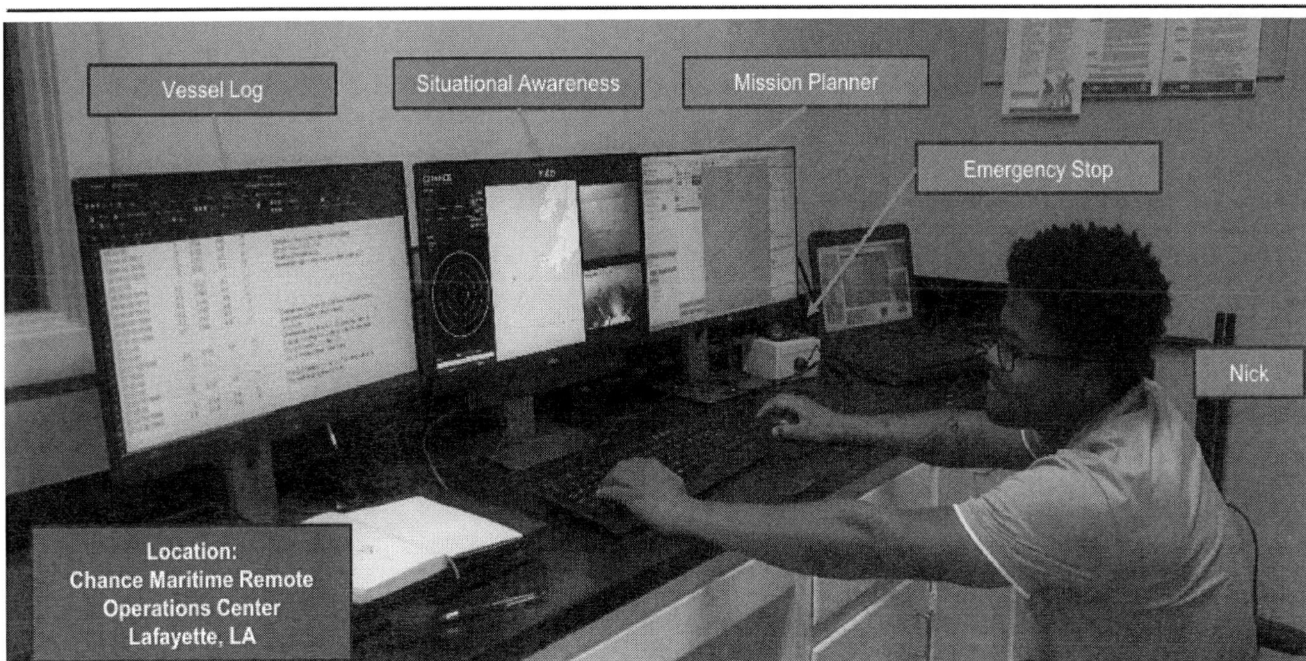

27-Sep-23

10

Operational Methodology

CHANCE
MARITIME TECHNOLOGIES

- Escort vessel used hand controller for launch and recovery on trailer
- Control handoff in harbor with constant communications to local team
- Route plotted as guidance, Navigation largely used heading hold rather than waypoint navigation to more easily avoid uncharted hazards (crab pots, etc.)

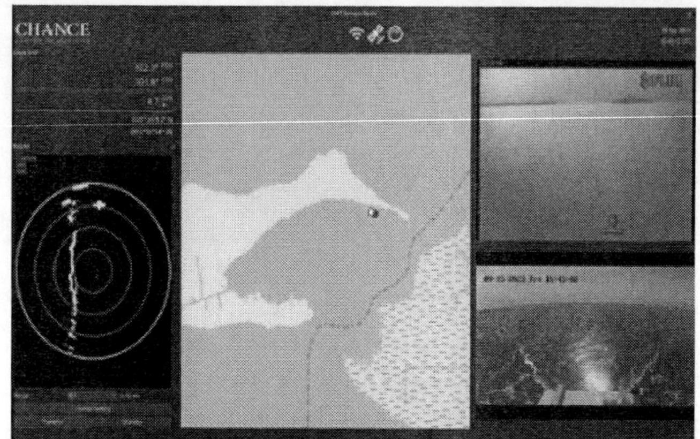

Data Collection & Delivery

CHANCE
MARITIME TECHNOLOGIES

- Vessel data (position, navigation, vessel systems) exfiltrated over Starlink
- Payload sensor data downloaded on recovery
    - Note: most payloads can be exfiltrated over Starlink
- Data uploaded to cloud and delivered to customer same-day

Salinity PSU
0 - 0.15
0.15 - 15.44
15.44 - 16.6
16.6 - 17.79
17.79 - 18.92

0 2.5 5 km

Lessons Learned

CHANCE
MARITIME TECHNOLOGIES

- Challenging environment: unprecedented shallow water conditions, crab pots
- Thermal & visual cameras critical for avoiding crab pots
- Radar critical for interpreting range to hazards and maintaining 360 situational awareness
- Starlink was remarkably reliable, supporting high data throughput back to our Remote Operations Center in Lafayette, LA

27-Sep-23 13

Next Generation

CHANCE
MARITIME TECHNOLOGIES

- Currently conducting endurance testing on next generation **MC-29** [3]
- Long endurance open ocean USV
- 20-30 day endurance
- Kilowatts of available payload power
- Climate controlled payload space, 19" rack, moon pool

27-Sep-23 14

Learn More

CHANCE
MARITIME TECHNOLOGIES

www.ChanceMaritime.com

linkedin.com/company/chance-maritime-technologies/

27-Sep-23

References

CHANCE
MARITIME TECHNOLOGIES

1. M. P. Scherer and T. S. Chance, "Enabling over the horizon operation on a line-of-sight uncrewed surface vehicle," *OCEANS 2023 Gulf Coast*, Biloxi, MS, USA, 2023.
2. Inland Protector Datasheet, https://chancemaritime.com/wp-content/uploads/2023/11/Chance-Tug-Datasheet-2023-09-20-Final-r2.pdf
3. MC-29 Datasheet, https://chancemaritime.com/wp-content/uploads/2023/11/Chance-MC-29-Datasheet-2023-09-20-Final-r2.pdf
4. LO-SPAT Program, http://www.lospat.org/

27-Sep-23

Fast Digital Inspection (FDI) for Integrity Management

Underwater Intervention, November 2023

Some Definitions - Apologies to Those Who Are Familiar With These Terms

CVI – Close Visual Inspection similar to a DVI – detailed/stationary or static inspection (BP)

Hull, propellor or specific area of a subsea structure

GVI – General Visual Inspection

Pipelines

GVI-1: A low or close pass inspection including visual and/or non-visual techniques
GVI-2: A high pass or standoff inspection including visual and/or non-visual techniques

Structures

Dynamic general visual inspection acquiring either mono or stereo imagery of (generally) vertical structures. GVI usually employs forward-facing camera(s). The option for mono or stereo is a product of inspection speed, coverage, and the outputs required for the inspection.

FDI – Fast Digital Inspection

Pipelines and Structures

A dynamic contactless inspection where all required inspection data can be collected with the vehicle underway.
ROV moving at 0.4 to 0.6 knots. Towed shallow water up to 4 knots. AUV – 3 to 5 knots.

CP – Cathodic Protection assessment – contact – touching the structure, contactless CP – field gradient assessment

OOS – Out of Straightness

DoB – Depth of Burial

Today's Agenda

1. The goals of an FDI
2. Deliverables and data available from FDI
3. What equipment, planning and execution elements are involved in an FDI
4. Pitfalls to watch for in planning and executing
5. Advantages of accurate subsea digital twins

FDI Stakeholders – We Are Trying to Get Them All Around the Table

For innovation to happen faster, the discussion must include all of the following:

Integrity software solution providers – Wood NEXUS IC, Aize Integrity Elements, EIVA?

Incumbent inspection community – CSWIP 3.4U is still structured around video

ROV/AUV contractors

Survey contractors

Field Gradient (contactless) CP solution providers

Imaging solution providers

Mux and navigation sensor manufacturers

Some Numbers

Late last year, we completed a major inspection campaign:

- Three deep water fields
- Six months of planning before field operations
- 419 structures inspected
- 200 structures fully processed onboard
- 8.4 days ROV at depth
- 18 days dock to dock

"We saved ~40% of the vessel days compared to the last time we did this inspection as a GVI."

Ray Smith, Survey and Inspection Manager, DOF Subsea in Perth

ZUPT

FDI Objectives & Goals

- Accurate Digital Twins
  - True representations of subsea assets – and how they change over time
  - Quantitative data for engineering decisions - not qualitative video
  - All data georeferenced – 3D CP heat maps layered onto the structure model
- Data and Deliverables Considerations:
  - Minimize "big-data" complexities
  - Smooth transition from image/video eventing to spatially correct model analysis
  - Accurate, automated change detection
  - Models/images available to all across the operators/contractor's enterprise
- Enhance Efficiency:
  - Reduce vessel days and remove the need for survey/inspection personnel offshore
  - Leads to a reduced carbon footprint
- Data Acquisition Innovation:
  - Deploy systems permanently on field resident ROVs/AUVs.
  - Automate data flow – automate data processing – reduce data volumes

ZUPT

PLANNING

Pre-Job Planning

A very clearly defined scope of work:

What are we going to inspect/model – and why

What are the required tolerances of the delivered products from the inspection/survey?

This leads to a very clearly defined list of deliverables: Who is the customer? What do they want?

Point cloud – binary or ASCII *.ply or *.obj

Mesh – binary or ASCII *.ply with the *.jpg (texture mapping file)

Pseudo video - *.mp4 – with video overlay generated from EXIF data within the images

All images *.raw – for now this is a Zupt proprietary format with everything we need saved All images used for processing models *.bmp (lossless) *.jpg (lossy) with EXIF populated

Real time precise navigation string $PZUPT, re-processed navigation string after Delph/Janus if available

INS post processed with tide corrections if available, note in real time everything is WGS84 – post processed can be in local reference frame.

What is the destination of the data?

Conventional metrology deliverable or Isometrics for metrology, or OOS or *.stl, *.stp?

IM deliverables Coabis, Nexus, Integrity elements, folder structures – moving from video to models?

Cloud locations (AWS, Azure, GCP) – data lake/repository access within your enterprise??

Project Planning – Details, Details, Details

What do we mean by details – just a few (of many) examples:

Depth – is it tide corrected – predicted tide or client provided tide

100% coverage of structures, even the underside of wellheads, drives lighting and instrument placement?

Clear definition of cloud locations (Azure/AWS/GC) for what deliverables?

Reliable access/permission to your cloud location for upload

Do you (the client) really want "all of the raw data"?

Does "raw" data include raw IMU, raw image, raw CP sensor data – or processed value "raw" data

Stereo delivers two of the same image – do you want both as a deliverable (doubles data storage)

What needs to be processed to the final deliverables offshore – what is sensible QC?

Strong coordination/PM between Client, ROV, Inspection, and Survey teams

ZUPT

Importance of Planning for Efficiency

As we get answers to the previous questions, we can then start to plan.

Total data acquisition time can be accurately calculated based on structure sequence, vessel/ROV capabilities (tether lengths, SIMOPS limitations) and structure types. Accurate field drawings and any shape files for structures greatly assists accurate planning. The outputs from this planning are the amount of raw data that will be collected, the number of vessel minutes per structure and deliverable processing time.

How accurate is the planning?

OPERATION	PLANNED	ACTUAL	DIFFERENCE
Field A	69.58 hours	82.75 hours	-13.17 hours
Field B	46.00 hours	49.77 hours	-3.77 hours
Field C	53.83 hours	68.65 hours	-14.82 hours

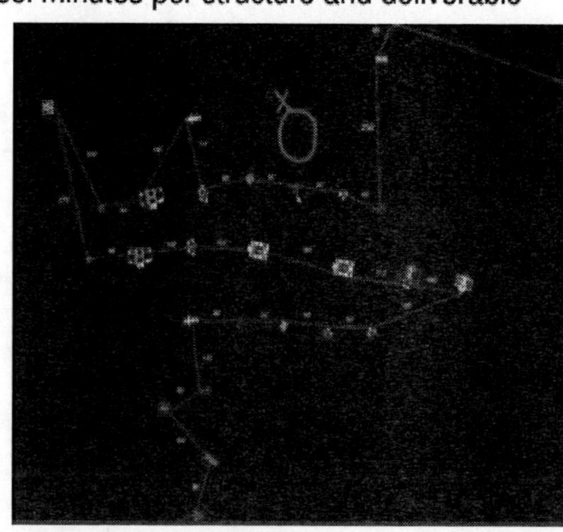

ZUPT

Sharing Space on the ROV – Productivity is Everything

The further an imaging solution is from the target – leads to a less accurate model with lower resolution. BUT – do we need the level of resolution available?

The further the imaging solution is away from the structure leads to greater coverage and higher productivity.

Field Gradient CP systems want to be well out in front of the ROV.

Imaging systems do not want anything in their field of view.

Lighting is critical – balanced illumination across the scene with offsets from imaging sensors, so that particulate matter is not as reflective, drive light locations and design.

Open water flow through other sensors determine their location.

Full 3D CAD layout prior to mobilization helps share the space tremendously.

ZUPT

Compromise

ZUPT

OPERATIONAL

Data Acquisition – A Lot Going On

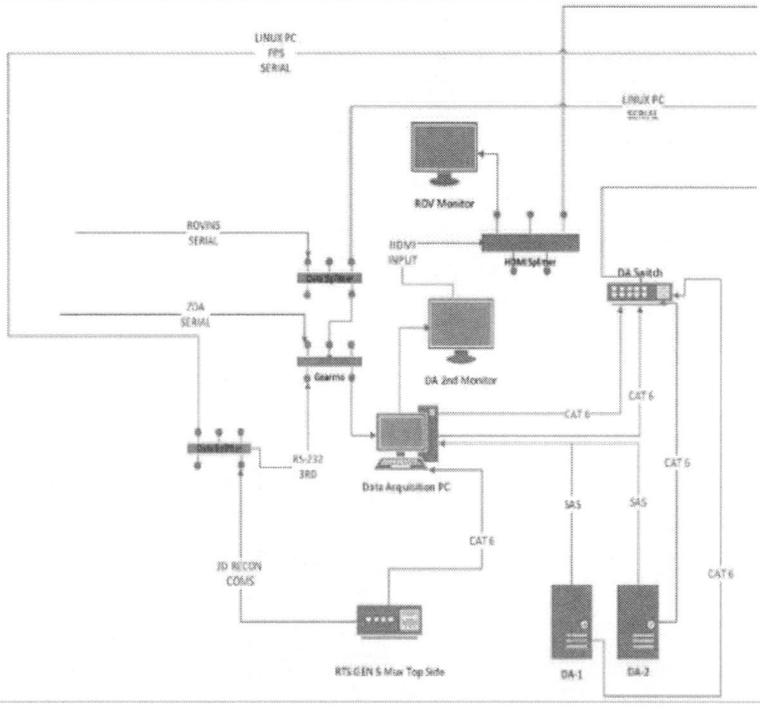

Reduce Mobilization Time Offshore

Ease of mobilization:

SAT prior to the equipment arriving onshore
- Integration between all FDI sensors?
- Difficult interfacing completed onshore
- Detailed data flow checked during SAT

Sensor and light location determined onshore

Topside:
- 10Gbps network racks/storage preassembled
- Processing stations rack-based and fully tested

ROV agnostic FDI Skid including:
- Mux
- 3D Modeling image sensors
- Contactless CP/Contact CP
- INS/DVL
- Depth
- CO2 Sensor?
- CH4 Sensor?
- PAH?

Do You Have the Data?

Real Time QA/QC

How do we *ensure* all the data is collected while ROV is on the structure?

Instrument Calibration

Sharing ROV time for all sensor calibrations (filed gradient and imaging)

How are we ensuring the quality of the data?

Offshore Storage

Many Terabytes of data

Redundant storage requirements

Data transfer validation – what tools?

Same Tool – Now "On The Job" Real Time QC

The goal is to get the 3D Recon system to a point where just the ROV crew can collect and QC data acquisition without the need for survey onboard. 3D Recon Painter is also being used in real time on the job to "paint the structure". Initially we were just painting the structure green if within our field of view and the SLAM flag is good. Now we are painting the structure with the sparse point cloud 3D Recon generates in real time to clearly identify coverage. This went live in Australia. ROV had this on the screen in front of them during data collection.

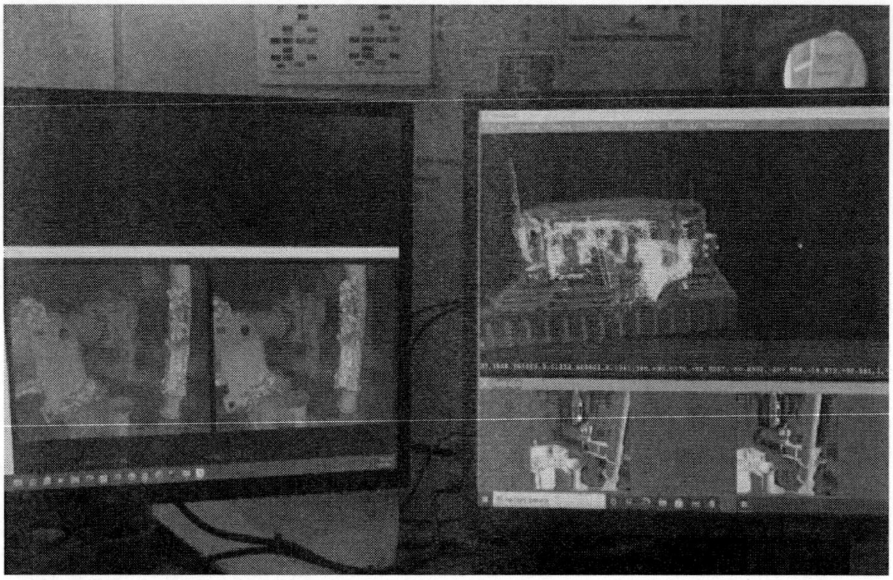

ZUPT

17

Data Acquisition Benefits

Our experience suggests that FDI can reduce campaign lengths by 40% compared to a conventional GVI.

Real-time CP "heat map" enabled through the precise structure relative position from imaging solution.

Real-time CP heat map only requires ROV to stop to contact CP where they matter. No need to revisit the field for this data collection

All eventing and anomaly documentation can be completed onshore.

ZUPT

18

DATA PROCESSING

Data Processing

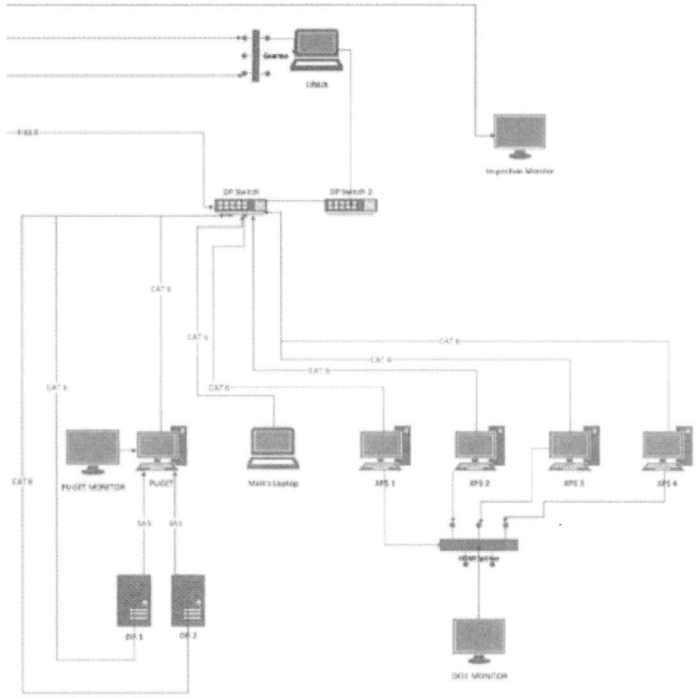

Processing Steps – Time Taken For 1 Hour Of Data Acquisition

PROCESSING STEP	DURATION FOR 1 HOUR ACQ
Convert Images Using 3D RECON-UTIL	
Generate .PZUPT string	Completed in real time
Generate video	1.2 hours
Generate .bmp	0.5 hours
Generate .jpg with EXIF	1.0 hours
Generate Metashape/Agisoft Project	
Open Agisoft Project generated by 3D Recon Util	No time
Run image alignment	2 hours
Clean up the point cloud	0.3 hours (data set dependent)
Build Mesh ("Sparse Mesh")	0.8 hours
Reduce Overlap	0.2 hours
Build Mesh ("Depth maps")	3.0 hours
Build Texture	1.0 hours
Export Camera Pose file and *.ply/*.obj	0.25 hours

Still Improving

- Time and GPU Intensive
  - Not a simple PC task, sometimes not even a Data Processing PC task
- Final deliverables are NOT generated in real time – but how soon are the deliverables required?
  - Onshore versus Offshore Processing and the personnel requirements
  - Bandwidth from Offshore
- What model resolution is good enough?
- Not all the data may be initially processed – Does the remaining raw data need to be stored - permanently?
- Can the end user manipulate a 15GB model?

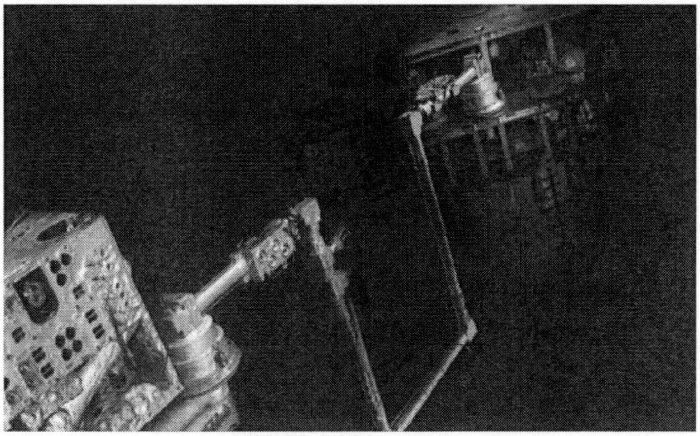

USE

Processed Data

Automated Change Detection

High-definition models, images, pseudo video with conventional overlay, color palletized mesh and point cloud (model), spatial CP heat map, all georeferenced

Most file formats available

Stored on the cloud for distribution throughout the organization

Again – decimation is valuable to ensure file size works for the intended use.

More Improvements

Optimized Planning:
Efficient methods to prevent wasted time offshore

Simplify the ROV interface, automate ROV trajectory around structures, data transfer to shore

FDI skid in design

Real Time QC:
Eliminate the need for additional personnel offshore

Real time QC methods to ensure all data is acquired

Standardized Deliverables:
Data that fits into industry standard inspection programs

Operators Drive Geographic Business Unit Alignment:
Standardize processes across global business units

ZUPT

25

Summary

FDI is the future of subsea inspection, but how do we get where we want to be in 5 years?

- FDI acquisition and processing technology is mature and readily available
- Automated real-time coverage QC will eliminate the need for survey personnel to be present offshore
- Conventional GVI deliverables, IN ADDITION to FDI deliverables, eliminates the concerns with integration with current inspection programs
- The "risk" of missing an anomaly using FDI instead of GVI is the same.

We need complete "buy-in" by the Operators, Engineering Contractors and the "inspection community" to enable this solution.

ZUPT

26

VideoRay
Underwater Intervention & Workboat 2023
DEFENDER

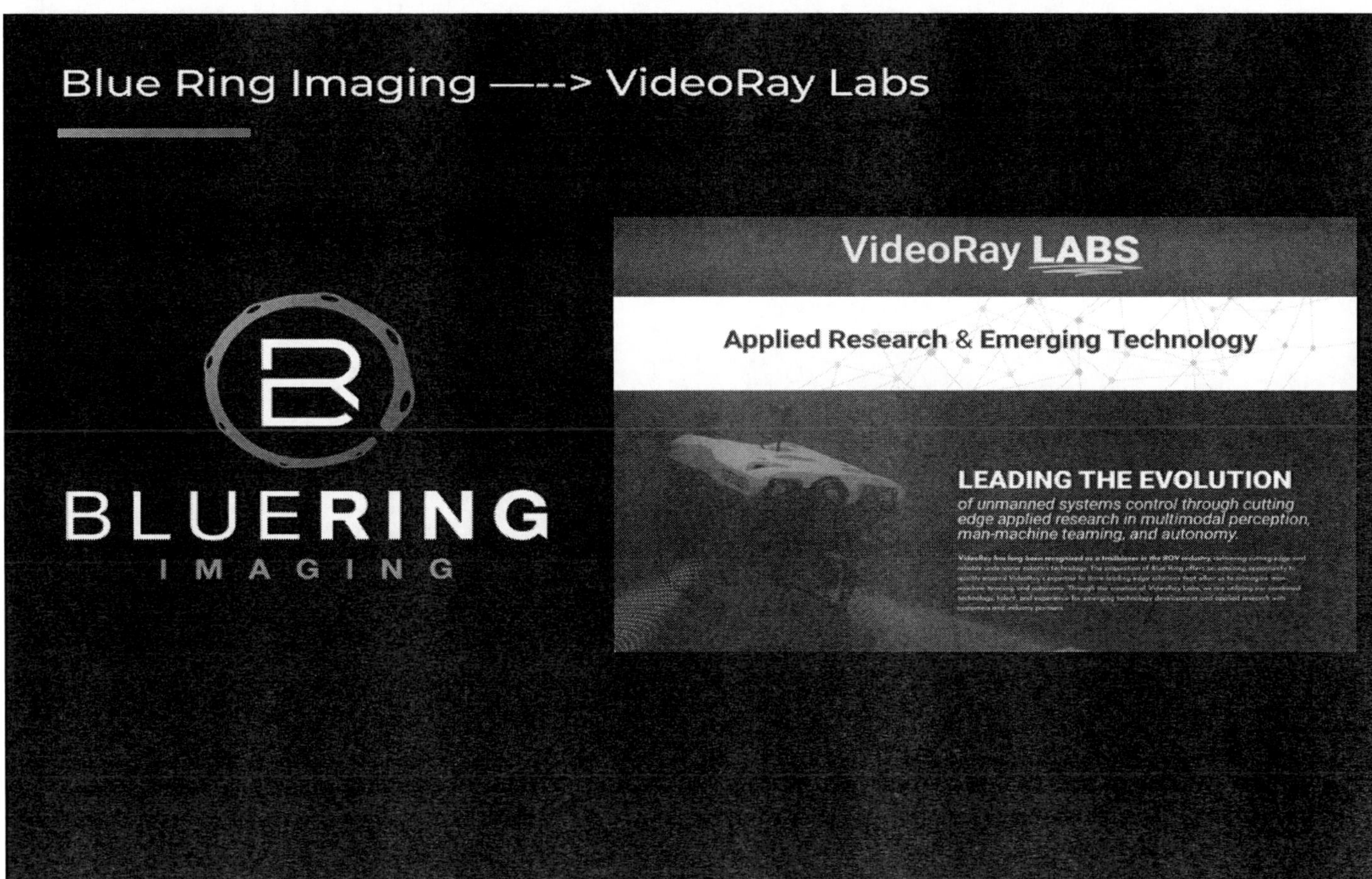
Blue Ring Imaging —--> VideoRay Labs
BLUERING
IMAGING
VideoRay LABS
Applied Research & Emerging Technology
LEADING THE EVOLUTION
of unmanned systems control through cutting edge applied research in multimodal perception, man-machine teaming, and autonomy.

“The Hub” for Maritime & Defense Companies in St. Pete

360° Camera Experience

Casey Bio

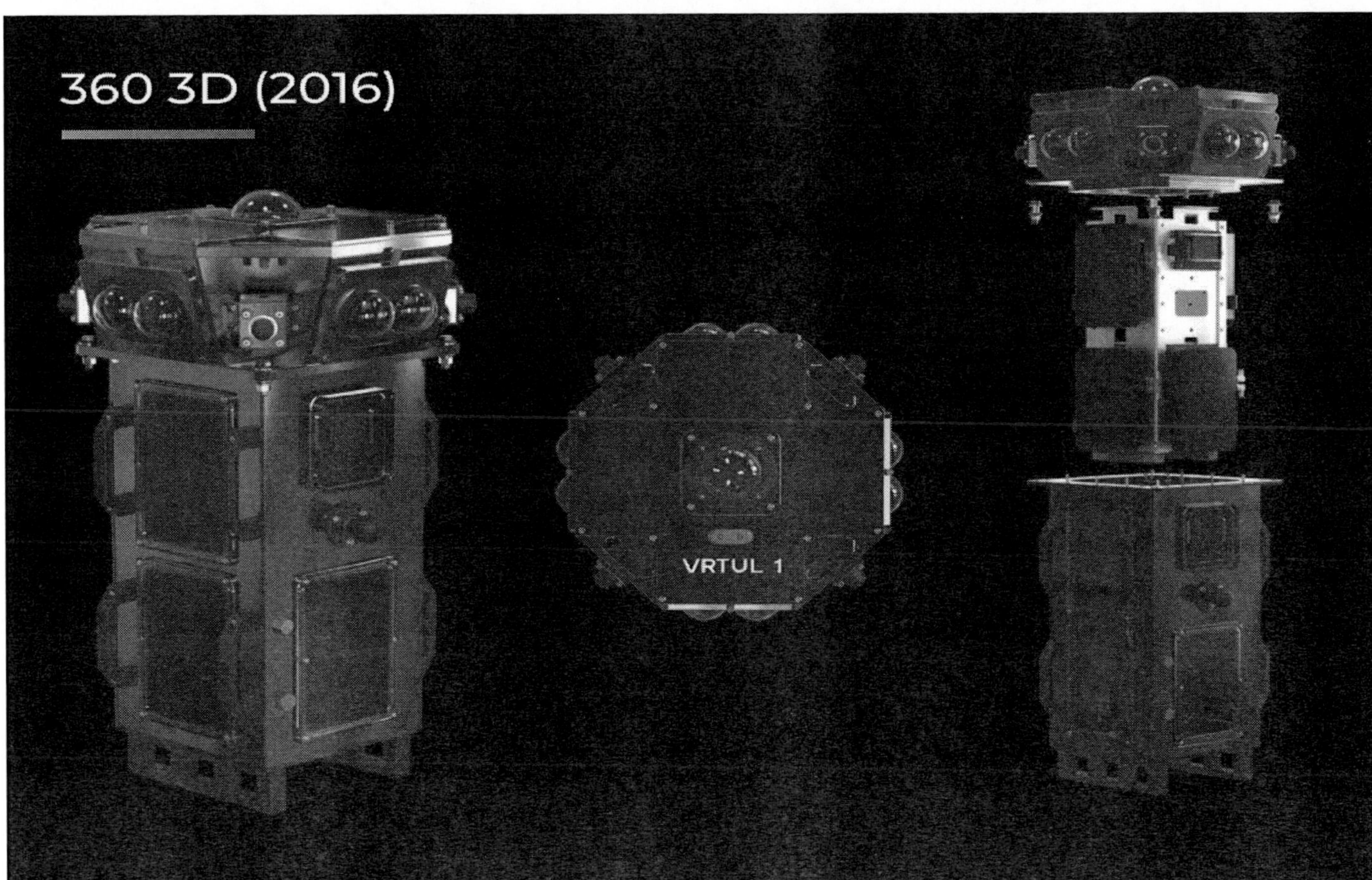
360 3D (2016)
VRTUL 1

360 3D (2017)

360 3D (2019)

Blue Ring Clients

Our 360 camera arrays have been used both for cinematic filming and 360 live-streaming

OCEANX
GOOD MORNING AMERICA
Google
Discovery CHANNEL
FACEBOOK
BBC
NATIONAL GEOGRAPHIC
FESTIVAL DE CANNES
Earthx2019
CONSERVATION SUMMIT JACKSON HOLE WILDLIFE FILM FESTIVAL
International OCEAN Film Festival
TRIBECA FILM FESTIVAL
OXFORD INTERNATIONAL FILM FESTIVAL

Blue Ring Achievements

Creator of the first 360 3D underwater camera system(s)

Developed first cinema class 360 video camera for a submarine (Seamagine)

Developed first underwater live streaming VR camera and produced the first underwater 360 live broadcast (Good Morning America, 2017)

Blue Ring Origin Story w/ MBARI (2018)

Whitepaper

In a 2021 white paper published by Monterey Bay Aquarium Research Institute they showed that when using Blue Ring products (OctoCom and OctoView) together that manipulator capture times in stereo VR ***decreased by 50%, for both rookie and experienced pilots.***

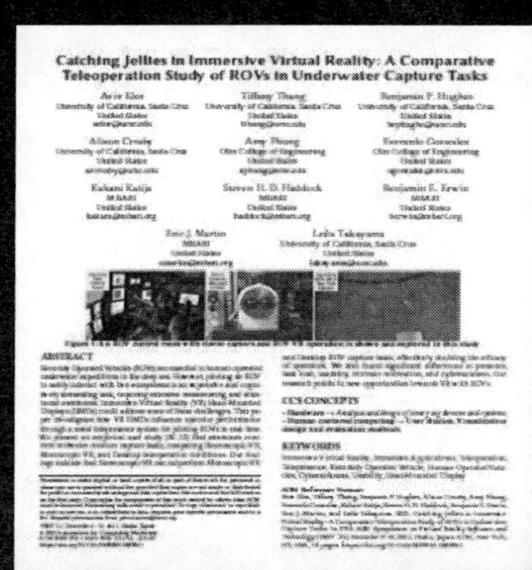

Catching Jellies in Immersive Virtual Reality: A Comparative Teleoperation Study of ROVs in Underwater Capture Tasks

ABSTRACT

OctoView Individual Benefits (eg Pilot)

OctoView now has 5 years of academic research, experimentation, and end user studies before its public release in 2023 working with MBARI, OceanX, and PMS 408.

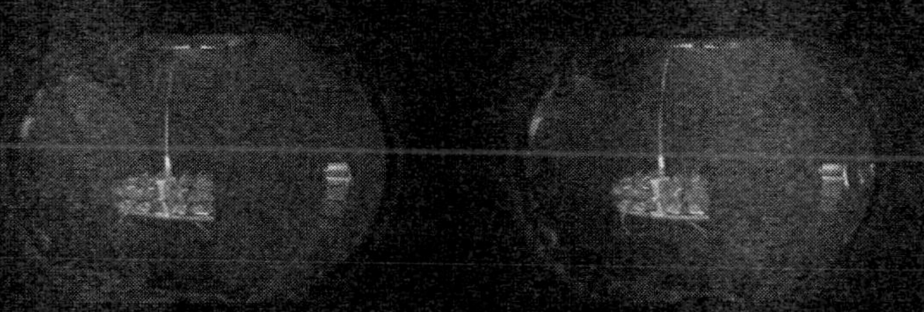

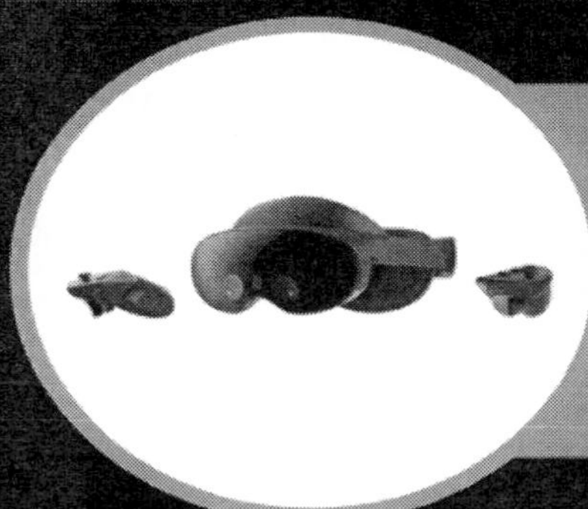

Demonstrated Results

- Simplifies complex tasks
- Decreases task completion times
- Decreases individual workload
- Increases UxS usability
- Improves pilot performance

OctoView Corporate Benefits & Business Outcomes

DoD and commercial customers are assessing, adopting, and evaluating OctoView for their pilots and fleet for the aforementioned reasons.

OCEANEERING® | Office of Naval Research ONR Science & Technology | OCEANX | PEO USC

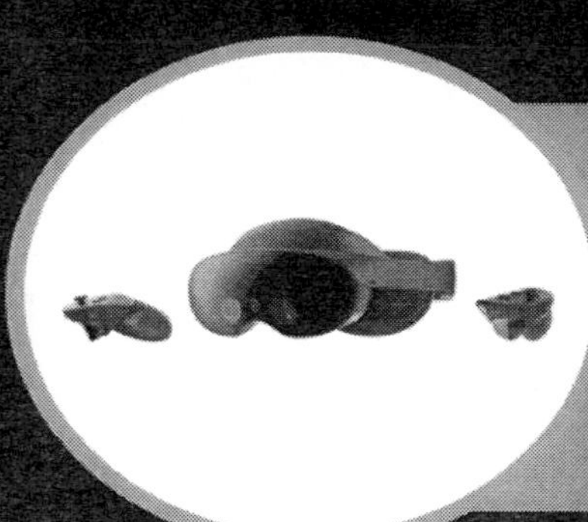

Benefits

- Enhance Man-Machine Teaming
- 360° Situational Awareness
- Improve Mission Execution
- Reduce Total Cost of Operations
- Enrich Team Collaboration with Multiple Headsets
- Improve Vehicle Safety

Continued Testing with MBARI, OceanX, PMS 408

2018 - Present

VideoRay Defender

Defender 1.0

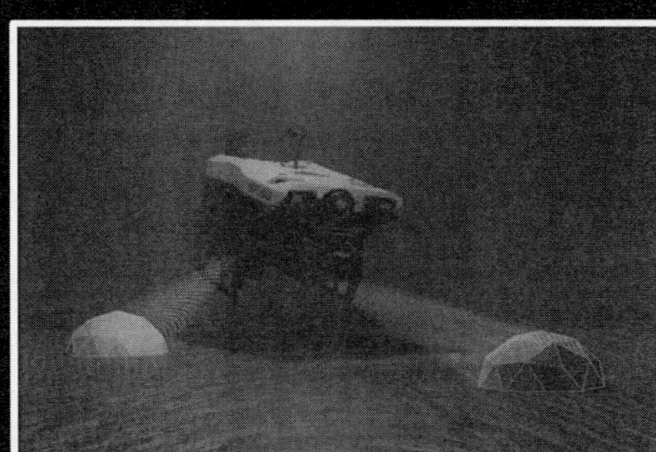

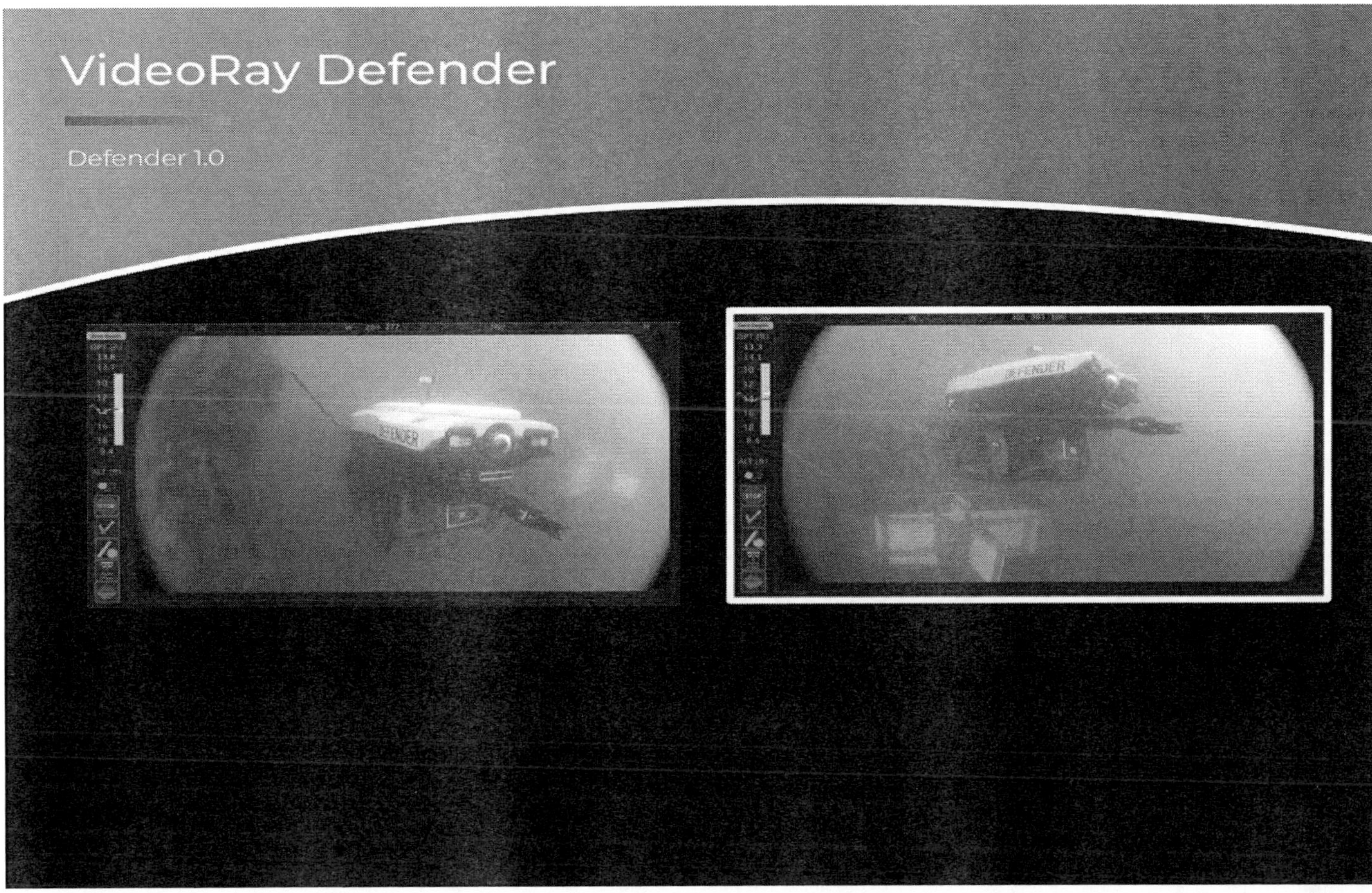
VideoRay Defender
Defender 1.0

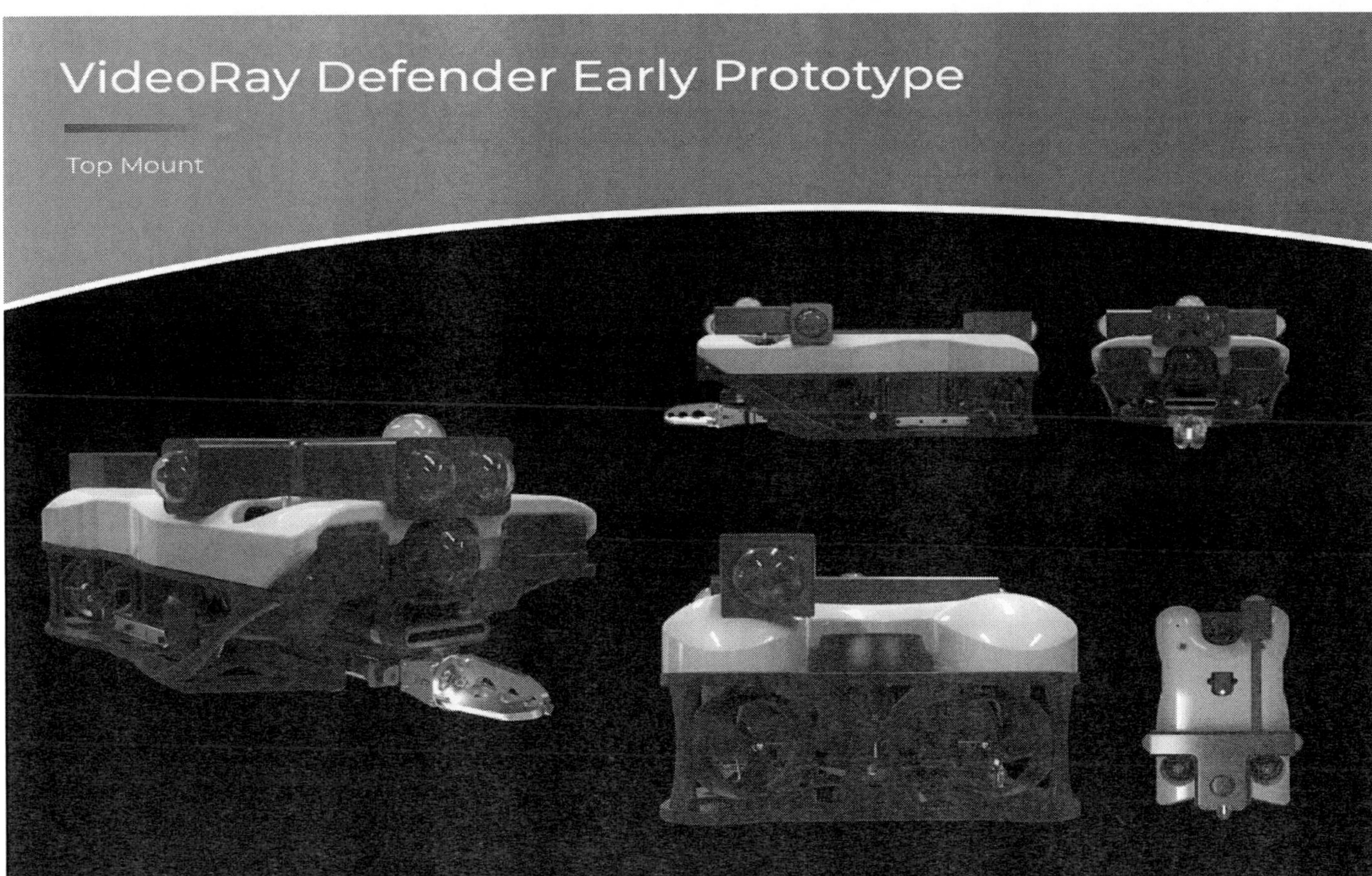
VideoRay Defender Early Prototype
Top Mount

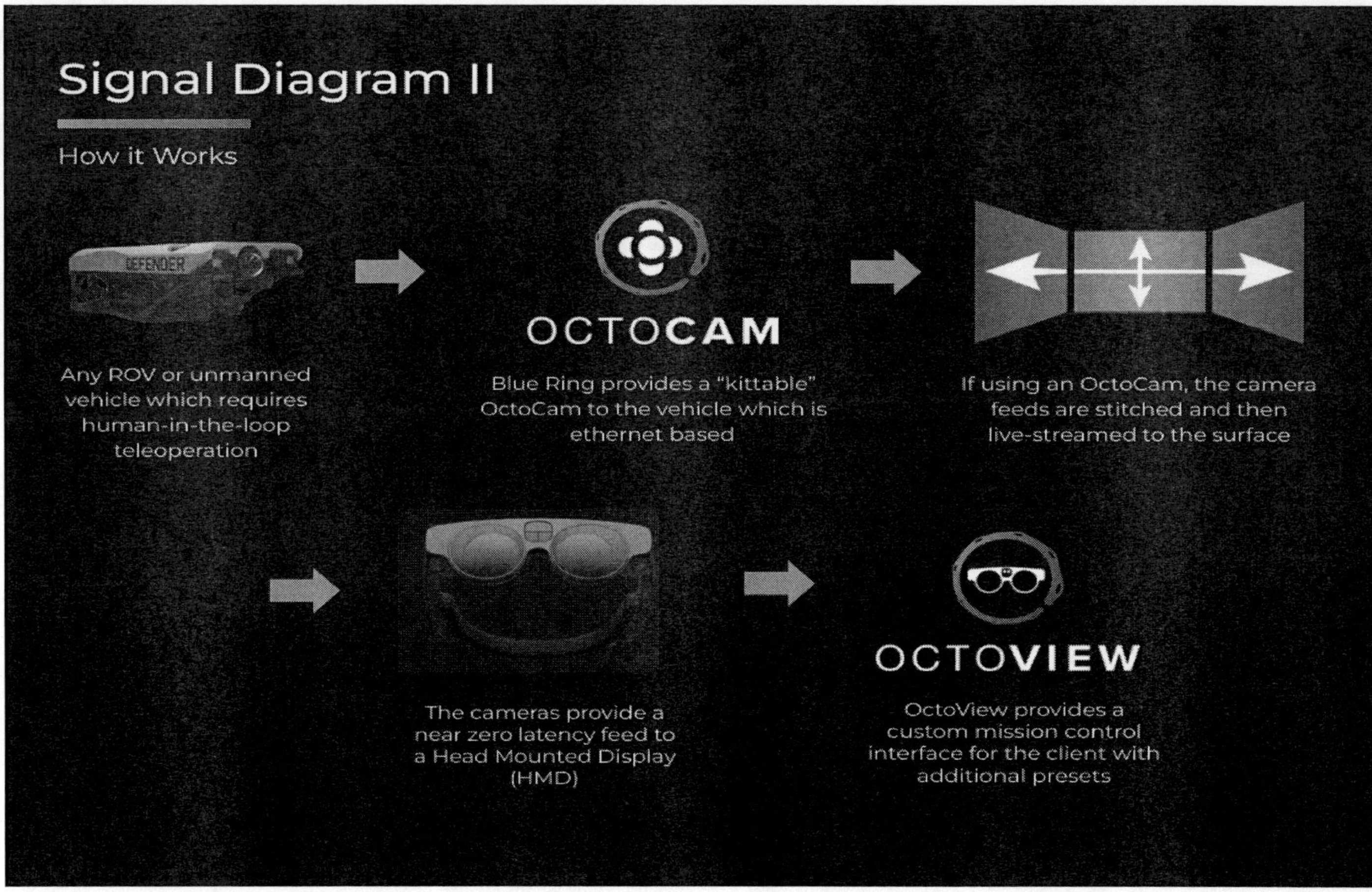
Signal Diagram II
How it Works
DEFENDER
Any ROV or unmanned vehicle which requires human-in-the-loop teleoperation
OCTOCAM
Blue Ring provides a "kittable" OctoCam to the vehicle which is ethernet based
If using an OctoCam, the camera feeds are stitched and then live-streamed to the surface
The cameras provide a near zero latency feed to a Head Mounted Display (HMD)
OCTOVIEW
OctoView provides a custom mission control interface for the client with additional presets

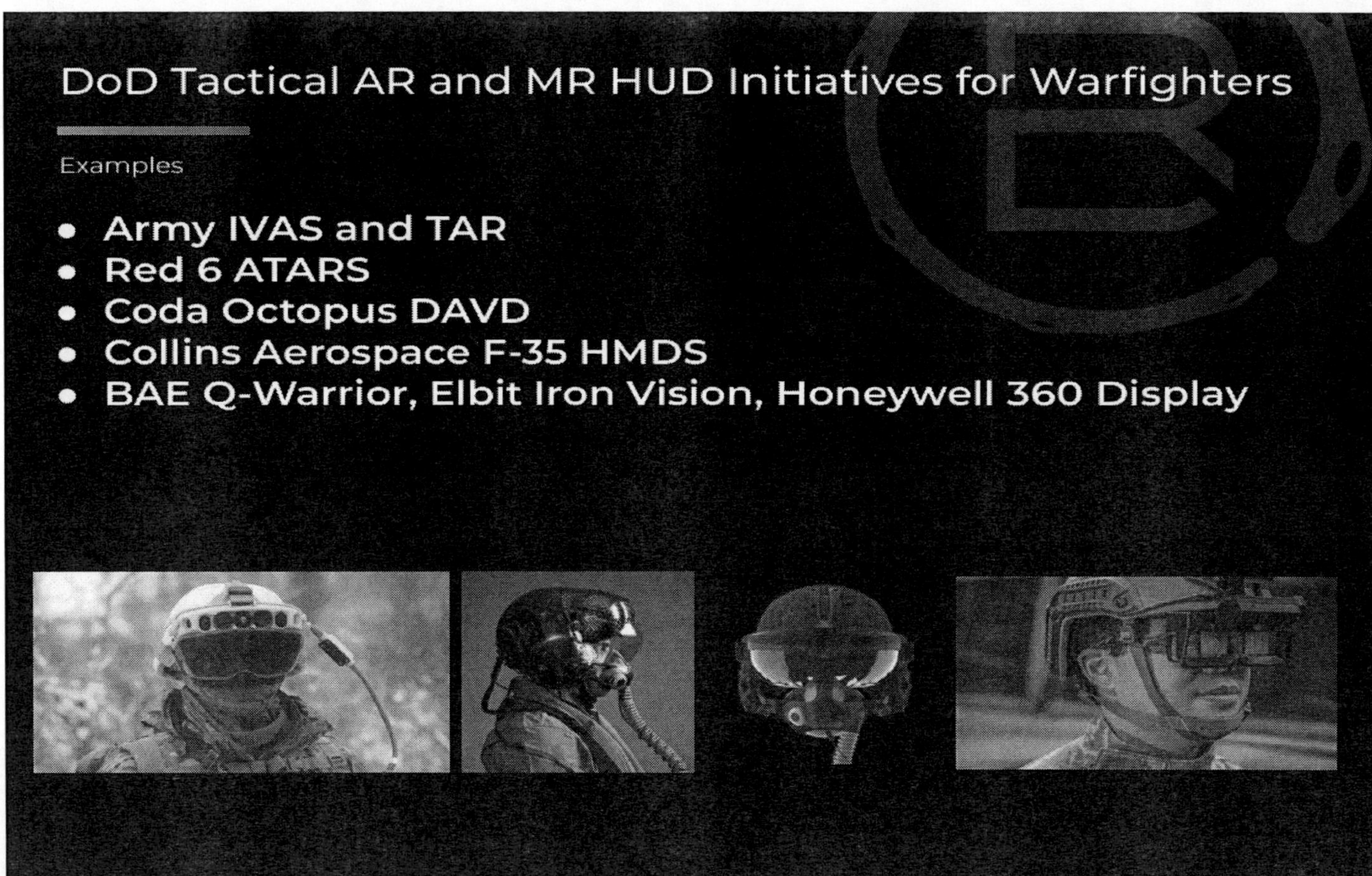
DoD Tactical AR and MR HUD Initiatives for Warfighters
Examples
• Army IVAS and TAR
• Red 6 ATARS
• Coda Octopus DAVD
• Collins Aerospace F-35 HMDS
• BAE Q-Warrior, Elbit Iron Vision, Honeywell 360 Display

OctoView for Embodied AI

OctoView is intended to integrate voice, haptics, and machine intelligence paired with 3D vision to develop the most intuitive, easy to use interface for controlling an ROV.

OCTOVIEW

OctoView Capabilities

Physical video walls are expensive and OctoView saves time, money, and space implementing "virtual video walls" for clients which can be deployed anywhere.

Our app is agnostic in every sense of the word including the vehicle, operating system, comms, headset, and end user device.

OCTOVIEW

OctoView Testimonial

*"I had a healthy skepticism of XR being useful when it came to ROV piloting. **But seeing is believing. I am now a full convert.**" (2021)*

Greg Diffendale
Ocean Exploration Trust ROV Pilot

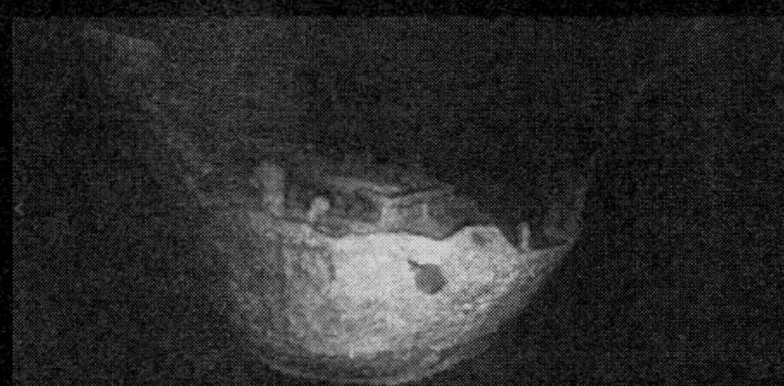

OctoView Testimonial

"I flew a lot last week with the headset on...
***The 180 3D was mind blowing.** There is a special bit of flying we do*
in the mid water where we have to catch animals in little containers.
It's usually one of the more challenging things any pilot has to do.
***I was able to catch this weird worm thing like it was nothing.**" (2019)*

- ***Ben Erwin***

MBARI ROV Pilot using Blue Ring Imaging ROV VR Technology

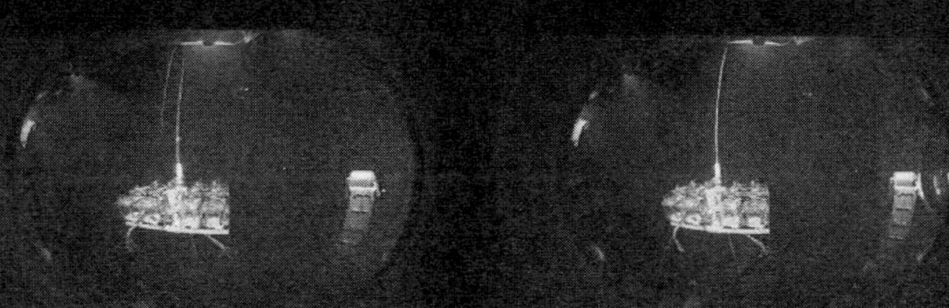

UNDERWATER INTERVENTION
Powered by WORKBOAT
The widest range of mission-critical interconnect technologies in the world
Best Practices
for Subsea Connector Selection
Eric Hawes
Glenair.

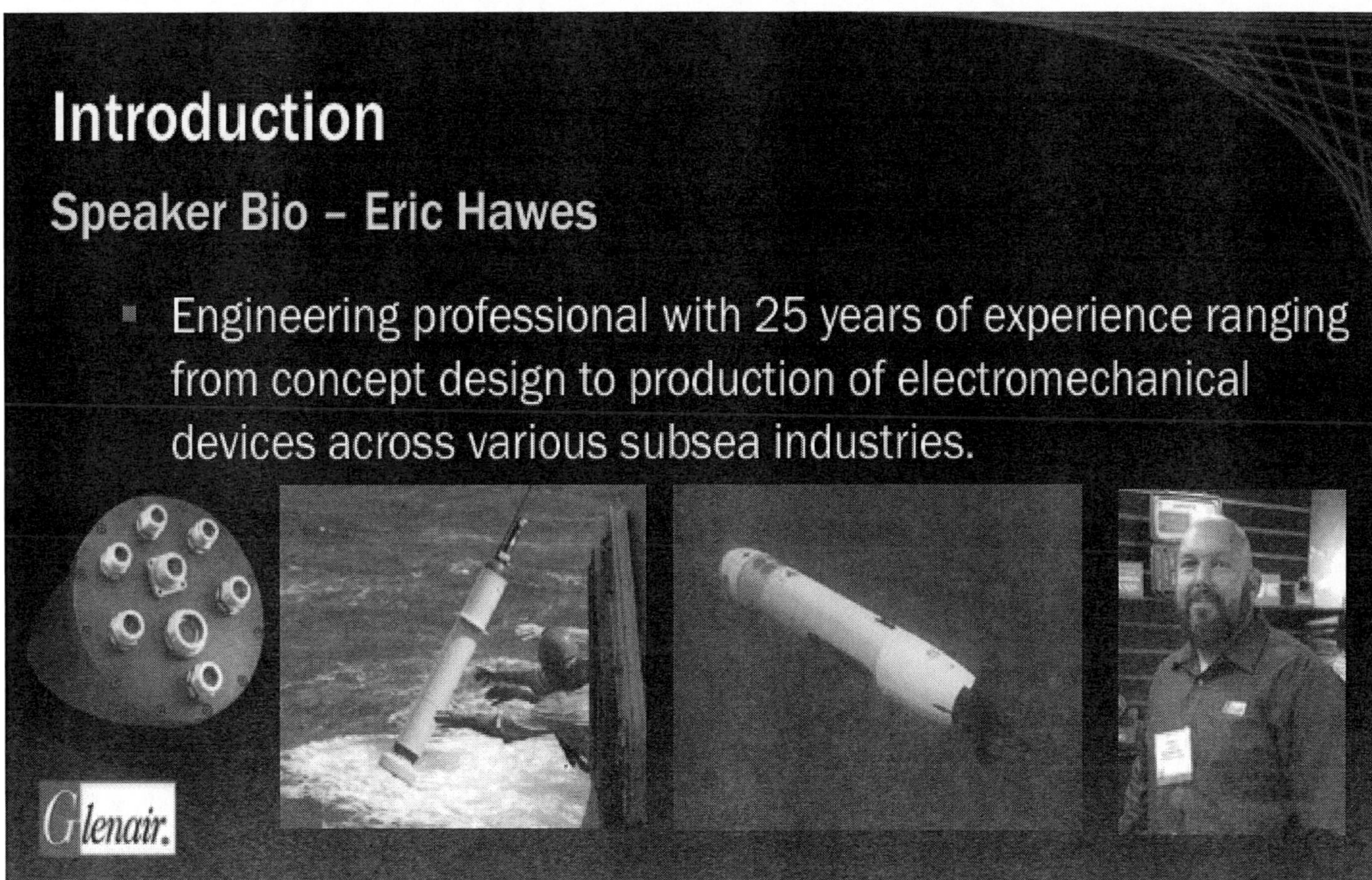
Introduction
Speaker Bio – Eric Hawes
▪ Engineering professional with 25 years of experience ranging from concept design to production of electromechanical devices across various subsea industries.
Glenair.

Best Practices for Subsea Connectors

Outline

- Design Requirements / Interface Scenario
- Mechanical Performance
- Electrical Performance
- Optical Performance
- Internal Interface
- Cost Tradeoffs

Glenair.

Preface Discussion

Connectors in general are a point of high failure rate and high stress concentration. Looking at your system diagram:

- Is a complete disconnect needed?
- Could the connector be a Feed-Thru / Penetrator?
- Can number of discrete contacts, data and power be reduced or combined to minimize size and number of penetrations?

Glenair.

Pressure Barrier

Penetrator and Feed-Thru

SeaKing

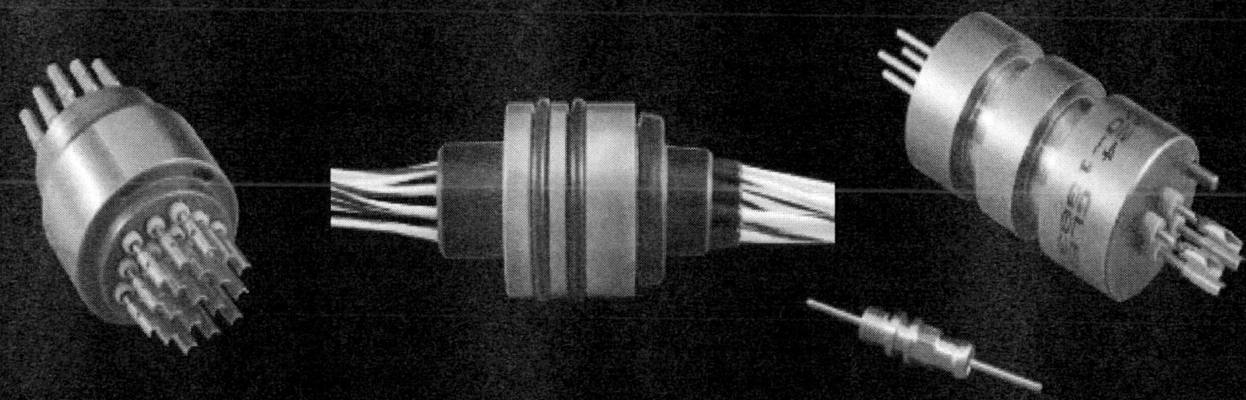

Hermetic Penetrator | Booted Hermetic Feed-Thru | High-Pressure Hermetic and Single-Pin Feed-Thru

Glenair.

Guts of a connector that directly interface your bulkhead

Design Requirements / Interface Scenario

Inputs needed in order to provide the best solution

- Depth Rating and Environment
- Connector Mate Type
- Transmission Type
- Transmission Distance
- Pressure Rating Type
- Load Bearing Cable System
- Subsea Cable Type
- Governing Body Specifications (API, NORSOK, MIL-SPEC)

Glenair.

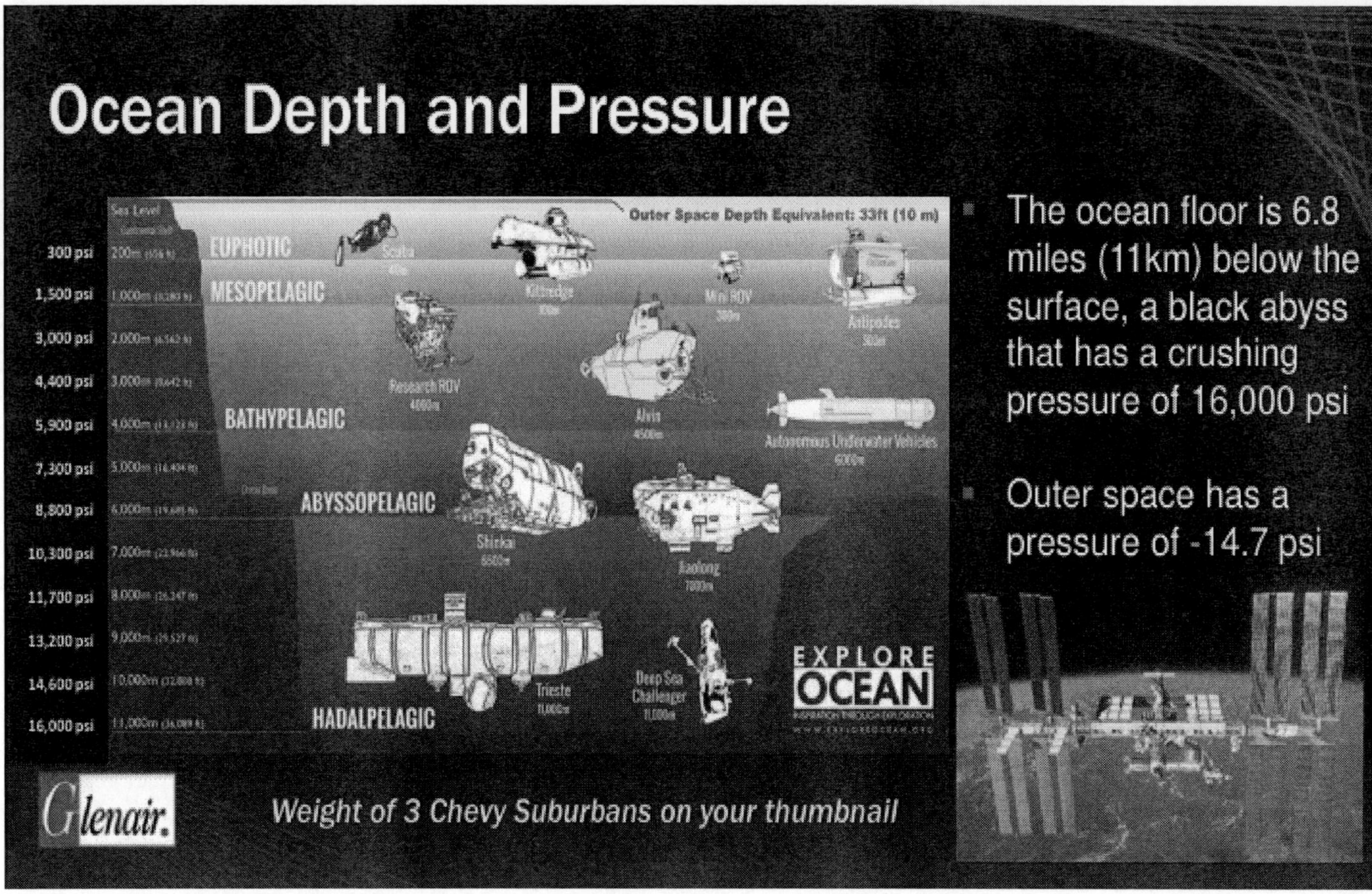

Ocean Depth and Pressure
Outer Space Depth Equivalent: 33ft (10 m)
EUPHOTIC
MESOPELAGIC
BATHYPELAGIC
ABYSSOPELAGIC
HADALPELAGIC
300 psi
1,500 psi
3,000 psi
4,400 psi
5,900 psi
7,300 psi
8,800 psi
10,300 psi
11,700 psi
13,200 psi
14,600 psi
16,000 psi
EXPLORE
OCEAN
The ocean floor is 6.8 miles (11km) below the surface, a black abyss that has a crushing pressure of 16,000 psi
Outer space has a pressure of -14.7 psi
Glenair.
Weight of 3 Chevy Suburbans on your thumbnail

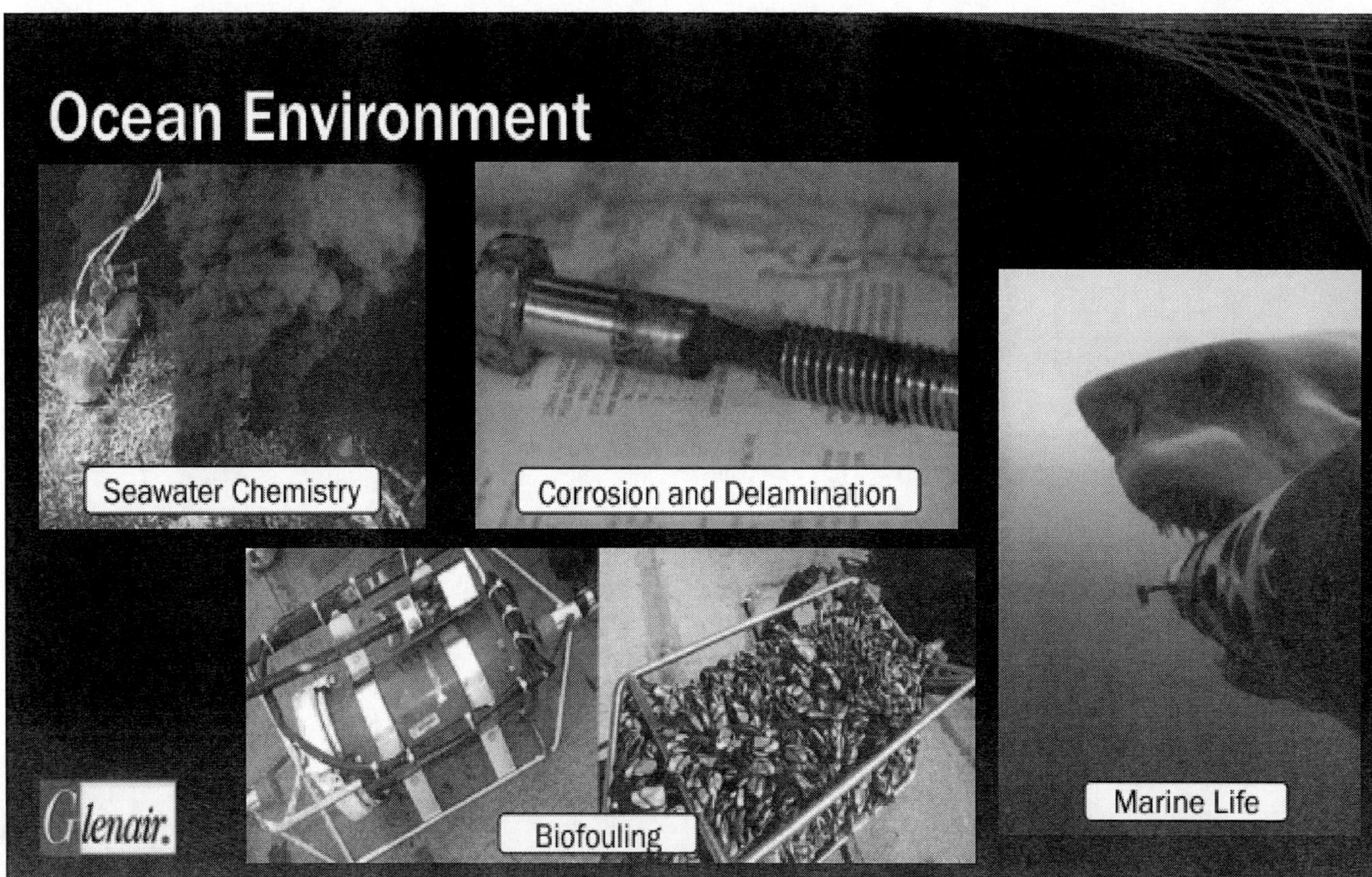

Ocean Environment
Seawater Chemistry
Corrosion and Delamination
Marine Life
Biofouling
Glenair.

Connector Mate Type

Dry Mate - Splash Mate - Wet Mate

- Dry Mate: Topside, Laboratory, Clean Environment
- Splash Mate: Wet Environment, Contacts designed to self squeegee or isolate connection
- Wet Mate: Complete Mate and De-Mate at Depth

Dry Mate

Splash Mate

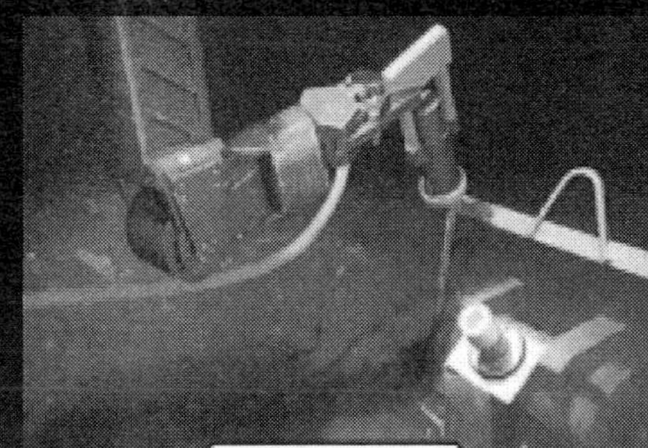

Wet Mate

Transmission Type

Data – Ethernet and other High-Speed Signal Protocols

- **Ethernet:** 10, 100, 1000, 10G-BaseT
- **Serial/Mil/Aero Data Bus:** RS-232,422,485, MIL-STD-1553,1760, CANBUS, ARINC-429,664, IEEE-1394 (Firewire)
- **Peripherals/Display/Video:** 50&75ohm COAX, USB 2.0/3.0 and SATA, DVI/HDMI/Displayport, SMPTE HD/3G-SDI, ARINC 818 Video, Serial Rapid I/O (sRIO), Fiber Channel, PCIe, SGMII, XGMII

Power

- Low Power: 5v-500v
- Hi Power: 1kv and above, DC, AC, High Voltage, High Current

Transmission Distance

How far do you need to push the data?

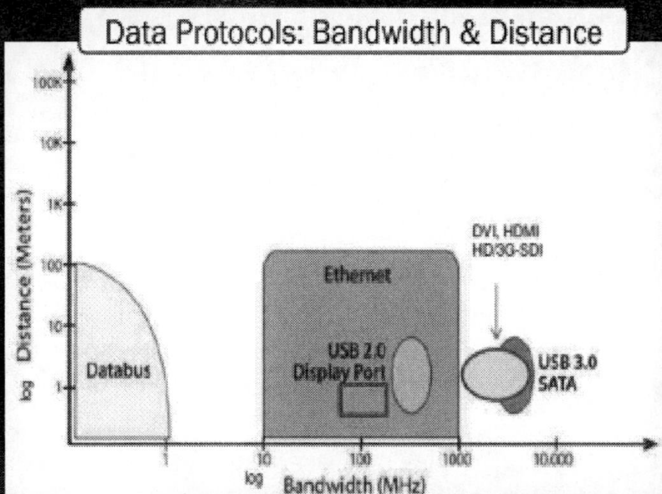

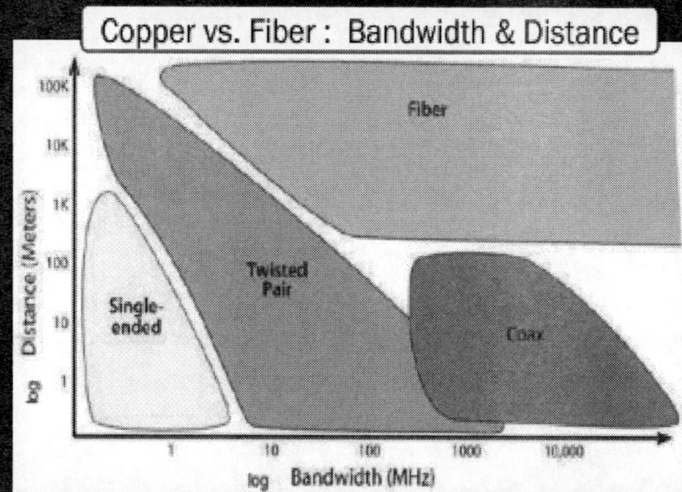

Glenair.

Pressure Rating Type

Where and to what extent do you keep the water out?

- Pressure barriers and location
  - Mated Pressure Rating vs Open Face Rated

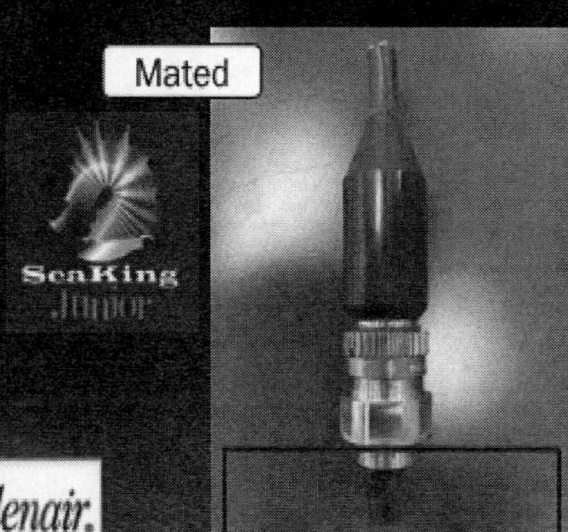

Glenair.

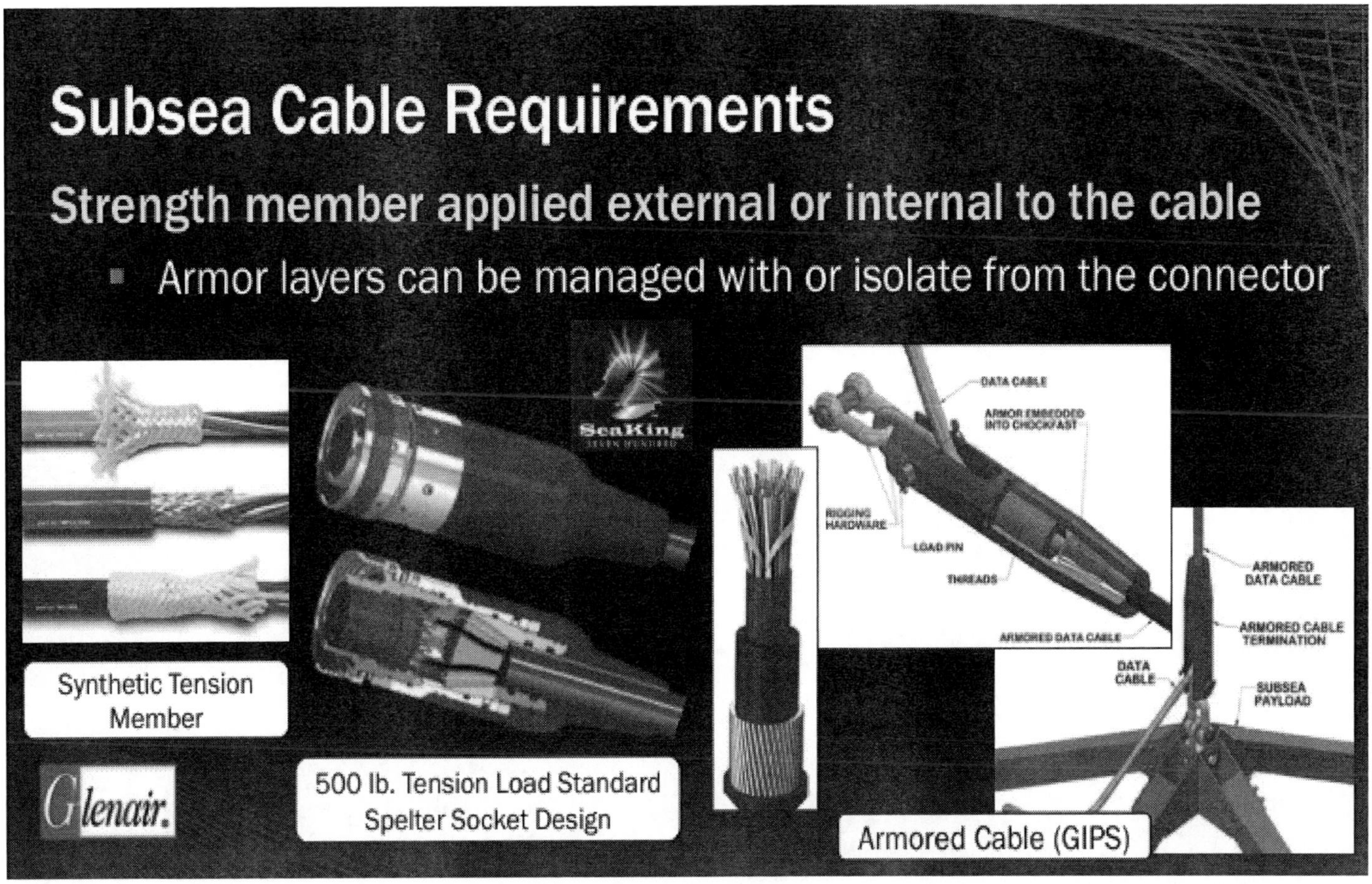
Subsea Cable Requirements
Strength member applied external or internal to the cable
Armor layers can be managed with or isolate from the connector
SeaKing
DATA CABLE
ARMOR EMBEDDED INTO CHOCKFAST
RIGGING HARDWARE
LOAD PIN
THREADS
ARMORED DATA CABLE
ARMORED DATA CABLE
ARMORED CABLE TERMINATION
DATA CABLE
SUBSEA PAYLOAD
Synthetic Tension Member
500 lb. Tension Load Standard Spelter Socket Design
Armored Cable (GIPS)
Glenair.

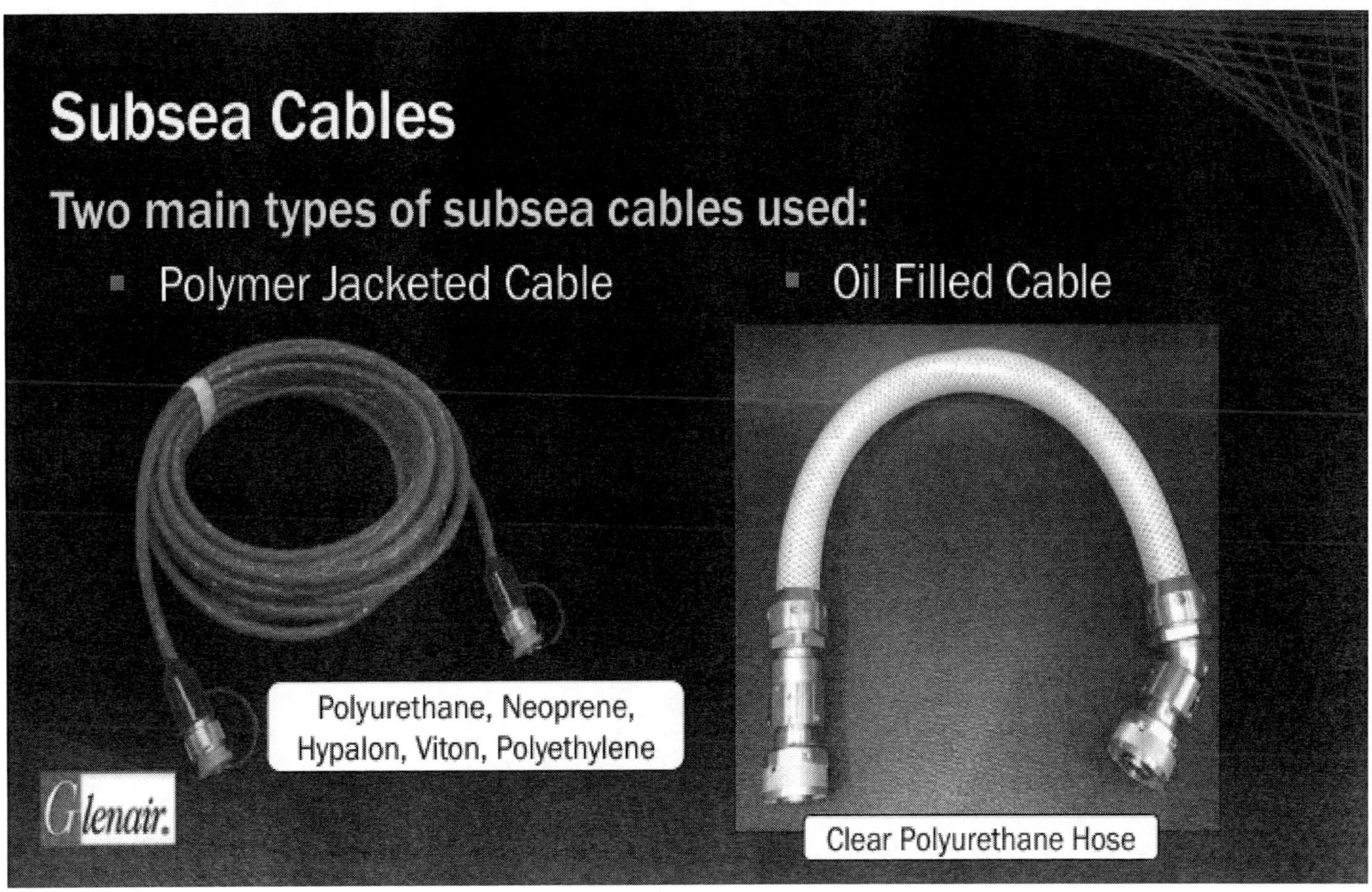
Subsea Cables
Two main types of subsea cables used:
Polymer Jacketed Cable
Oil Filled Cable
Polyurethane, Neoprene, Hypalon, Viton, Polyethylene
Clear Polyurethane Hose
Glenair.

Water Blocked Molded Cable

- Conventional molded cables use water blocking (anti-capillary) compound to fill voids between the conductor strands and block sea water ingress and fluid migration along the cable
- Terminated ends rely on overmolded elastomeric chemical bonds for water blocking

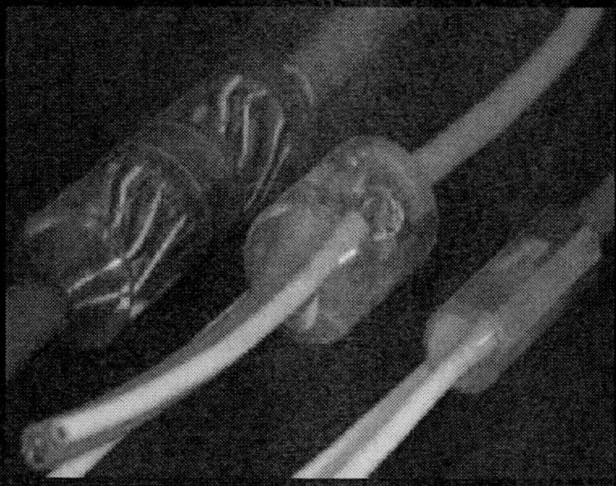

Polyurethanes absorb water approximately .3% to 1% by weight

Glenair.

Pressure Balance Oil Filled (PBOF)

- Many subsea applications utilize electrical and optical Pressure Balanced Oil Filled (PBOF) Connector and Cable Assemblies.
- PBOF cables use a compensating fluid-filled elastomeric tube as a conduit for the electrical and fiber optic lines.
- The elastomeric conduit allows the ocean pressure to transfer to the compensating fluid, equaling the pressure differential.
- PBOF technology has been widely utilized in ocean science exploration, drilling systems, production control systems, towed sonar arrays, submarines and ROVs

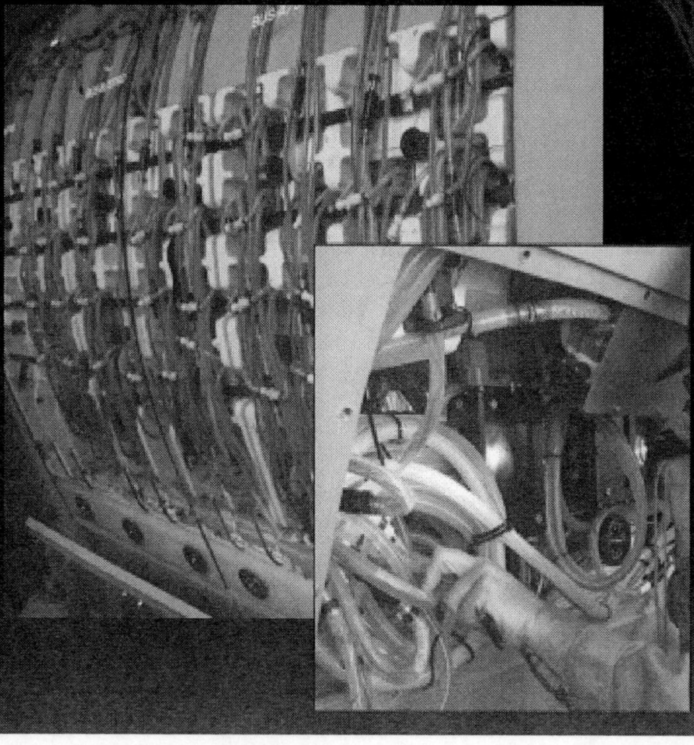

Glenair.

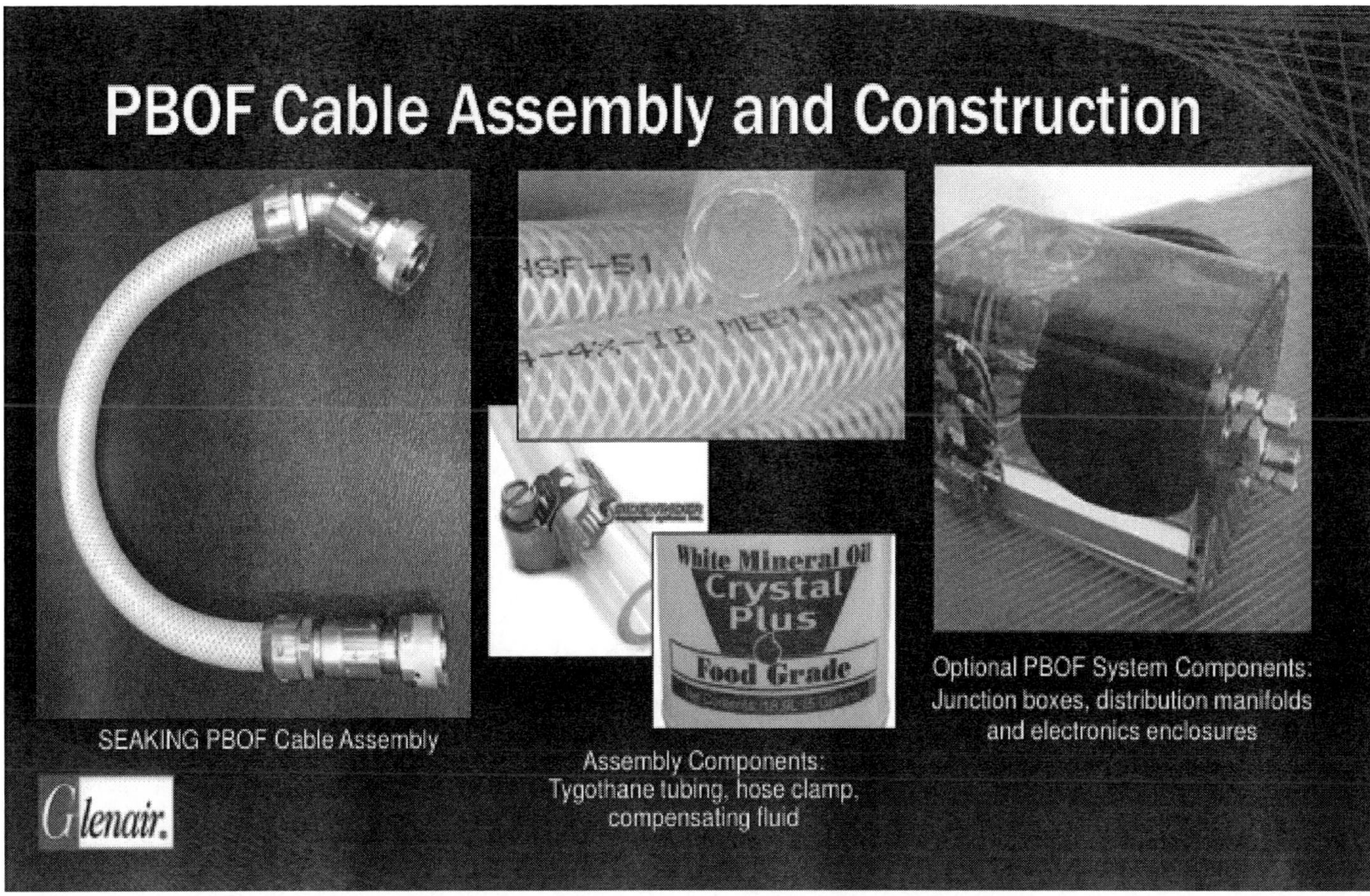
PBOF Cable Assembly and Construction
White Mineral Oil
Crystal Plus
Food Grade
SEAKING PBOF Cable Assembly
Assembly Components:
Tygothane tubing, hose clamp,
compensating fluid
Optional PBOF System Components:
Junction boxes, distribution manifolds
and electronics enclosures
Glenair.

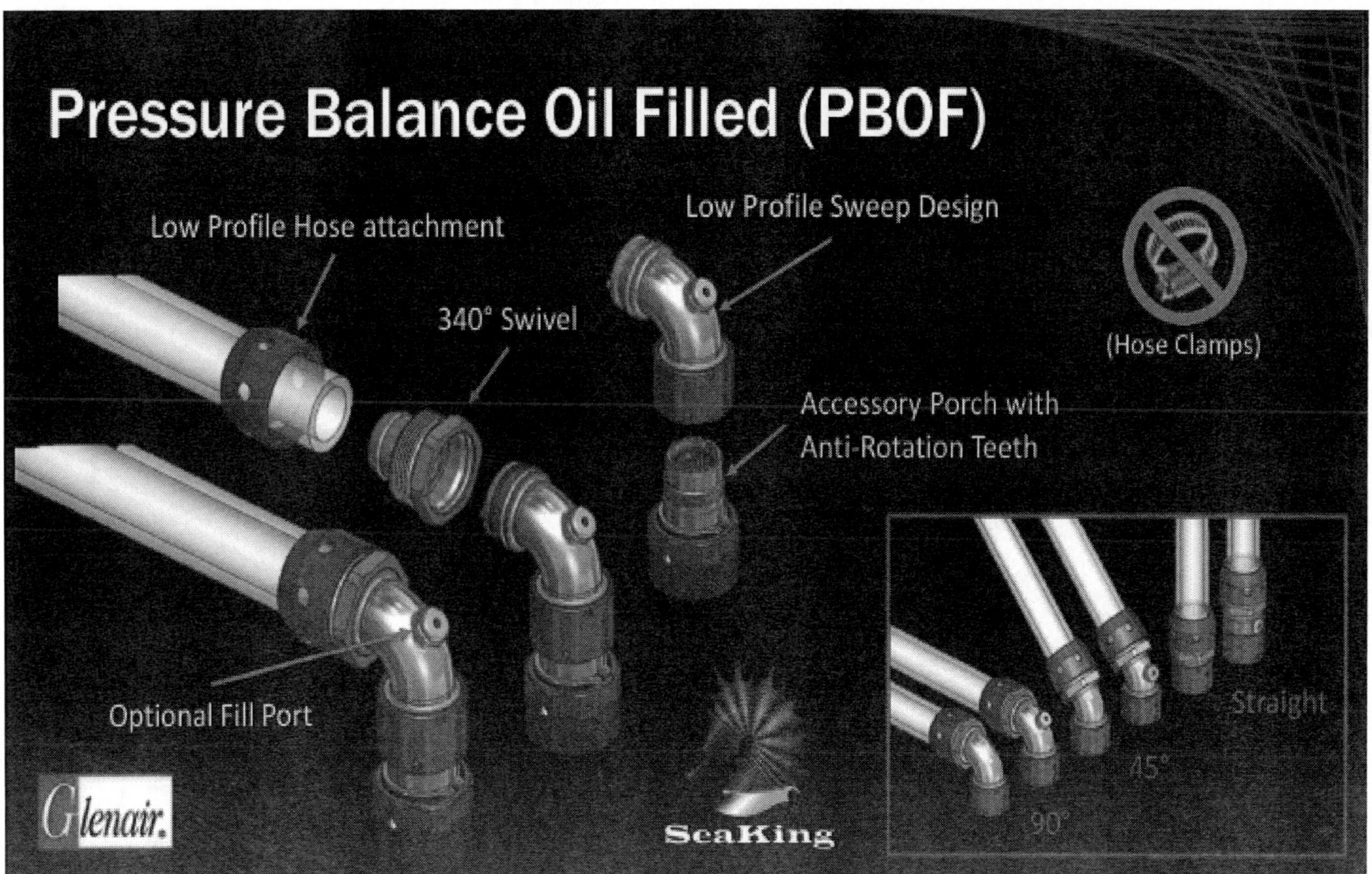
Pressure Balance Oil Filled (PBOF)
Low Profile Hose attachment
Low Profile Sweep Design
340° Swivel
(Hose Clamps)
Accessory Porch with
Anti-Rotation Teeth
Optional Fill Port
Straight
45°
90°
Glenair.
SeaKing

PBOF vs. Molded Cables

PBOF

- Pressure insensitive
- Easily reconfigurable/repairable
- All pressure loading is transferred from cable to connector insert face
- Continues to function if outer cable jacket is compromised
- Visible inspection of cable internals
- Minimal Tension Loads
- Shorter procurement times

Molded Cables

- Pressure loads directly on cable bundle
- Reduced Size
- Cable shares pressure load with molded connector assy
- Potentially lighter weight
- No oil-filling process
- Can be designed to handle high tension loads
- Polyurethane absorbs water overtime
- Potential long lead time for custom cable

Glenair.

Mechanical Performance

Push back the crushing pressure of ocean depth

Balancing material strength with:

- Material compatibility of bulkhead substrate (metallic)
- Material bonding characteristics (molded cable)

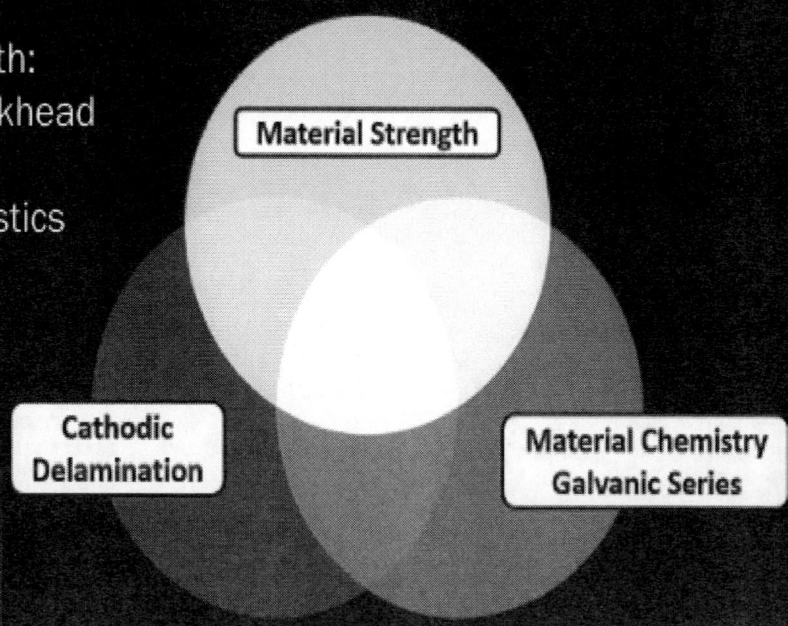

Glenair.

Mechanical Performance

Material Strength - Classical Solutions

- Metals
  - 316L Stainless Steel
  - Titanium Grade 2 & 5, 6AL4V, Grade 23 ELI
- Marine Bronze
  - UNS C46400 – Naval Brass
  - UNS C63000 – Nickel Aluminum Bronze
  - UNS C95500 - Aluminum Bronze
- Polymers
  - GRE – Glass Reinforced Plastic

Glenair.

Galvanic Compatibility

- Galvanic Series lists metals by nobility
- When dissimilar metals are in contact in the presence of an electrolyte (sea water) they have a voltage potential and the less noble metal will exhibit galvanic corrosion.
- Rate of corrosion varies based on the dissimilarity of the metals and areas exposed to the electrolyte

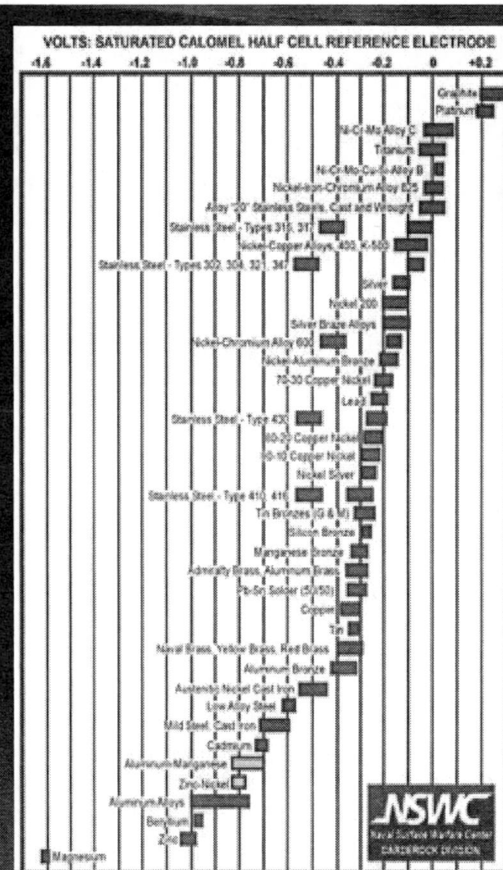

Glenair.

Cathodic Delamination

Erosion of Metal to Polymer Bonds

- In the presence of a galvanic reaction an elevated pH environment can occur
- This pH environment can attack the primer layers and delaminate the bonding material

Cathodic Delamination of Marine Hardware: Causes and Mitigation Strategies

Thomas Ramotowski
NAVSEA Warfare Center Newport
Sensors and Sonar Systems Department (Code 15)
1176 Howell Street, Newport, RI, 02841
ramotowskits@npt.nuwc.navy.mil

Glenair.

Anti-Cathodic Delamination Ring

SeaKing

- Non metallic ring electrically isolates the polymer bond on the plug connector

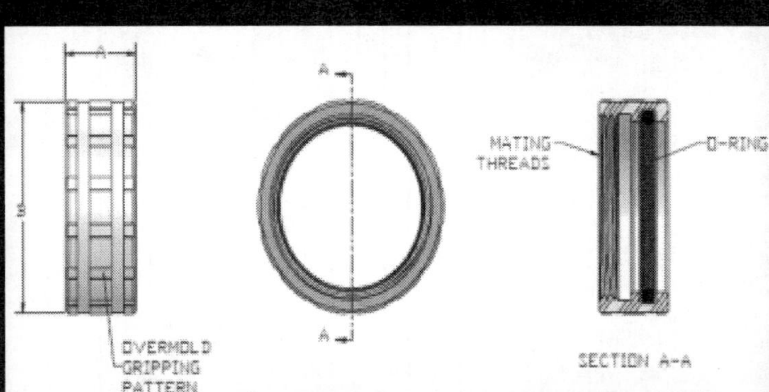

Glenair.

Mechanical Performance

Material Advances

- Fiber Reinforced Composites
  - PEEK – PolyEtherEtherKetone (glass fiber filled)
  - PEK – PolyEtherKetone (glass fiber filled)
  - Ultem / PEI - PolyEtherImide
- Exotic Metals
  - 15-5 PH & 17-4 PH - Martensitic, Chromium-Nickel-Copper, Precipitation-Hardened Stainless Steel
  - 2205 – Duplex Stainless Steel
  - Inconel 625 & X-750 – Nickel-Based Superalloy
  - Monel 400 – Nickel Copper Alloy
  - Nitronic 50 & 60 – High Strength Austenitic Stainless Steel

SeaKing
PEEK

Glenair.

Mechanical Performance

Fiber-Reinforced Composites

- Fiber-reinforced composites (FRC) have found widespread use in interconnect products across a variety of industries with excellent track record of performance and reliability
- FRCs offer a variety of benefits:
  - Reduced weight
  - Reduced magnetic signature
  - Superior corrosion resistance
  - Large operating temperature range
  - Excellent chemical resistance
  - Non-conductive
  - "Galvanically Agnostic"
  - High-speed manufacturing
- Challenges can be mitigated by applying best practices in material selection and component design

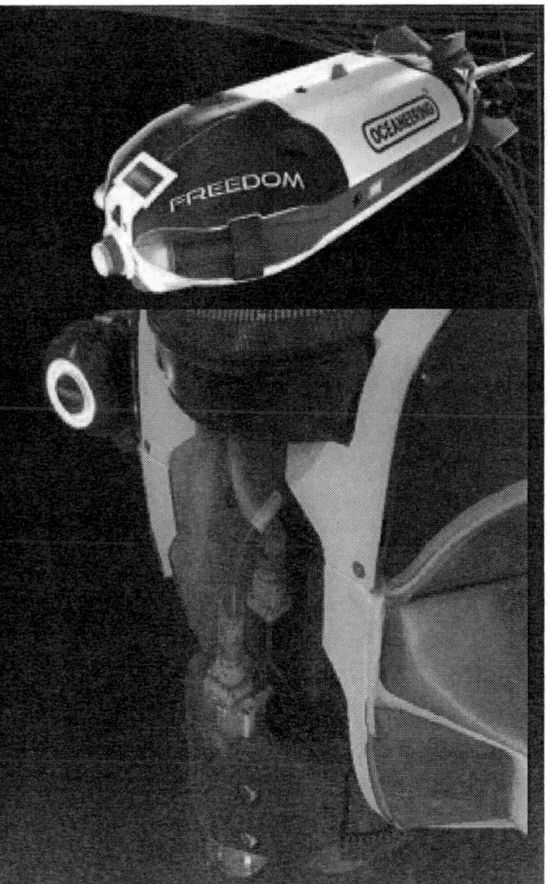

Glenair.

Weight Reduction

- Why is weight reduction important?
  - Reduces power consumption of ROV/AUVs
  - Increases number of vessels that can be used for deployment and intervention
  - Improves ergonomics of diver-manipulated objects
- Connectors are often overlooked as a potential source of weight reduction
- Let's look at a deployment frame with 50 mating pairs of dry-mate connectors
- Weight in Air of a SeaKing 700, Size M, Mated Pair
  - 316 Stainless Steel Shells: (.89 lbs [.40 kg] each) x 50 = 44.3 lbs [20.1 kg]
  - PEEK Shells: (.34 lbs [.15 kg] each) x 50 = 16.9 lbs [7.6 kg]
- A weight savings of 27.4 lbs [12.4 kg], just by changing connector material!

316 Stainless Steel
(Density = .289 lb/in^3)

SeaKing

PEEK – 30% GLASS FILLED
(Density = .055 lb/in^3)

SeaKing PEEK

Glenair.

Electrical Performance

Transmission Information Needed:

- Power - Voltage and Current, AC or DC, Phase
- Signal - Ethernet, USB, Coax, Twisted, Shielded
- Electrical testing performance and to which standard
- Adhere to recommended contact/wire sizing
- Recommend not splitting power across multiple contacts
- Recommend not grounding shields to shell

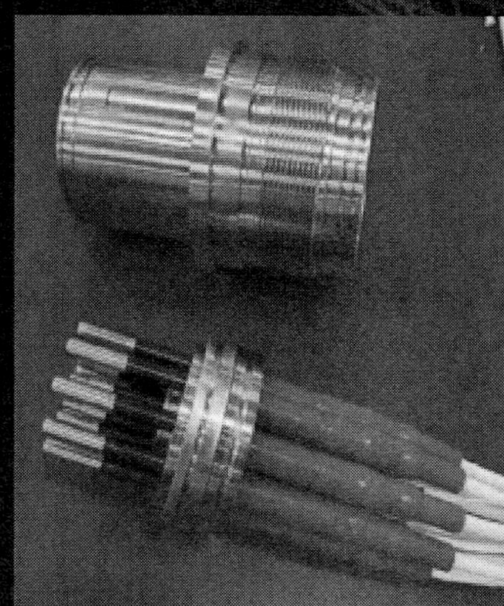

Glenair.

Contact Specifications				
Contact Size	#22	#20	#16	#12
Amps	3	5	10	17
AWG Accomodation	22	20	16	12

SeaKing™ Performance Specifications		
Pressure Rating	Plug: 10,000 psi, mated condition Receptacles: 10,000 psi mated and open face	per ISO 13628-6
Electrical	600 VDC typical	per MIL-STD-202, Method 301
	5 GOhm insulation resistance at 500 VDC	per MIL-STD-202, Method 302

The Advantages of Fiber Optics

Five Reasons Fiber Optics play a growing role in Subsea

1. Reduced Size and Weight: Relative to copper, optical fiber is smaller and lighter
2. EMI Immunity: Light is not subject to electromagnetic interference, RFI or voltage surges.
3. Huge Bandwidth Over Long-Distances: Fiber can transmit infinitely more data than bulky copper cable
4. Spark/Arc Immunity: Fiber media is totally electrically isolated, no risk of spark or short-circuit from a damaged cable
5. Enhanced Security: Light pulses, unlike electrical signals, are almost impossible to intercept or monitor

Glenair.

Optical Performance

Solution Selection

- Matching Existing Fiber
- Single Mode or Multi Mode
- Bend Tolerant Fiber
- Loss Budget
  - Insertion Loss
  - Back Reflection
- In-Board Termination
  - SC, FC, ST, LC or None

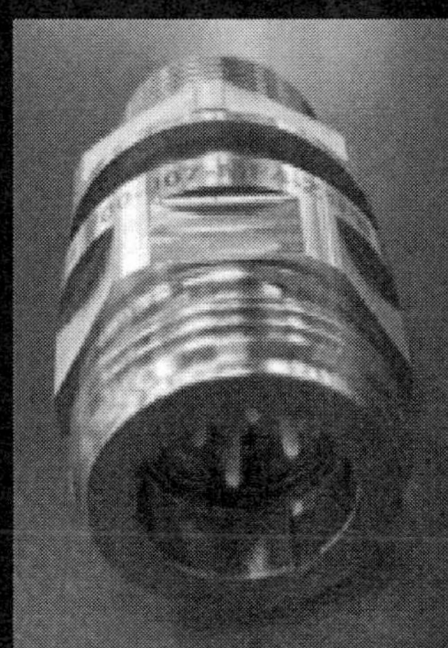

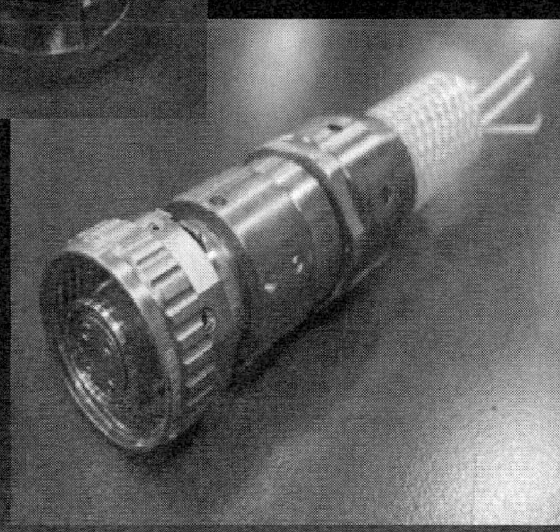

Glenair.

Internal Interface

Wire Routing

- Pigtail Wires
- PC Tails / PCB Interface
- Flex Circuit
- Bulkhead feed-thru / In-Board Disconnect

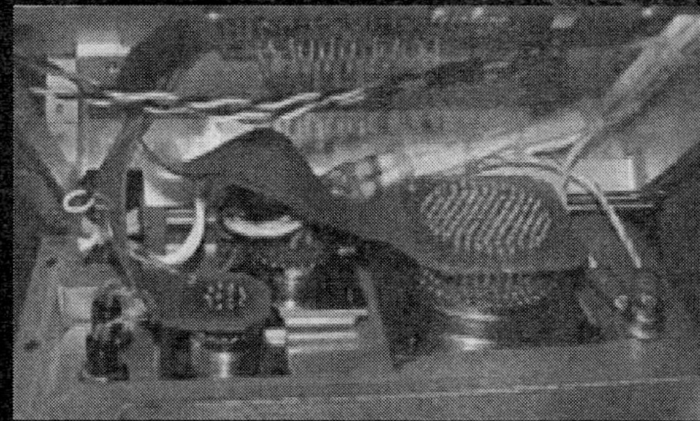

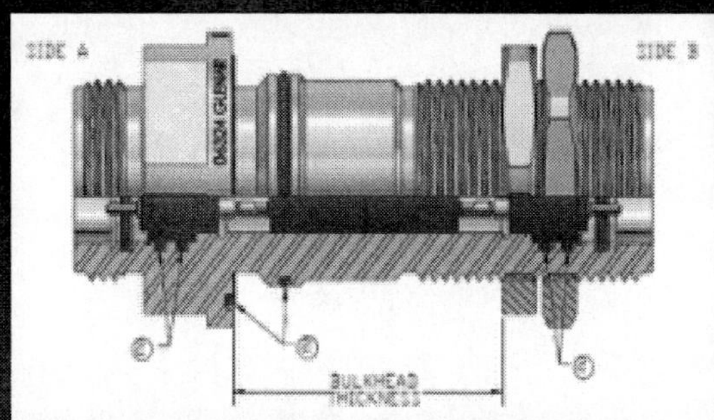

Glenair.

Why Choose Flex?

Flex assembly vs. round wire assemblies

- Less weight and bulk
- Virtually eliminates wiring errors
- Total installed cost is less
- Tightly assembled electronic packages
- Dynamic (Flex in application)
- Direct in 3 different axis

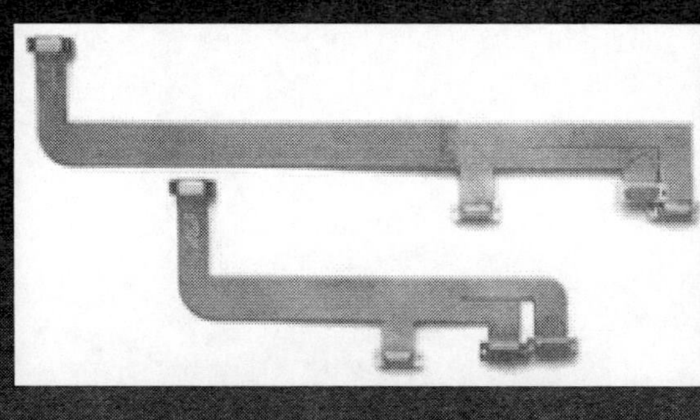

Glenair.

Cost Tradeoffs

Performance and Cost Drivers

- Number of Seals / Environmental Barriers
- Number of Mate / De-Mate Cycles
- Maintenance and Serviceability
- Cost of Failure
- Cost of Replacement

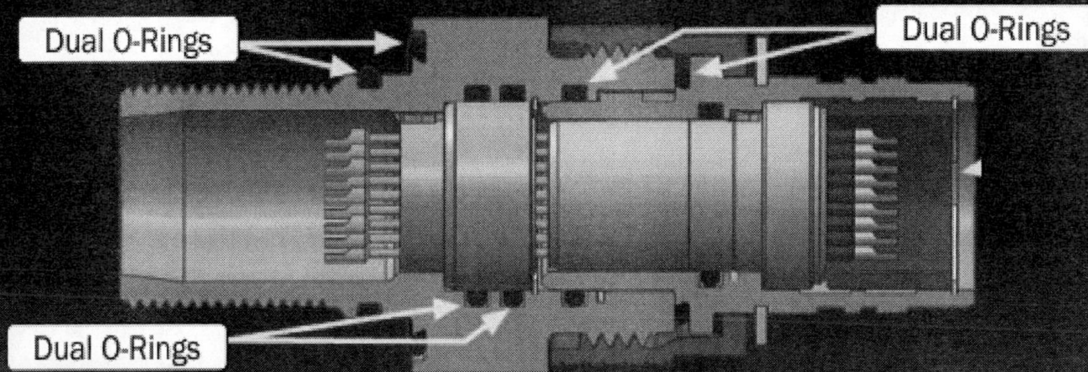

Glenair.

UNDERWATER INTERVENTION

Powered by: WORKBOAT

For further questions and discussions please visit me at Glenair Inc. – BOOTH: UI-3856

SeaKing PEEK

Glenair.

MICRO-PSI

SeaKing

VOYIS

Illuminate the Unknown

Increasing Perception of Inspection-Class ROVs with Optical Sensors

Autonomously capture high-accuracy, 3D digital representations of your environment in real-time while reducing survey costs and carbon footprint.

WELCOME

Agenda

1. About Voyis
2. Inspection Challenges
3. Voyis Optical Sensor
4. Integrated Solutions
5. Case Studies

Russell Giles

4+ years of experience working in various Civil Industries including mining, deep foundations, construction & consulting

About Us

IN A NUTSHELL

We are building hardware that allow you to see the depths like you see the surface.

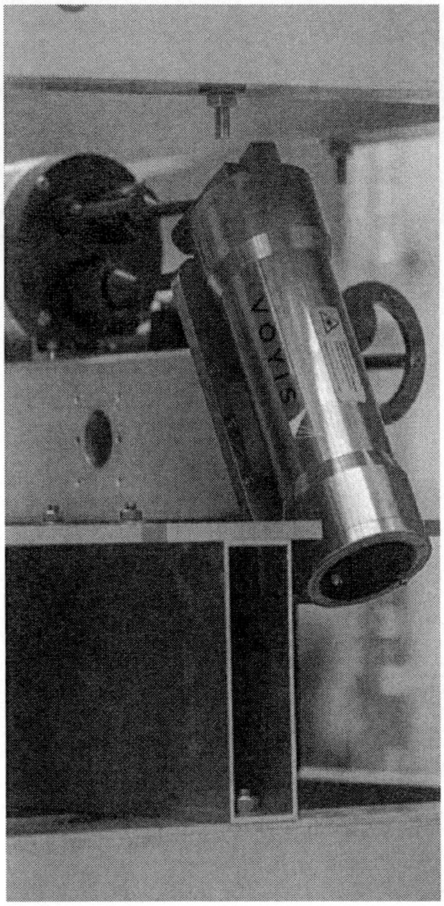

OVERVIEW

Our Products

Standalone

Underwater Laser Scanners
Our high-resolution Insight laser scanners can be used for dynamic or static surveys and inspections.

Imaging Systems
Observer cameras paired with our Nova LED panels for crisp, evenly illuminated stills images.

Solutions

Recon AUV Payloads
A complete, power-efficient payload to expand your platform's capability.

Perception ROV Skids
Configurable inspection-class or WROV skids that suit your needs.

CIVIL INFRASTRUCTURE

Inspection Challenges

- Turbid water and current create difficult inspection conditions
- Gaps in data collections due to current optical sensors
- Poor quality data leads to increased processing time and low accuracy models
- Difficult for repeatable modeling and data collection of critical infrastructure.

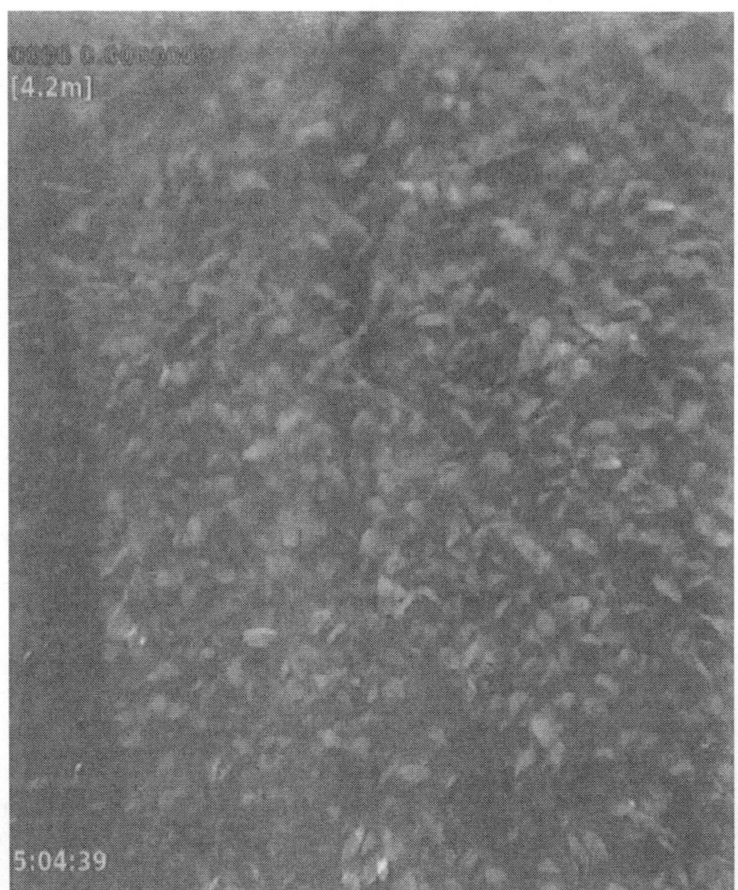

Discovery Systems

Imaging Systems

IMAGE ENHANCEMENTS

Real-Time Image Correction

Raw — Undistort & Debayer — Colour & Lighting

Automated image correction with no input parameters

Captured on a REMUS 600 AUV

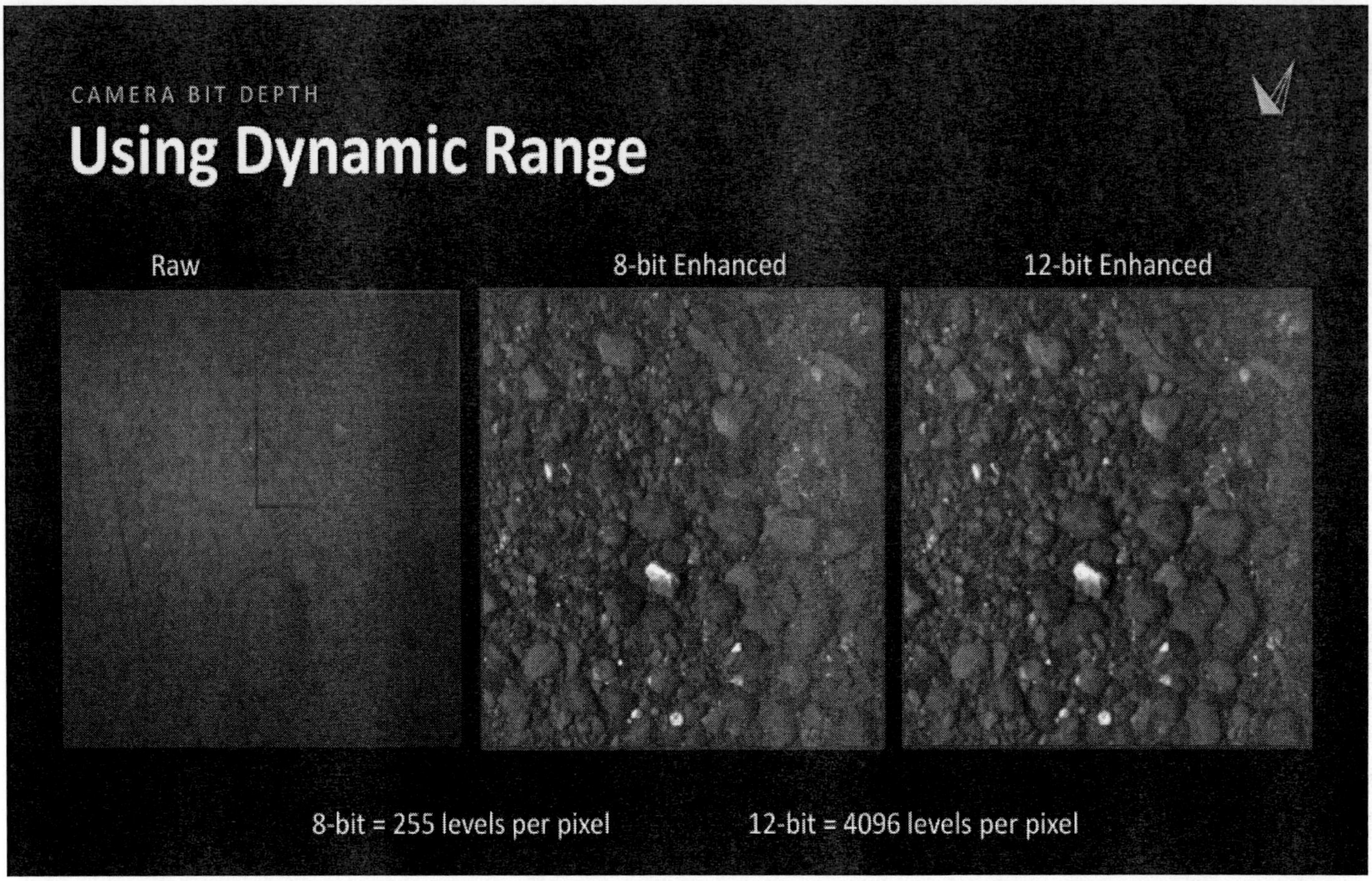
CAMERA BIT DEPTH
Using Dynamic Range
Raw
8-bit Enhanced
12-bit Enhanced
8-bit = 255 levels per pixel
12-bit = 4096 levels per pixel

Discovery & Nova Mini
The first smart ROV Piloting Camera with low latency vehicle piloting video and 3D inspection with powerful edge-computing. Delivers detailed video, simultaneous crisp stills images, and onboard IMU data to support accurate photogrammetry.
Discovery Stereo & Nova Mini
Compact Stereo Camera solution with edge-computing for real-time 3D modelling, supporting inspection of vertical targets and enhancing general visual inspections. Delivering video, stills images, and real-time disparity maps to advance autonomous operations.

DISCOVERY VISION SYSTEMS
Core Advantages
REAL TIME DATA
Low Latency Piloting
Edge Image Correction
High Res Video
Crisp Stills
Onboard Point Clouds
ENABLE AUTONOMY
DDS Architecture
High Quality Data Onboard
Enable Station Keeping and
Autonomous Manipulation
3D MODELING
Onboard IMU
Factory Stereo Calibration
Edge Image Correction with
Imbedded Metadata
Global Shutter Imaging

Video
Discovery Recordings for Inspection Purposes

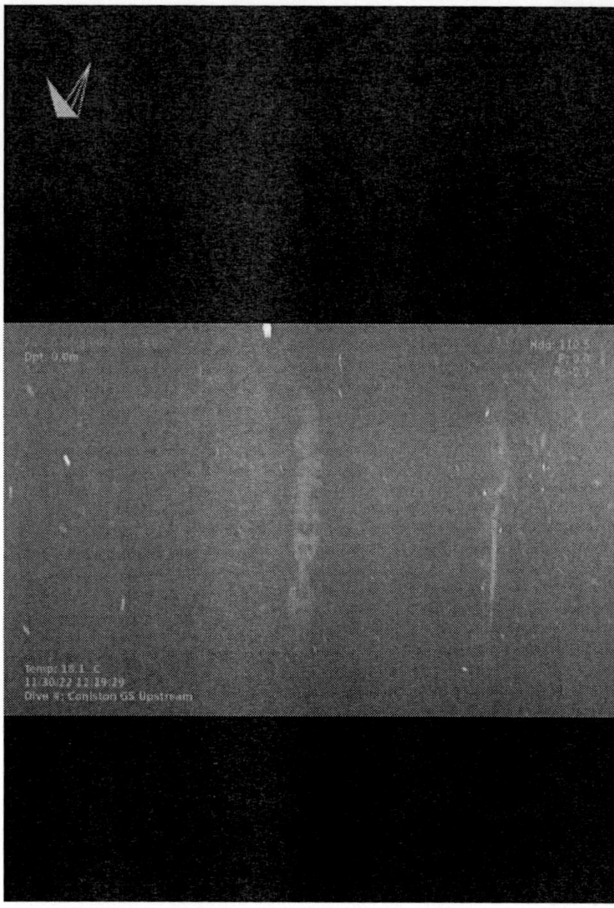
Dpt: 0.0m
Coniston GS Upstream

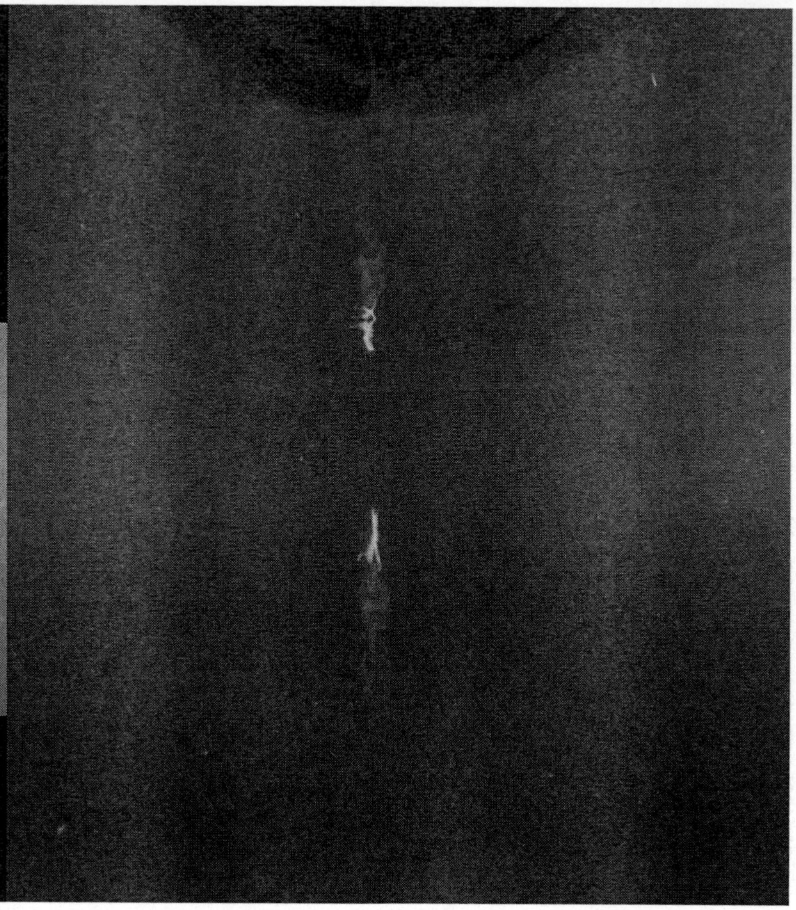

Stills

Discovery Stills for Inspection Purposes

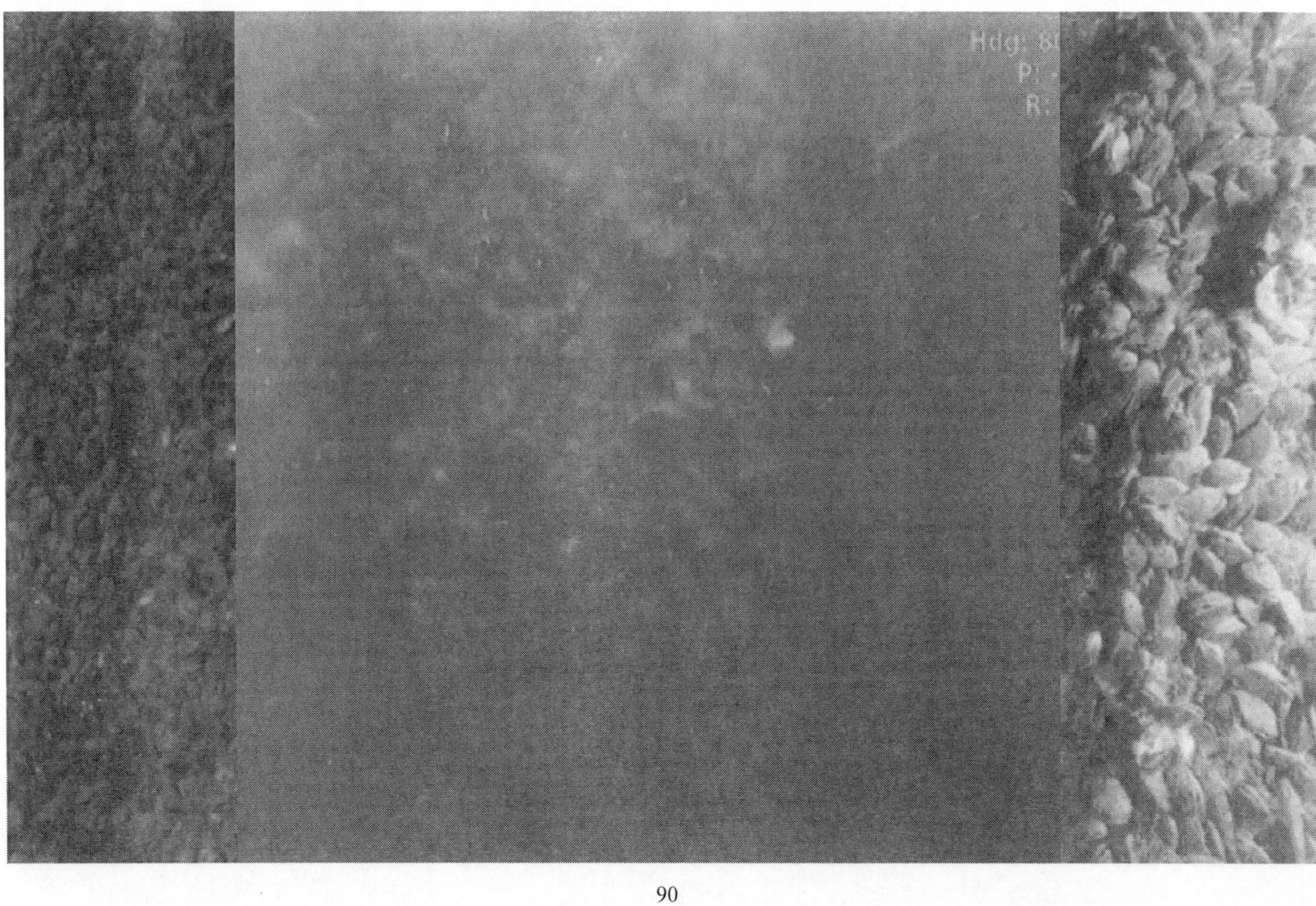
Hdg: 8
P:
R:

3D Models

Created from Discovery Vision Systems

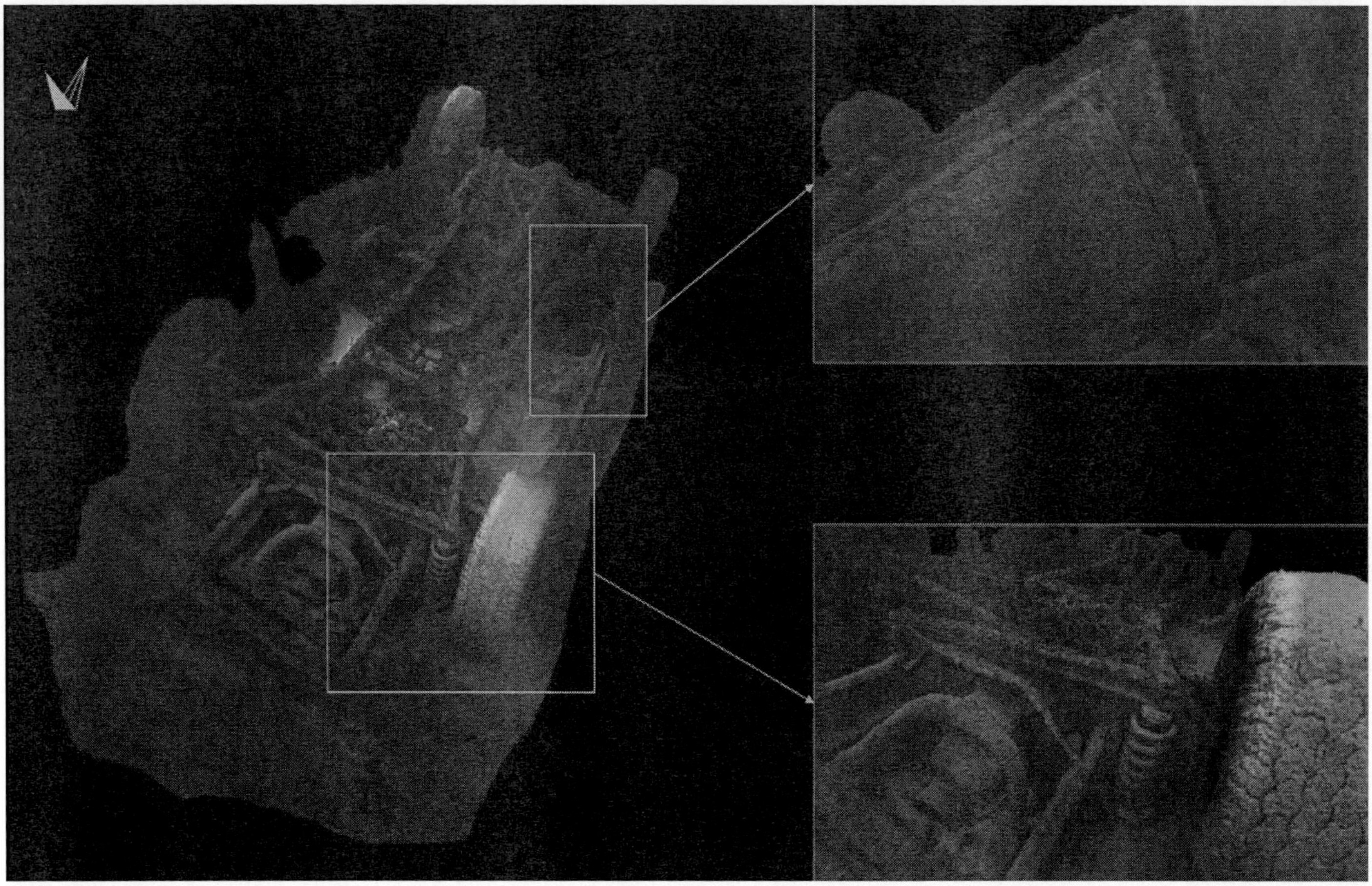

Courtesy of DOF Brasil

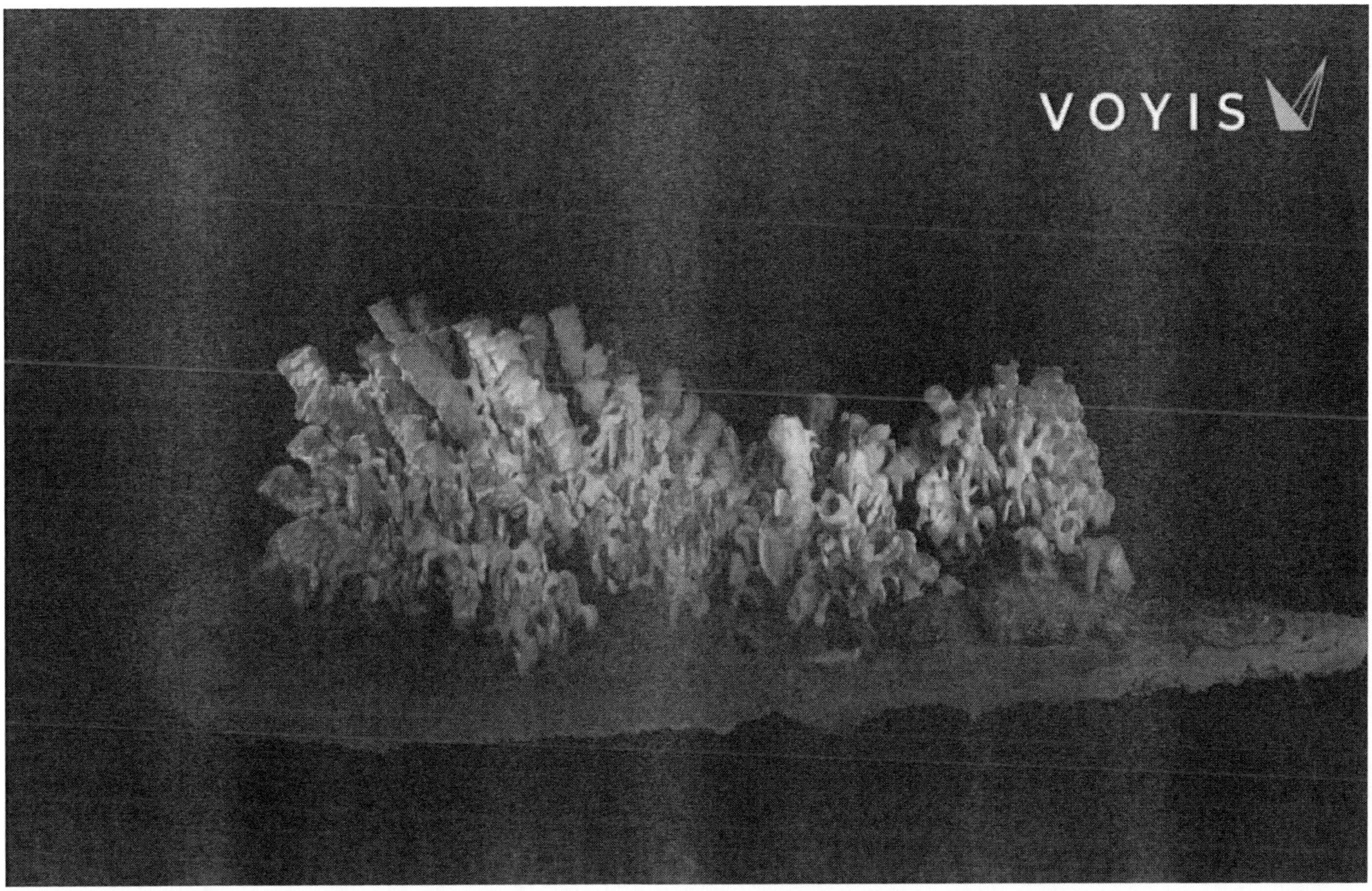

Integrated Solutions

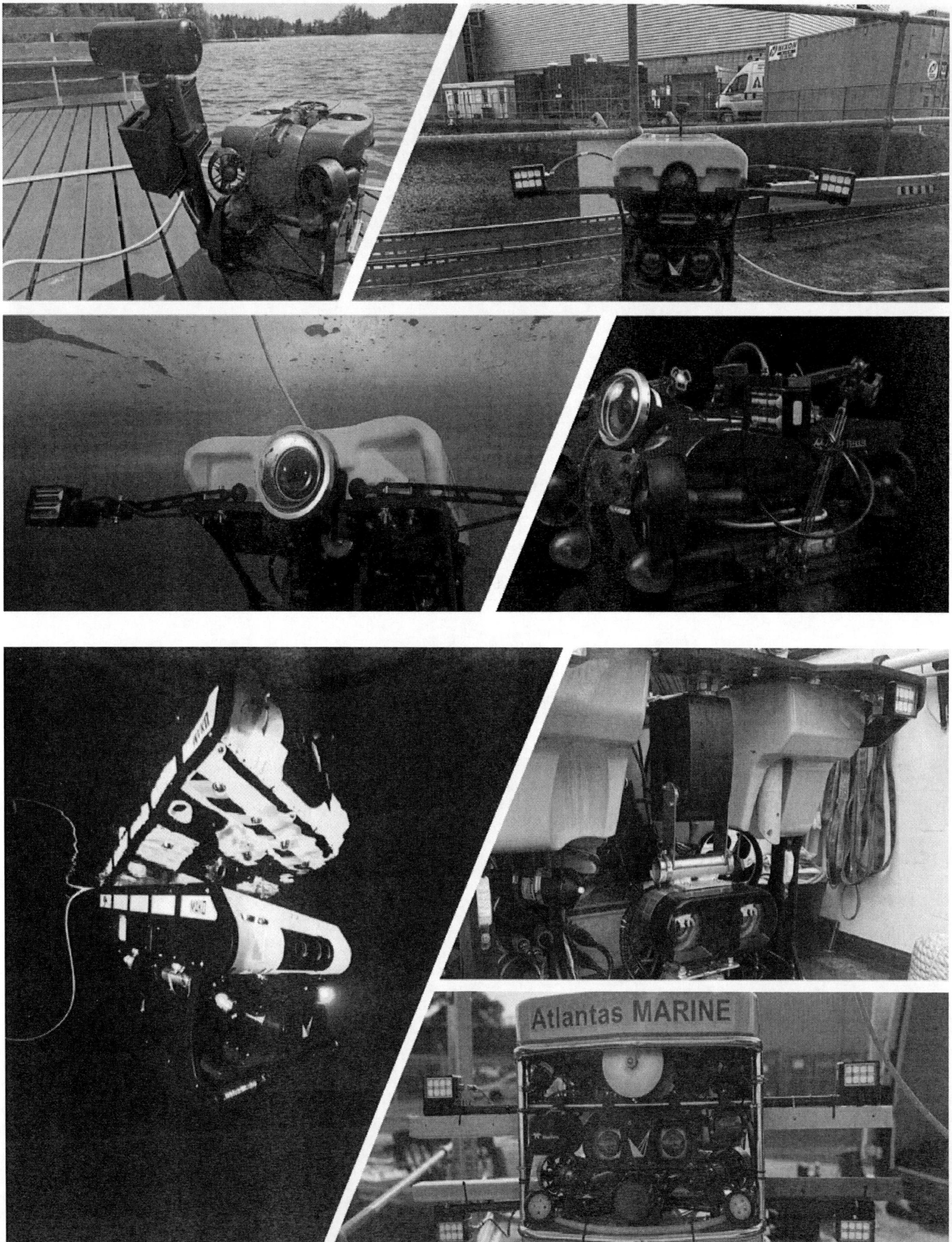
Atlantas MARINE

Case Studies

CASE STUDY

OPG– Auburn Facility

Location:	Auburn Hydro Electric dam
Depth:	8m
Vehicle:	VideoRay Defender ROV
Equipment:	Discovery Mono Camera & Perception Skid
Objectives:	Complete 360 scans of trash racks, stop logs, concrete walls for a condition assessment

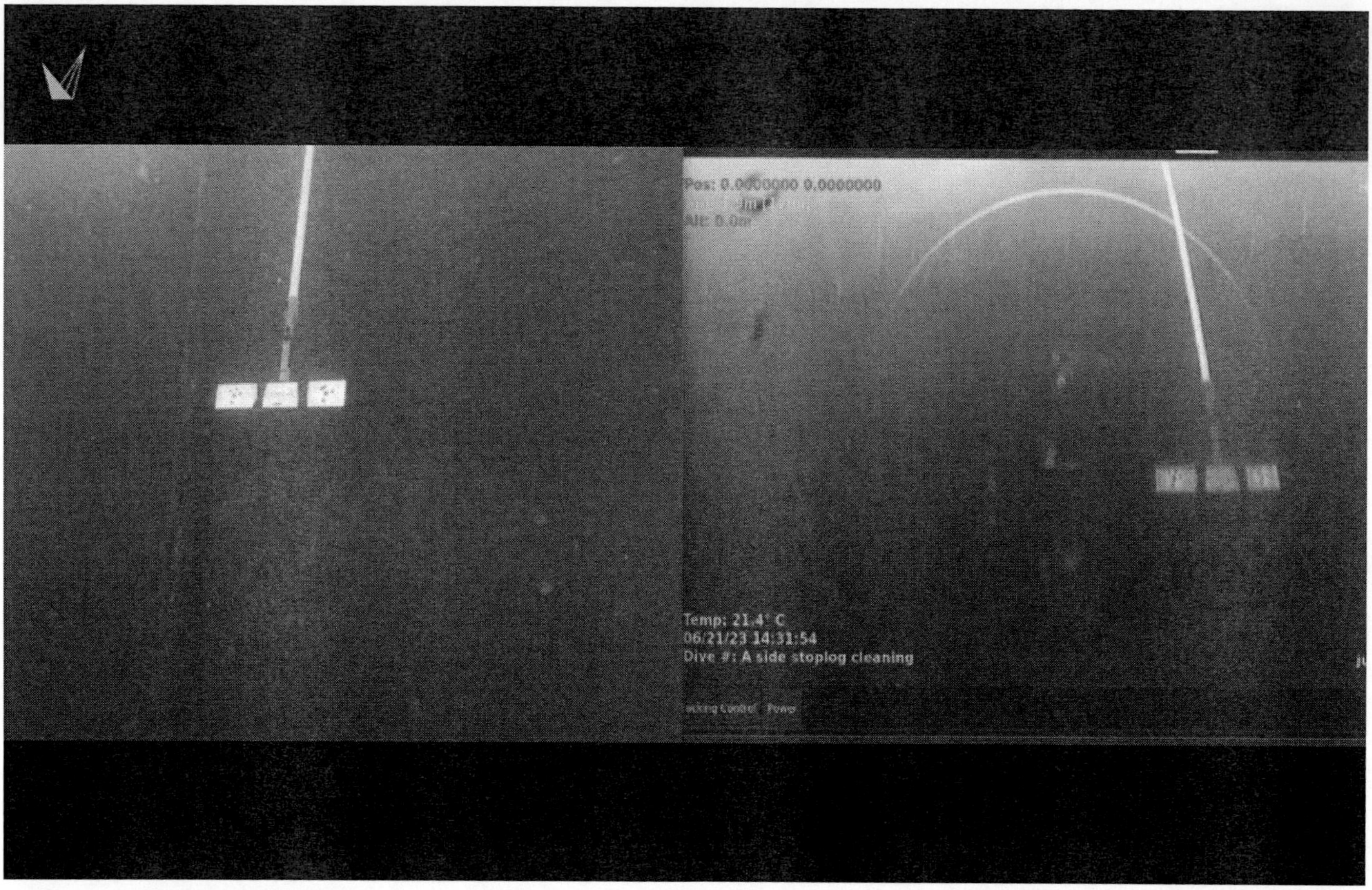
Pos: 0.0000000 0.0000000
Alt: 0.0m
Temp: 21.4° C
06/21/23 14:31:54
Dive #: A side stoplog cleaning

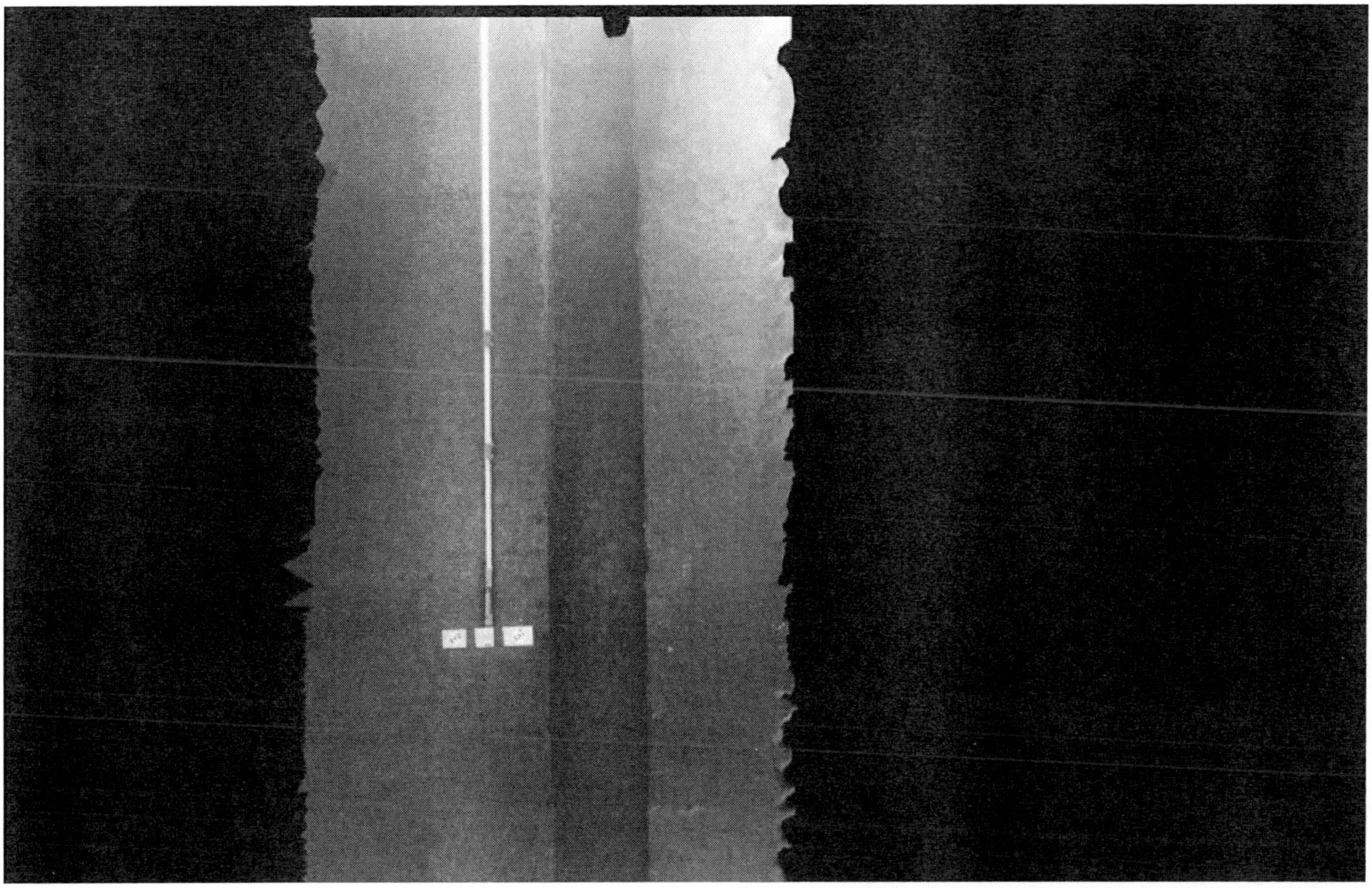

CASE STUDY

OREC Demonstration

Location:	Blyth, UK
Depth:	10m
Vehicle:	VideoRay Defender & Ocean Modules V8
Equipment:	Discovery stereo Camera
Objectives:	Condition assessment of Monopile, anodes & CPS

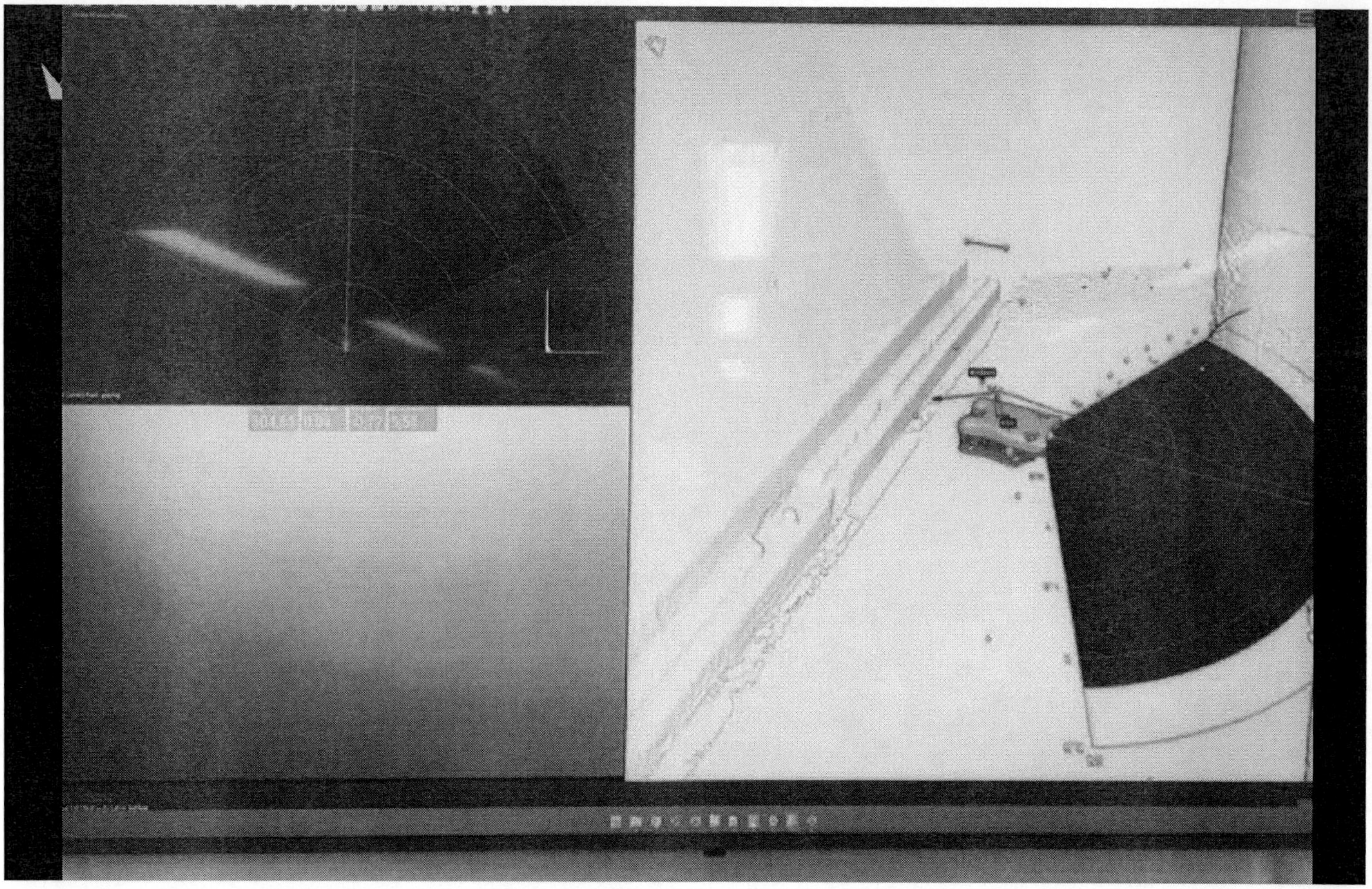

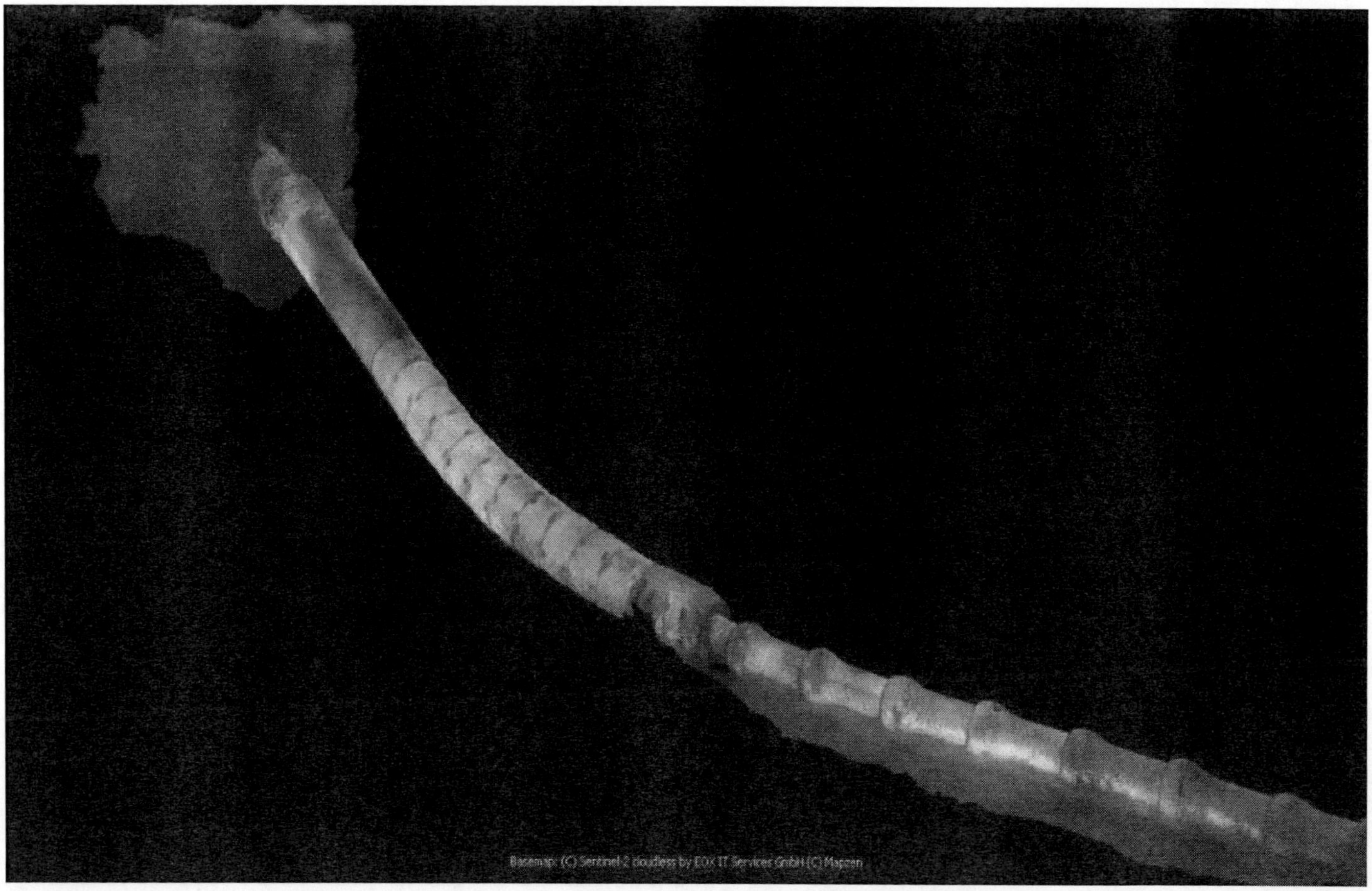
Basemap: (C) Sentinel-2 cloudless by EOX IT Services GmbH (C) Mapzen

THANK YOU

Questions?

Russell Giles

4+ years of experience working in various Civil Industries including mining, deep foundations, construction & consulting

VOYIS

UNDERWATER INTERVENTION

MTS marine technology society
Opportunity runs deep™

Advancing Shallow Water and Surf Zone Operations with the Bayonet AUGV A Technical Overview

David Pearson, CMarTech
Technical Solutions Architect
Greensea IQ

Nov 30, 2023

Greensea IQ

Agenda

- Greensea IQ Update
- Very Shallow Water & Surf Zone Environment
- Traditional Methods in VSW/SZ
- Bayonet AUGVs
- Real-world Applications and Case Studies
- Future Opportunities
- Q&A

Corporate Update

On Sep 1st 2023, Greensea Systems, Inc. (Greensea) merged with Armach Robotics, Inc. (Armach) and Bayonet Ocean Vehicles, Inc. (Bayonet). Armach and Bayonet were both founded by Greensea and had operated as independent companies. Seeing an opportunity to create a greater impact in ocean robotics, the founder and CEO of all three companies, Ben Kinnaman, decided to merge the companies.

This singular entity provides our customers greater efficiency and access to technology and solutions.

3

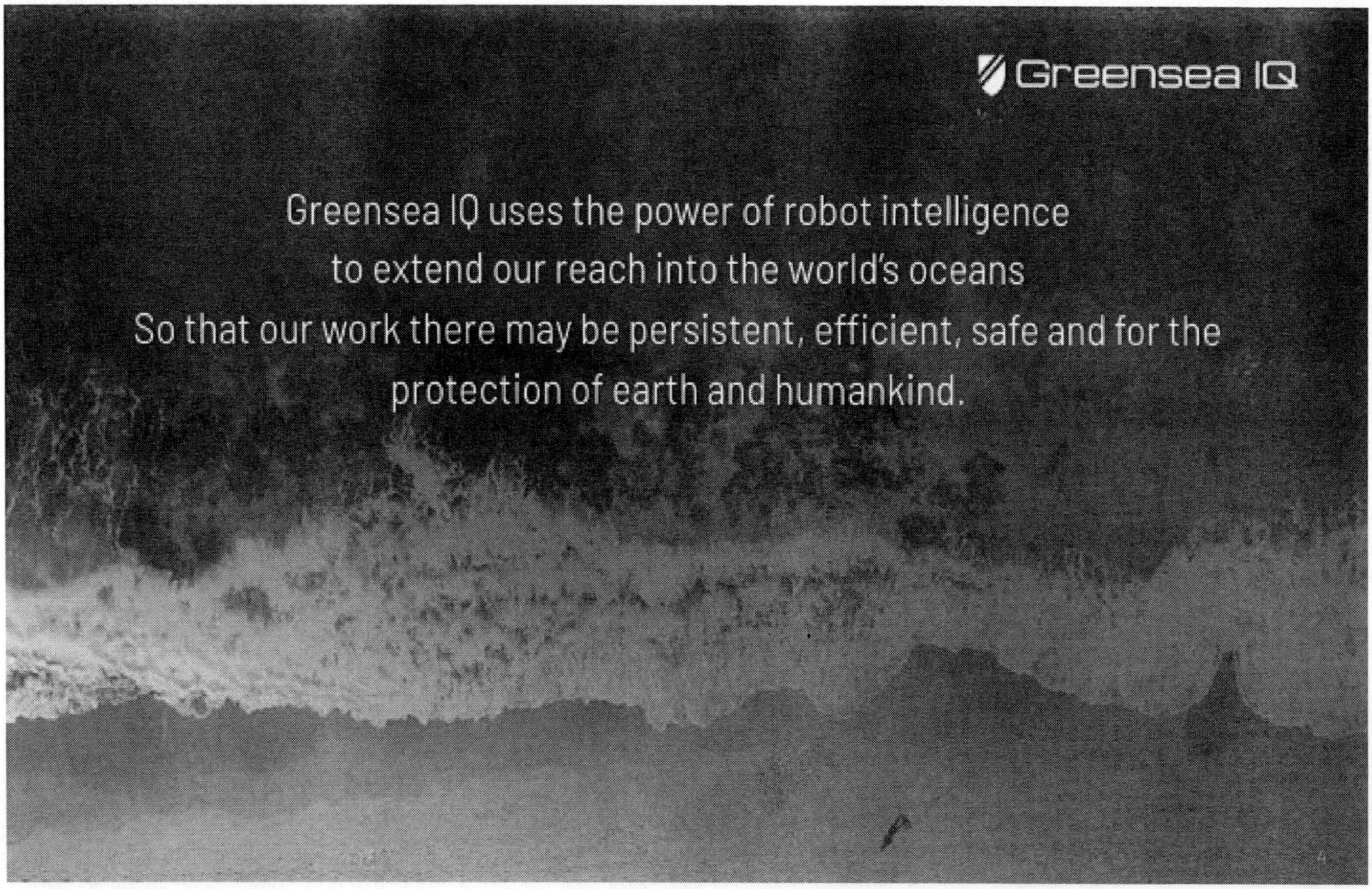

Greensea IQ

KEY MARKETS

Defense

Defending those who defend us.
Pushing the advanced capabilities of OPENSEA into new and existing platforms to create intelligent systems that increase the safety and effectiveness of the operators in the SOF, EOD, and MCM communities.

UXO & EOD

Solving a worldwide problem.
Using UXO industry-led collaboration to field a safer and more rapid clearance survey work solution with Bayonet AUGVs and an APEX sensor sled.

Offshore Renewables

Enabling a new era of sustainable energy.
Offering intelligent integrated systems powered by OPENSEA and connected remotely through Safe C2 to meet the demand for unmanned platforms to perform survey, inspection, and maintenance offshore.

Ship Maintenance & Performance

Reducing carbon emissions and fossil fuel use in maritime transportation.
Intelligence-driven service leveraging advanced robotic technology and hull data collection to deliver an always clean hull through an exclusive service program for vessel owners and operators.

Science

Understanding climate change and its impact on the world's oceans.
Uniting research institutions on vehicle development in the ocean using OPENSEA and GSIQ integrated systems to rapidly advance next-generation vehicle development and deployment.

Greensea IQ

Surf Zone Operations

Environment Complications & Existing Methods Limitations

6

Surf Zone Operations

Greensea IQ

Dynamic Nature: Constantly changing water depths, wave action, and sediment movement make surf zones a challenging environment.

Critical Area: Surf zones are the interface between open ocean and shoreline, crucial for maritime navigation, habitat, and natural resource exploitation.

Technical Complexity: Traditional surveying methods in the surf zone require complex hardware, specially trained crews, and extensive logistical planning.

High Operational Costs: The intricacies of managing heavy equipment and specialized personnel contribute to increased costs for surf zone operations.

Safety Concerns: The variable and often hazardous conditions in the surf zone create additional safety risks for human operators.

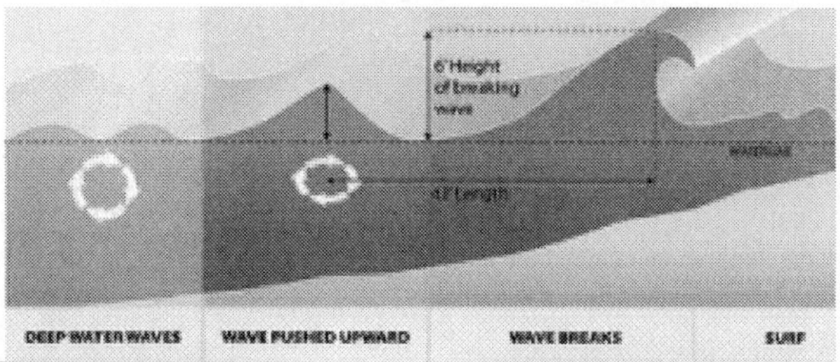

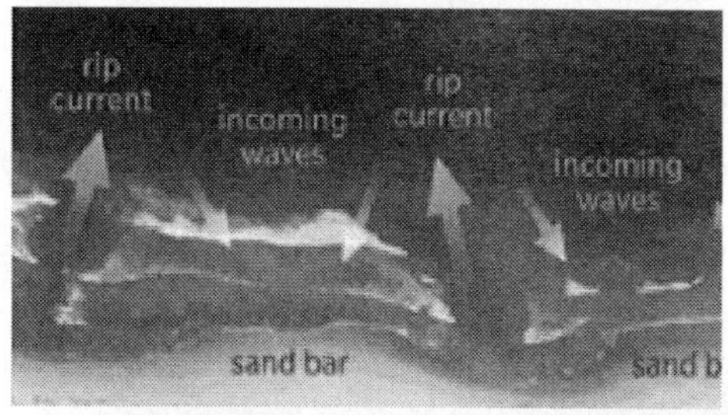

Greensea IQ

Traditional Methods - Manned

Human Factors & Safety
Risks to Equipment: High
Risks to Life: VERY HIGH

System Capabilities
Data Collection Efficiency: MED
Data Quality: MED
System Reliability: MED
System Availability: LOW (weather or personnel limited)

System Costs
Capital Cost: \$\$-\$\$\$\$
Operational Cost: \$\$\$

Difficult to Impossible get sensors near seafloor.

Greensea IQ

Modern Methods - UxV

Human Factors & Safety
Risks to Equipment: MED
Risk to Life: Very LOW

System Capabilities
Data Collection Efficiency: HIGH
Data Quality LOW: MED
System Reliability:MED-HIGH
System Availability: HIGH (weather limited)

System Costs
Capital Cost: $
Operational Cost: $$

Difficult to Impossible get sensors near seafloor.

9

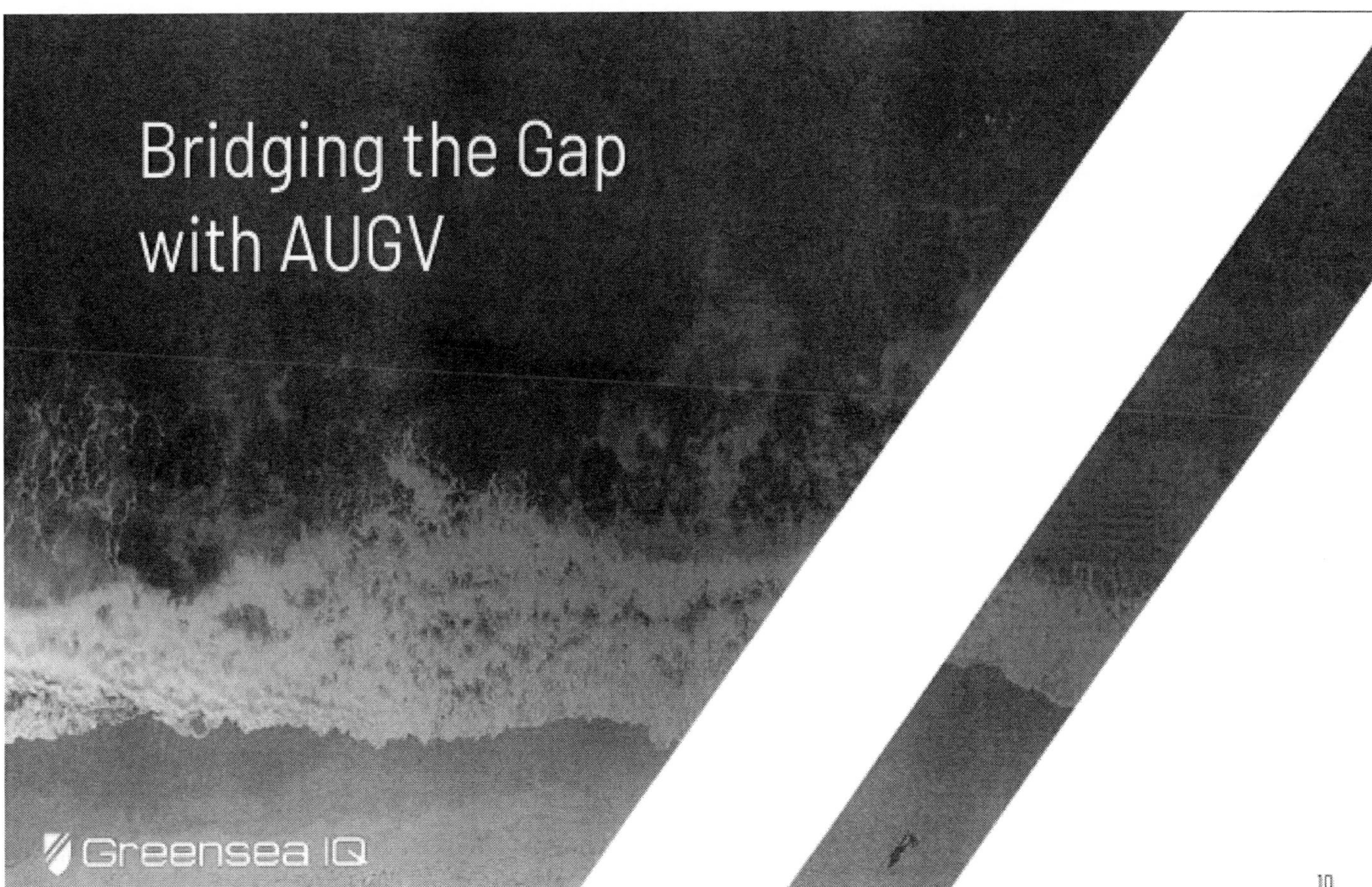

10

Greensea IQ

GSIQ - Bayonet AUGVs

Human Factors & Safety
Risks to Equipment: LOW
Risk to Life: Very LOW

System Capabilities
Data Collection Efficiency: MED
Data Quality: HIGH
System Reliability:HIGH
System Availability: HIGH

System Costs
Capital Cost: $$
Operational Cost: $$

Able to reliably navigate sensors at seafloor in the SZ.

11

Greensea IQ

Bayonet 150, 250 & 350

SPECIFICATIONS	150	250	350
DIMENSIONS	111.8 x 91.4 x 25.4 CM	121.9 x 121.9 x 38.1 CM	182.9 x 152.4 x 45.7 CM
OPEN DECK SPACE	66.0 x 30.5 CM	66.0 x 30.5 CM	152.4 x 71.1 CM
WEIGHT (WITH BATTERIES)	131.5 KG	176.9 KG	272.2 KG
DURATION	~6 HRS	~6 HRS	~4 HRS
RANGE (DRY/WET)	6/8 NM	6/8 NM	4/6 NM
DECK CAPACITY	90.7 KG	113.4 KG	158.8 KG
NAVIGATION POSITIONING	Greensea IQ GS4 INS		
COMMAND AND CONTROL SOFTWARE	OPENSEA, optional Greensea IQ User Interface Software: Workspace, Professional Workspace, or EOD Workspace		

Battery Packs Per System		
Qty 2 + Payload	Qty 2-4 + Payload	Qty 2-8 + Payload
LiFePO4: 1 kWh | Li-ion: coming soon		

Bayonet 150
- Low profile & high stability
- Two-person portability
- Suitable for small boat deployment

Bayonet 250
- Carries heavy payloads
- Two-person portability

Bayonet 350
- Carries the heaviest payloads
- Can be disassembled for transport

12

Greensea IQ

Operational Success

B250 AUGV Accolades to Date, Commercial & Defense

Operational Availability:
> 50 Days of Operation

- EOD/MCM & UXO System Development and Defense Sales.
- Commercial Operations we provide full-service Robot-as-a-Service or Lease.

Surfzone Capability:
>6 ft Breaking Waves

- Survived Tens of Thousands breaking waves under operation.
- Consistently proven stability and navigational accuracy.

System Reliability:
> 200 Nmi Distance Traveled

- Top speed of 1.5 knots and nominal speed of 1 knot.
- Rapidly swappable batteries to keep operational cadence.

System Mobility & Rapid Deployment

- Proven launch and recovery from boat and beach deployments.
- High system mobility using pickup trucks.

13

Real-world Applications

Case Studies

Greensea IQ

Beach & Surfzone Survey

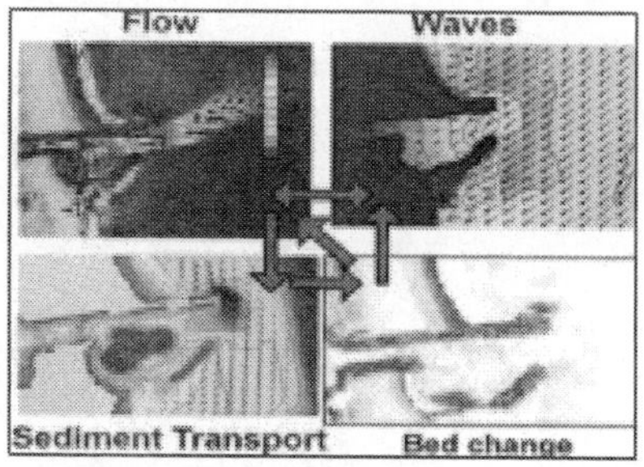

Situation

Monitoring of Sediment transport and Coastal Erosion is critical to support coastal economies and infrastructure.

Complication

Climate change is directly contributing to humanitarian emergencies from heatwaves, wildfires, floods, tropical storms and hurricanes and they are increasing in scale, frequency and intensity.

Solution == More Data

Proactive vs Reactive?

High Veracity In-situ observations coupled with Predictive Models for Storm Damage can be leveraged to actively manage and optimize the design and operation of coastal infrastructure.

15

Greensea IQ

RaaS Operational Efficacy

Amphibious Beach Reclamation Survey

- ➢ Shippable Around the World in a 48" x 40" crate
- ➢ Rapid Site Mobilization
- ➢ Increased Speed & Accuracy of Data Collection
- ➢ Proven & Intuitive Operator Interface
- ➢ Increased Safety & Operational Weather Windows

- ✓ Reduction of Operational Logistics
- ✓ Minimizes Data Validation & Verification
- ✓ Minimize Training
- ✓ Flexible Contracting Terms

16

Greensea IQ

EOD/MCM & UXO & Buried Objects

Situation

The surf zone represents a particularly challenging environment for Explosive Ordnance Disposal (EOD), Mine Countermeasures (MCM), and Unexploded Ordnance (UXO) operations.

Complication

Traditional methods of detecting and neutralizing ordinances in the surf zone are not only costly but also often yield incomplete or unreliable results, raising safety and security concerns.

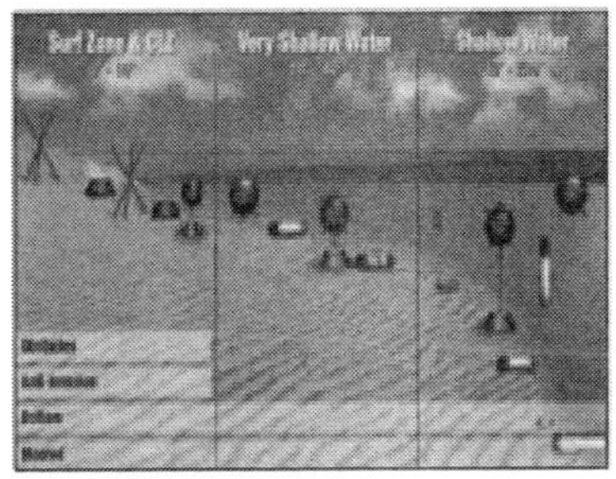

Solution == Safe Autonomy

Proactive, Safe, and Trusted autonomous systems can be leveraged to autonomously detect localize and clear underwater hazards effectively at a safe standoff distance.

Image from "21st Century U.S. Navy Mine Warfare". Program Executive Office Littoral and Mine Warfare – Expeditionary Warfare Directorate – US Navy 2009

17

Greensea IQ

EOD/MCM - Intelligent Autonomy & Perception

Equipped with OPENSEA Edge the B250 can be deployed for EOD applications. Assisting operators with AI enhanced Sonar & Optical Object Detectors to Classify and intervene on Mine-Like-Objects.

Mapping all relevant sonar contacts for Obstacle Detection and Avoidance.

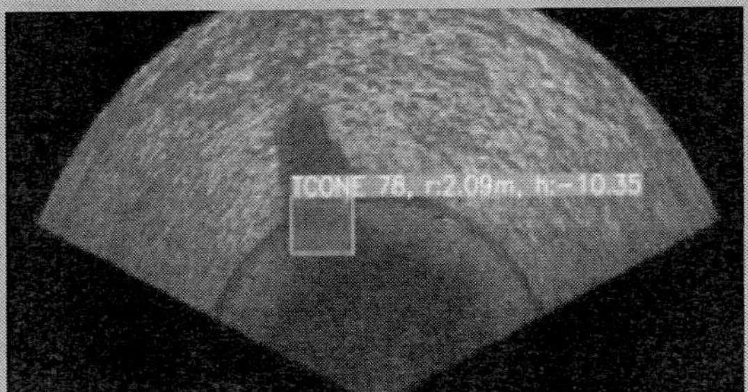

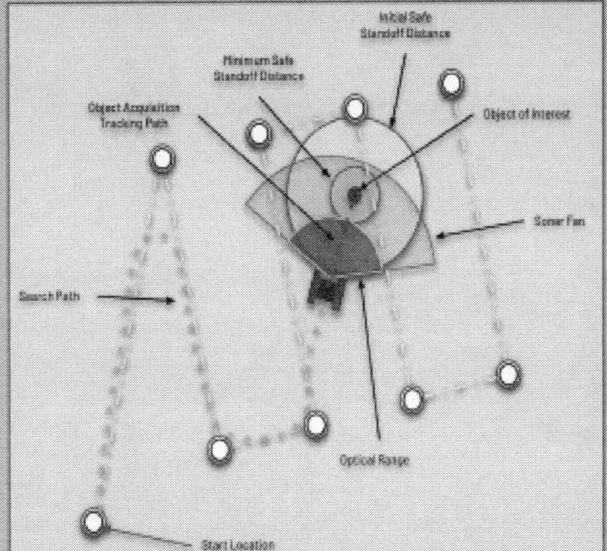

Greensea IQ

UXO & Buried Objects

Demonstrating the power and flexibility of OPENSEA to rapidly integrate and deploy a new sensor suite for the task of identifying buried UXO in the surf zone while showcasing the capabilities of the B250 AUGV.

- Operating over 8 days
- Covering >18 square kilometers
- Depths of up to 4 meters
- Swell of 1.5 meters
- Breaking wave upto 2 meters

Bayonet 250 provided a stable and reliable platform for conducting the survey. This included a towed sled equipped with White River Technology's (WRT) Marine APEX 3D EMI sensor, which was seamlessly integrated into OPENSEA on the B250.

Greensea IQ

Marine APEX 3D EMI - DAGCAP Accredited

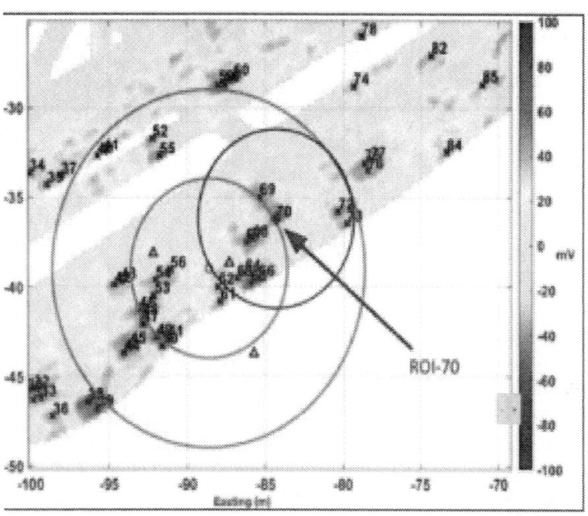

EMI Heat Map (GEOTIFF) & Target IDs

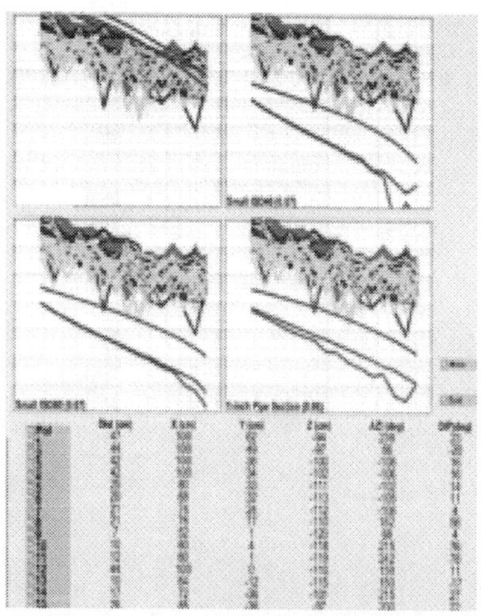

Classification Regression Models of known Targets

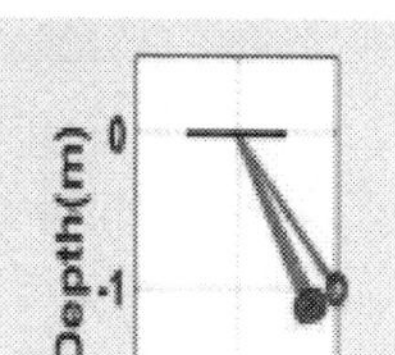

Percentage TOI Detected	100%
Clutter Rejection Achieved	>89%
TOI* Correctly Classified	96%
Northing/Easting Error (cm)	6.0 cm
Depth Error (cm)	2.5 cm

Depth and Position Measurements

20

Greensea IQ

Opportunity - Subsea Export Cables Survey

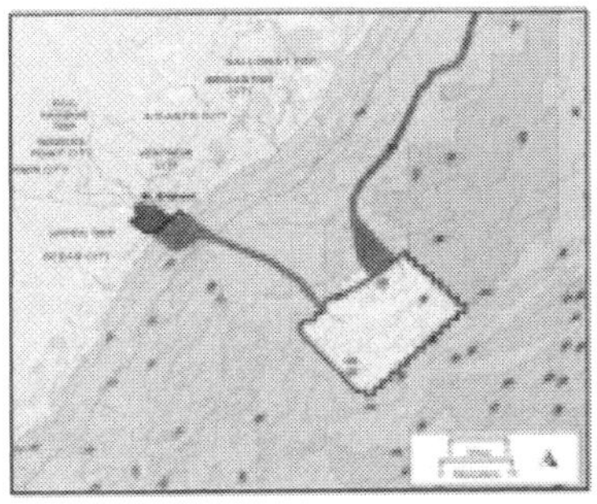

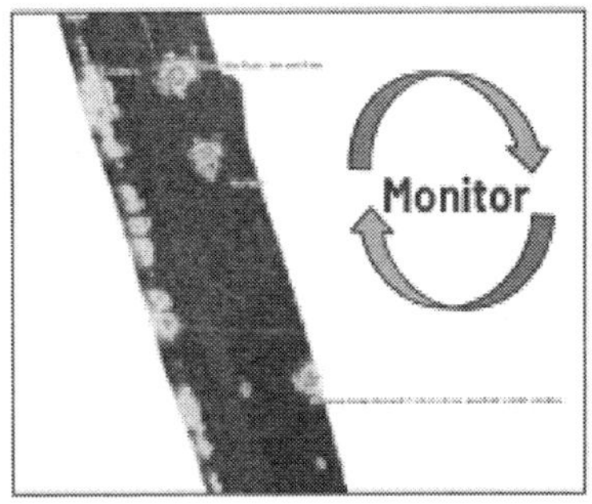

Situation

Increasing offshore wind energy is a positive shift toward sustainable power sources.

However, 80% of European offshore wind farm insurance claims have been cable-related.

Complication

Post-installation surveys and monitoring is an increasing concern for field longevity.

Complete survey thru the surf-zone with one system is near impossible today.

Solution == Lower Barrier for Operational Logistics

Proactive Not Reactive.

With frequent post installation route surveys, we can enhance field longevity, ensure infrastructure security and maintain energy independence.

Applications Overview

- **EOD MCM**
- **UxO & Buried Object Detection**
- **Dredge & Nurishment Operations**
- **Geophysical Survey**
- **Hydrographic Survey**
- **Environmental Monitoring**
  - Surf Zone
  - Estuaries
- **Custom Applications**
  - Aquaculture
  - Post-Storm Survey
  - Radiation Survey
  - If your sensor fits, Bayonet can collect your data!

Greensea IQ

COMPREHENSIVE SOLUTIONS

EXTENSIVE RANGE OF PAYLOADS | ADVANCING SHALLOW WATER & SURFZONE OPS

23

Greensea IQ
Intelligent Ocean Solutions

THANK YOU

+1-802-434-6033
sales@GreenseaIQ.com
GreenseaIQ.com

24

Greensea IQ

Citations

[1] A. Spicer Bak, P. Durkin, B. Bruder, M. J. Saenz, M. F. Forte, and K. L. Brodie, "Amphibious Uncrewed Ground Vehicle for Coastal Surfzone Survey," *Journal of Surveying Engineering-asce*, vol. 149, no. 4, Nov. 2023, doi: https://doi.org/10.1061/jsued2.sueng-1381.

[2] H. Francis and P. Traykovski, "Development of a Highly Portable Unmanned Surface Vehicle for Surf Zone Bathymetric Surveying," *Journal of Coastal Research*, vol. 37, no. 5, 2021, doi: https://doi.org/10.2112/jcoastres-d-20-00143.1.

[3] Holly Beck, "How To Read Waves: Lesson 1 – Types of Breaks - Surf With Amigas," *Surf with Amigas*, May 19, 2016. https://surfwithamigas.com/how-to-read-waves-lesson-1/

[4] M. Barnes, "CRAB 2.0 – DLBA's Modern Take on a Coastal Research Classic," *dlba-inc.com*, Jun. 26, 2023. https://dlba-inc.com/news-awards/crab-2-0-dlbas-modern-take-on-a-coastal-research-classic/ (accessed Sep. 29, 2023).

[5] Wang Dandi, S. Xing, Y. He, J.-Y. Yu, Q. Xu, and P. Li, "Evaluation of a New Lightweight UAV-Borne Topo-Bathymetric LiDAR for Shallow Water Bathymetry and Object Detection," *Sensors*, vol. 22, no. 4, pp. 1379–1379, Feb. 2022, doi: https://doi.org/10.3390/s22041379.

25

Greensea IQ

Backup Slides

26

Greensea IQ

WRT mAPEX Specs

Sensor Array:

- 3 Pulsed EM Transmitters
- 6 Triaxial High-Gain Receivers
- Wet-mateable connectors / cables

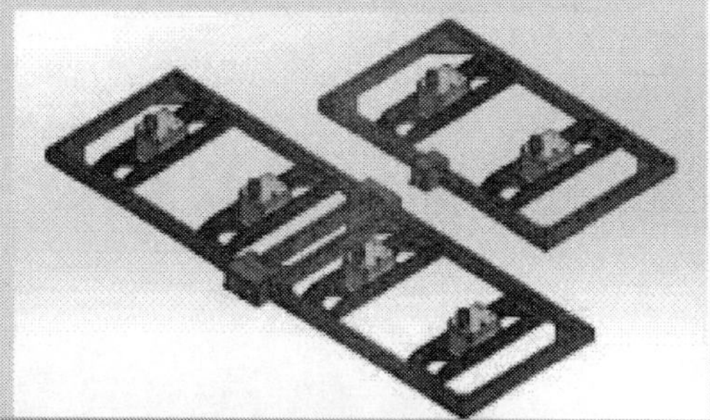

Feature	Specifications	Notes
	Towed Sensor Array	
Dimensions (LxWxH)	122.0cm X 160.0cm X 12.8cm	48"L X 63"W X 5"H
Weight	29.3 kg (64.4 lbs)	sensor array only
	Electronics Pressure Vessel	
Dimensions (DxL)	20.5cm X 46.0cm (8" X 18.1")	
Weight	9.25 kg (20.4 lbs)	
Depth Rating	100 ft seawater	
Power Specifications	22 A-hrs, Approx. 100 W	
Power Supply	2 x 13.2VDC LiFeO4 Batteries	Lithium iron supplies
Endurance	4 hours, 10 line-km	depends on conditions
Sampling Rate	10 Hz (full 3D sample)	
Data Interface	10BASE-T Ethernet (UDP)	
Locating Accuracy	Approx. 2% Depth (5cm XYZ to 2m)	AGC to 11 x Tgt. Diam.
Data Work Products	Geotiff, KML, Target Decision List	
Software	EMClass v3.16 (DAGCAP approved)	

Percentage TOI Detected	100%
Clutter Rejection Achieved	>89%
TOI* Correctly Classified	96%
Northing/Easting Error (cm)	6.0 cm
Depth Error (cm)	2.5 cm

27

Greensea IQ

B-250 Size & Weight

- LxW: 48" X 48"
- Height:
  - 11" Wheel
  - 15" Deck & Tracks
- Chassis Platform
  - 18" Width Clearance
  - 10" Ground Clearance
  - 12" X 26" Deck Plate
- Base Vehicle: 300 lbs (dry), 160 lbs (wet)
  - 250 lbs of Payload
  - 370 lbf drawbar (tow force)

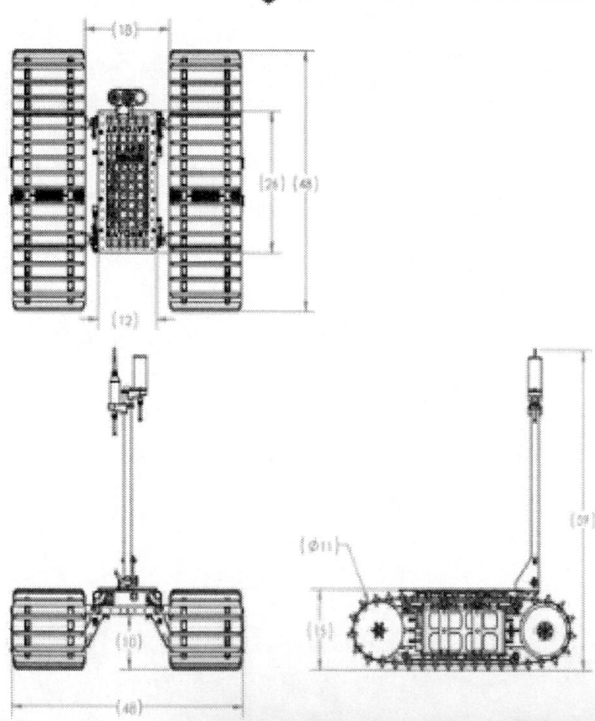

28

Innovative Strategies Toward
Preparing a Workforce
for the Blue Economy
Hans W. Van Sumeren
UNDERWATER
INTERVENTION
Powered by WORKBOAT
NOVEMBER 30, 2023
Northwestern
Michigan College

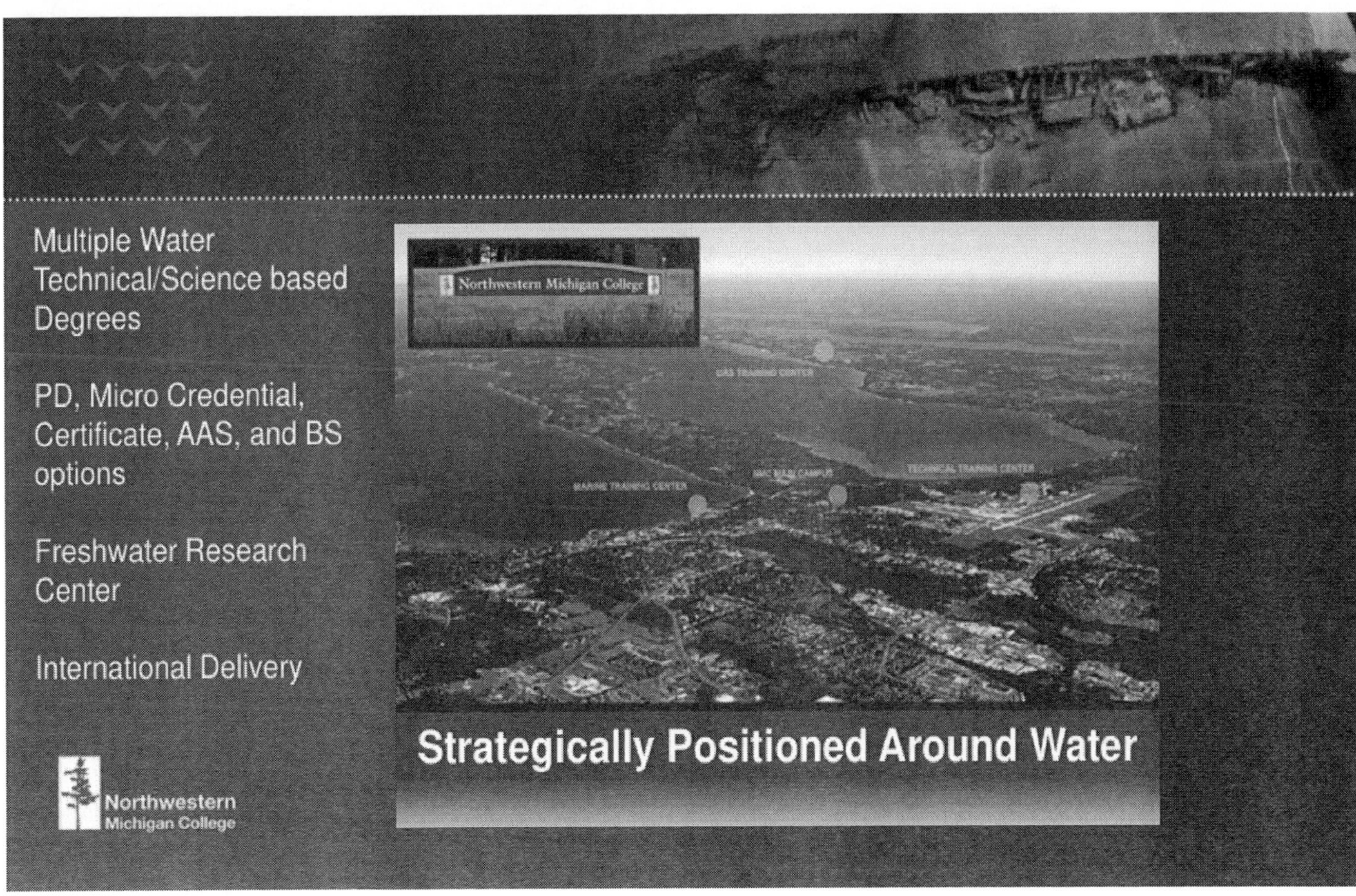
Multiple Water
Technical/Science based
Degrees
PD, Micro Credential,
Certificate, AAS, and BS
options
Freshwater Research
Center
International Delivery
Northwestern
Michigan College
Northwestern Michigan College
MARINE TRAINING CENTER
TECHNICAL TRAINING CENTER
Strategically Positioned Around Water

Distinctive Programming supporting global markets of need

Vision:

National leader in academic, professional development, and technical training for the geospatial/hydrospatial environments

Northwestern
Michigan College

Future Focused Education toward Innovation and Transformation

Traditional – Who We Have Been	Future – Who We Are Becoming
• Place Based	• Content Based
• Time Based	• Competency Based
• Knowledge Transfer	• Synthesis and Application
• Public Good	• Private Good
• Long Development Cycles	• Short Development Cycles
• Risk Averse	• Risk-taking
• Reactive	• Proactive
• Access	• Success

Northwestern
Michigan College

Key metrics for success

- Development of standards and competencies for a technical workforce
- Development of micro-credentials and stackable certificates
- Development/Delivery of PD programs for upskilling current workforce
- Strong connections with industry through advisory boards and connected training

Northwestern
Michigan College

Key Challenges

- Enrollment and attraction of new learners
- Addressing the technical workforce challenge – 10:1
- Job Classification and titles – lack of awareness
- Traditional pockets within the institution remain
- Defining the competencies for Microcredentials

Northwestern
Michigan College

Microcredentials

WORKFORCE DEVELOPMENT

What is needed?

- Innovative and flexible pathways
- Recognition of skills - microcredentials
- Partnerships

Northwestern Michigan College

Ability Levels	
Foundational (Learning focus)	Validated Experience: • 30 – 40 hours • Fundamentals of the technology • Beginning concepts • Uses/application • Basic knowledge demonstrated
Intermediate (Learning and competency demonstration/practicum)	Validated Experience: • Months/crosswalk • Technology applications • Intermediate concepts • More complex uses • Intermediate level knowledge demonstrated • PD at work and application
Advanced (Competency mastery through demonstration)	Validated Experience: • Practical experience (months to years) • Capstone Experiences – demonstration of abilities O Platforms O Sensors O Data

Northwestern Michigan College

- Fill Gaps
- Multiple pathways
- Flexible timeframes
- Fosters equity and inclusivity

Marine Technology (BSMT)
Mapping Micro-credentials to curriculum

Northwestern Michigan College

Technical Specialties

- ➢Sonar Systems
- ➢Data Processing
- ➢ROV/AUV Platforms
- ➢Fluid Power Systems
- ➢Project Management
- ➢Internship
- ➢GIS
- ➢CADD
- ➢Field Surveying Fundamentals
- ➢UAS Surveying Applications
- ➢Robotics/programming
- ➢Electronics

- 100% employment
- Industry driven curriculum
- Comprehensive technical skills
- Strong industry support
- Industry standards and credentials of value
- Diverse career opportunities
- Limited Competition

Water Quality and Environmental Technology
Credentialing the Technician

Northwestern Michigan College

Technical Specialties

- ➢ Site Assessment and Remediation
- ➢Groundwater Monitoring
- ➢Aquifer Sampling
- ➢HAZWOPER 40
- ➢Water Quality
- ➢Internship
- ➢GIS
- ➢CADD
- ➢Field Surveying Fundamentals
- ➢UAS Surveying Applications
- ➢Electronics
- ➢Bio, Chem, Geology

- Local, regional, national, international needs
- Industry driven curriculum
- Broad technical and science based skills
- Strong industry support
- Aging workforce and strong technician need
- Little competition

Strategy: Academic On-ramps
Northwestern
Michigan College
Survey Technician Training Program
PROGRAM OVERVIEW
This survey technician training program is intended to provide the fundamental skills required to become a survey technician and is designed to provide an overview of the technologies and their applications used by the survey and civil engineering industry.
TECHNICIAL TRAINING PROGRAM
> Fundamental skills development
> 80 hours of instruction
> Application focused
> Academic on-ramp
> Tuition cost covered
5-WEEK PROGRAM
Begins MARCH 2023
Completion APRIL 2023
Northwestern Michigan College
LOCATIONS
CORE SKILL SETS
Participants will receive by completion of the program:
> Survey credential (NMC)
> Job readiness training
> OHSA safety
> Construction site orientation
> Construction drawings
> Surveying basics
> Survey methods
> Equipment identification
> Interpreting maps and contour lines
PROGRAM REQUIREMENTS
18 years or older
High school or GED
Interest in a high-wage, rewarding career
Leica Geosystems
• Industry partners
• Attracting new talent
• Easing transition to academic program

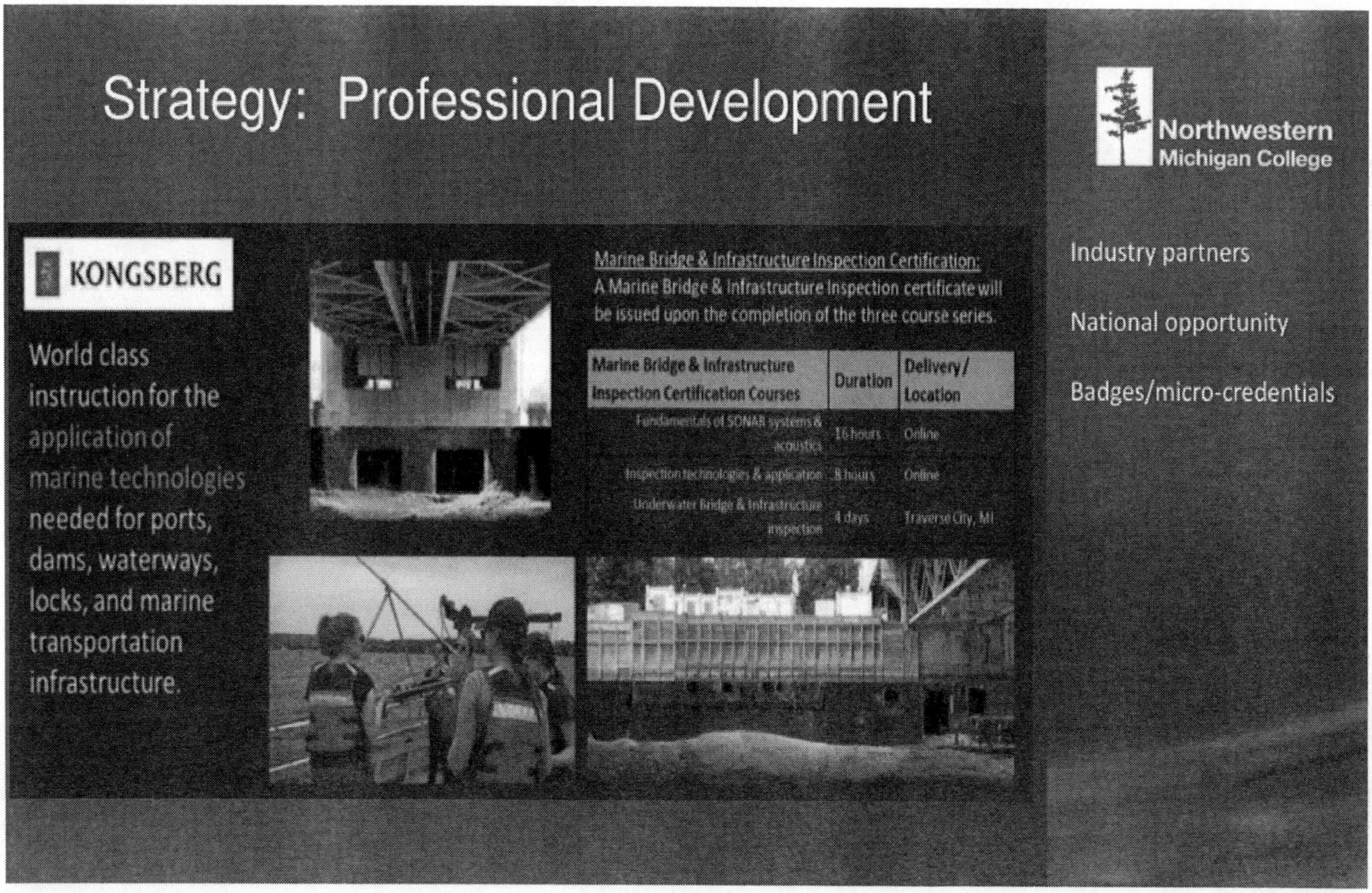
Strategy: Professional Development
Northwestern
Michigan College
KONGSBERG
World class instruction for the application of marine technologies needed for ports, dams, waterways, locks, and marine transportation infrastructure.
Marine Bridge & Infrastructure Inspection Certification:
A Marine Bridge & Infrastructure Inspection certificate will be issued upon the completion of the three course series.
Marine Bridge & Infrastructure Inspection Certification Courses | Duration | Delivery/ Location
Fundamentals of SONAR systems & acoustics | 16 hours | Online
Inspection technologies & application | 8 hours | Online
Underwater Bridge & Infrastructure inspection | 4 days | Traverse City, MI
Industry partners
National opportunity
Badges/micro-credentials

Freshwater Research & Innovation Center
Coming January 2027

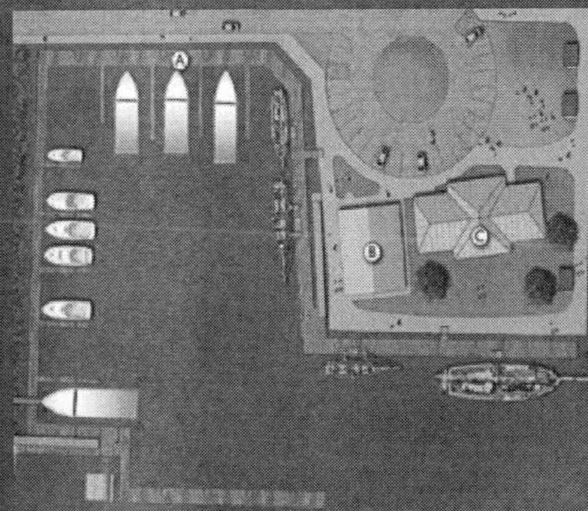

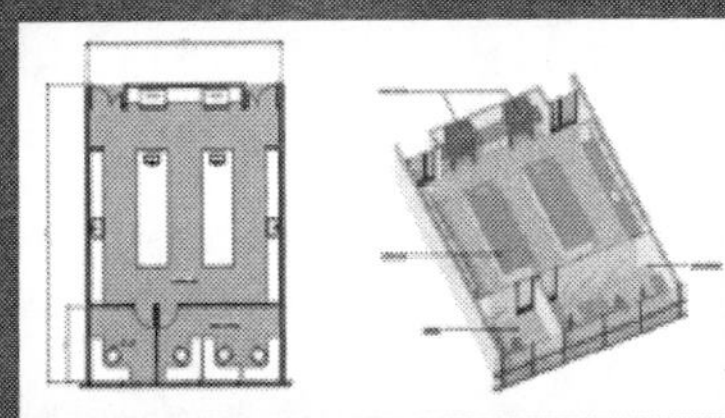

Northwestern
Michigan College

Key Takeaways

- Competency and content based learning can be applied multiple ways – all at the same time
- Applications and synthesis of learning modules
- Flexibility
- Industry partnerships
- Softskills and team based field experiences

Northwestern
Michigan College

Future
Focused
Education

Northwestern
Michigan College
Thank you.
hvansumeren@nmc.edu
WWW.NMC.EDU/WATER

SUNFISH

Field Demonstrations of Precision Mapping and Control by the SUNFISH® AUV in Support of Marine Archaeology

Kristof Richmond[1], Tyler Haden[1], Vickie Siegel[1], Miriam Alexander[1], Jason Gulley[2], Tamara Adame[3], Taylor Heaton[4], Kelly Monteleone[5], Rosita Worl[5]

[1]Sunfish, Inc. [2]University of South Florida [3]ProTec Dive Centers Institute [4]Cornell University [5]Sealaska Heritage

Our Submerged Past

Exploration Objectives:

- Submerged ancient caves
- Migration to the Americas
- Indigenous oral history

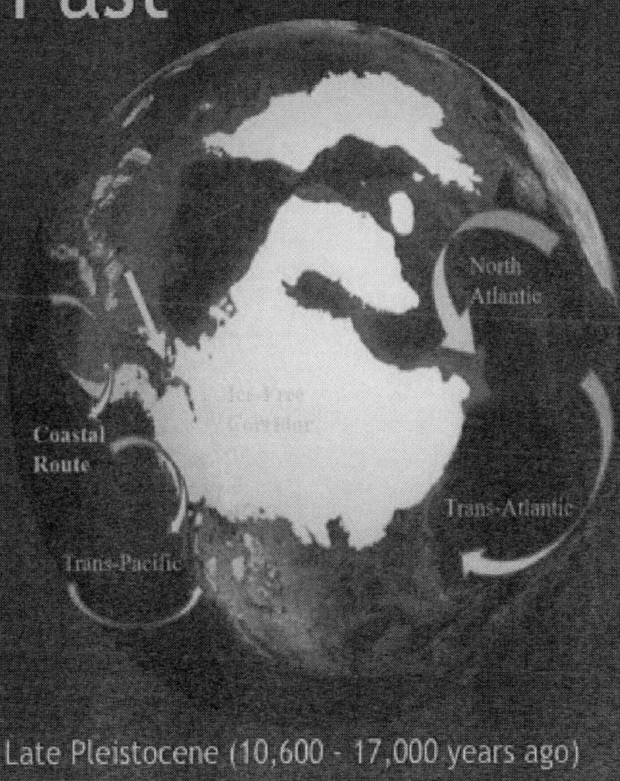

Late Pleistocene (10,600 - 17,000 years ago)

Tech to do Marine Archeology

UUV

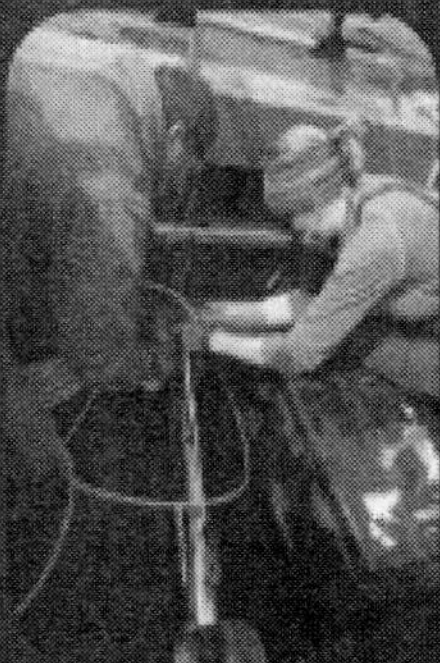

Archeology

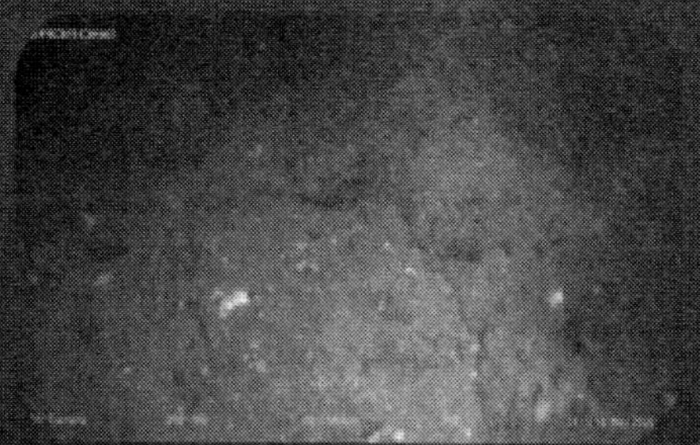

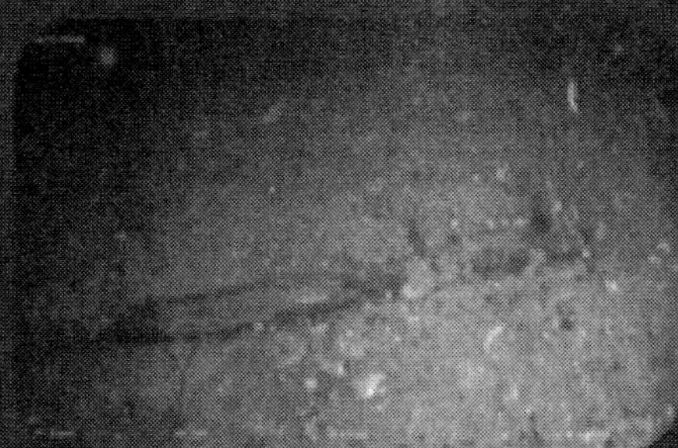

Exploration Plan

Year 1

- Find Ancient Caves
- Use ROV and SS

Year 2

- Enter Ancient Caves
- Use **SUNFISH AUV**
- Find History

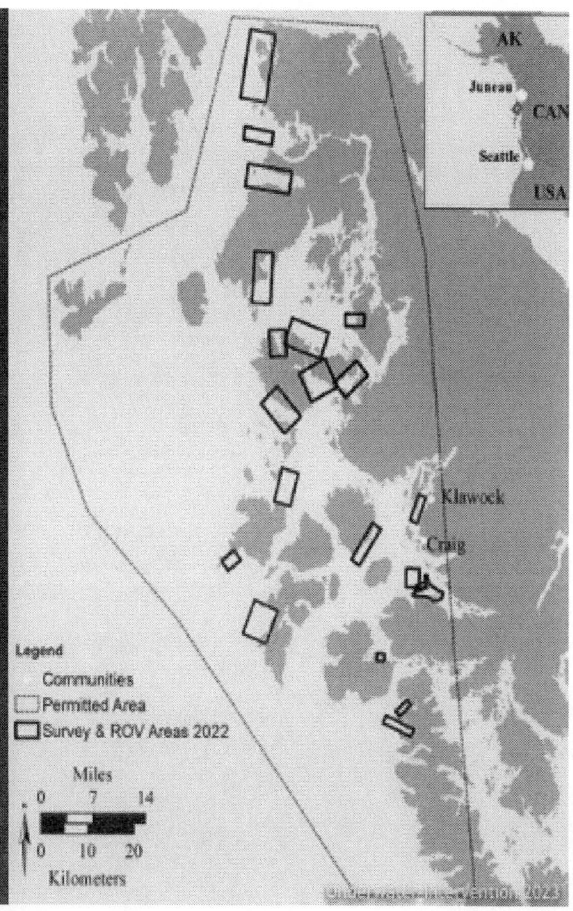

Platform

SUNFISH AUV

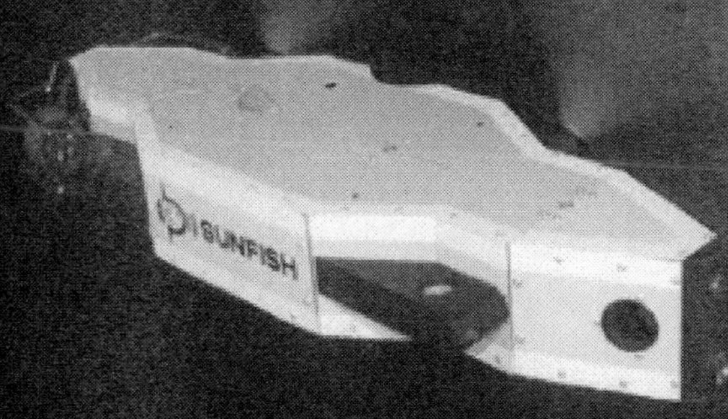

- 6-DOF control
- Person-portable
- 3D sonar SLAM
- Autonomous exploration
- Sonar and visual maps

Platform

SUNFISH AUV

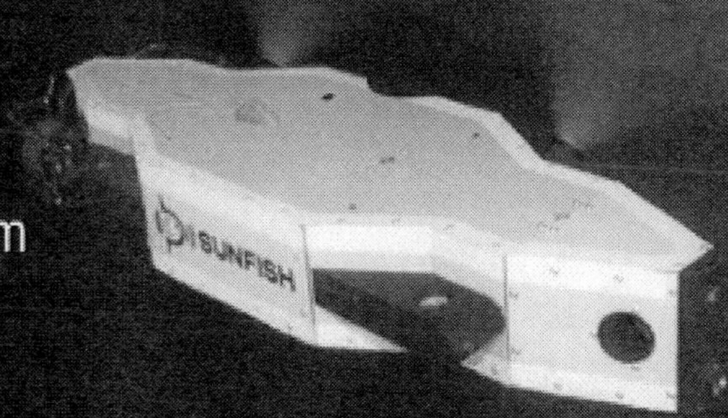

- Dimensions:
  - 1.64 m x 0.47 m x 0.22 m
- Weight: 60 kg
- Depth Rating: 200 m
- Range: ~12 km (15 hr)

Platform

SUNFISH AUV

- FOG IMU
- DVL
- Multibeam
- CT + D
- HD video + lights
- Strobed stereo cameras
- Acoustic Modem / USBL
- Water Sampler
- Nitrate

Platform

SUNFISH AUV

Spectrum of autonomy

- Fully Autonomous
- Scripted
- Supervised
- Remotely Operated

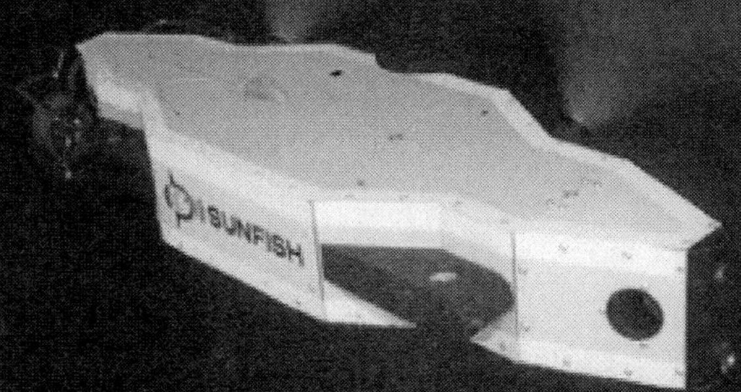

Sediment Sampling

Use vehicle maneuverability to actuate sampling.

AUV Objectives

1. Confirm targets from initial survey
2. 3D mapping and imaging
3. Identify sediments
4. Sample with divers and SUNFISH

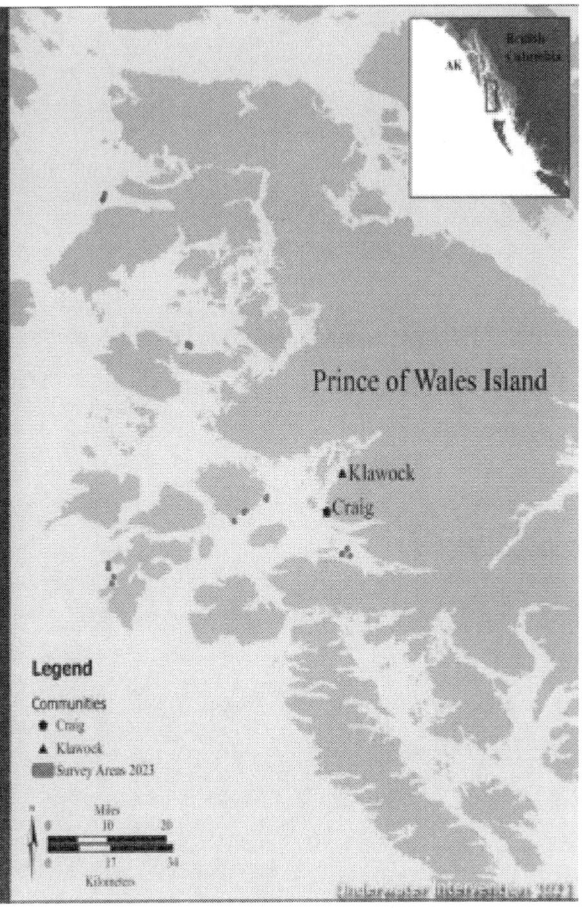

Field Logistics
© 2023 Sunfish, Inc.
Underwater Intervention 2023

Field Logistics
© 2023 Sunfish, Inc.
Underwater Intervention 2023

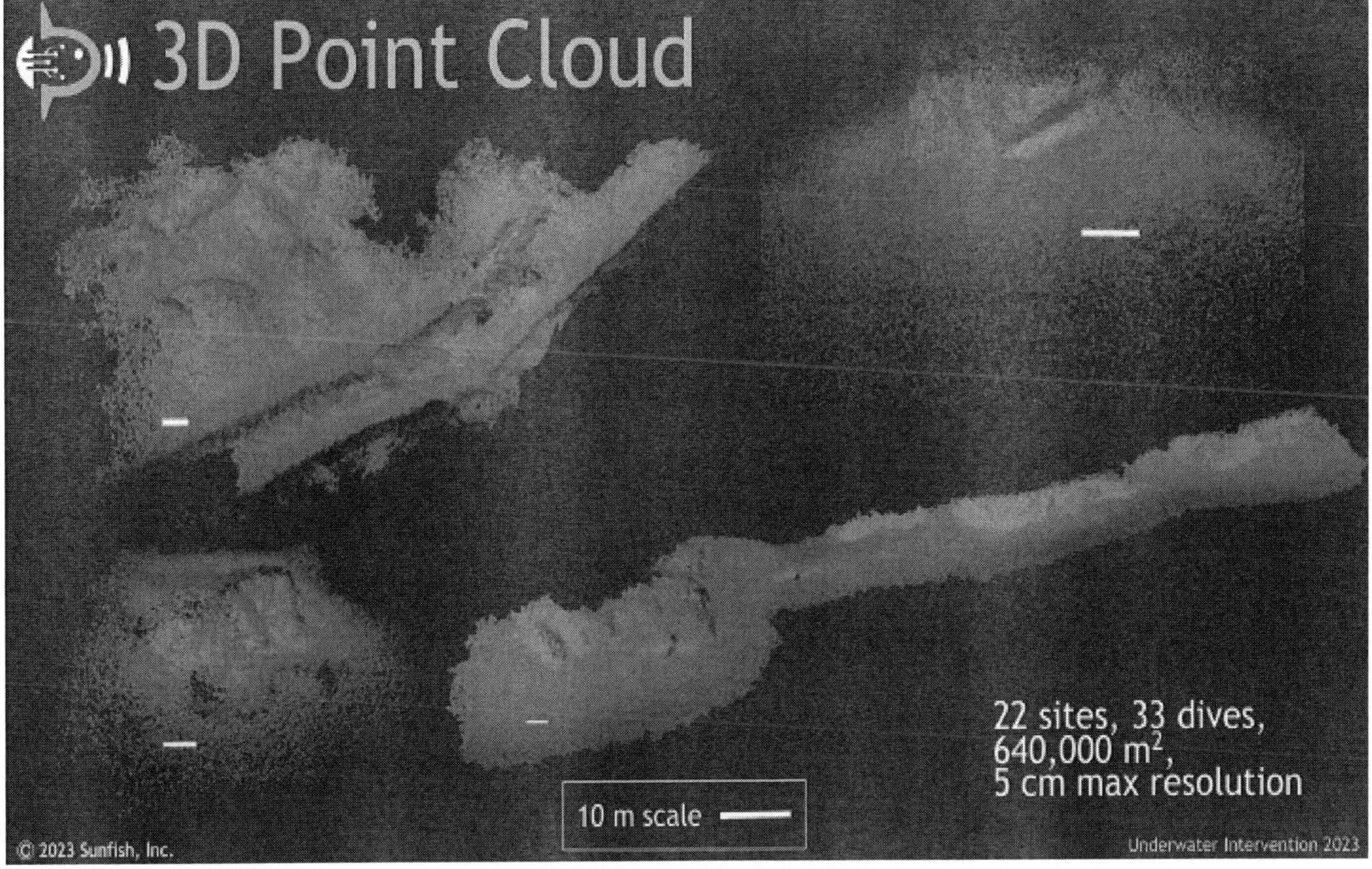
3D Point Cloud
22 sites, 33 dives, 640,000 m2, 5 cm max resolution
10 m scale
© 2023 Sunfish, Inc.
Underwater Intervention 2023

3D Exploration
© 2023 Sunfish, Inc.

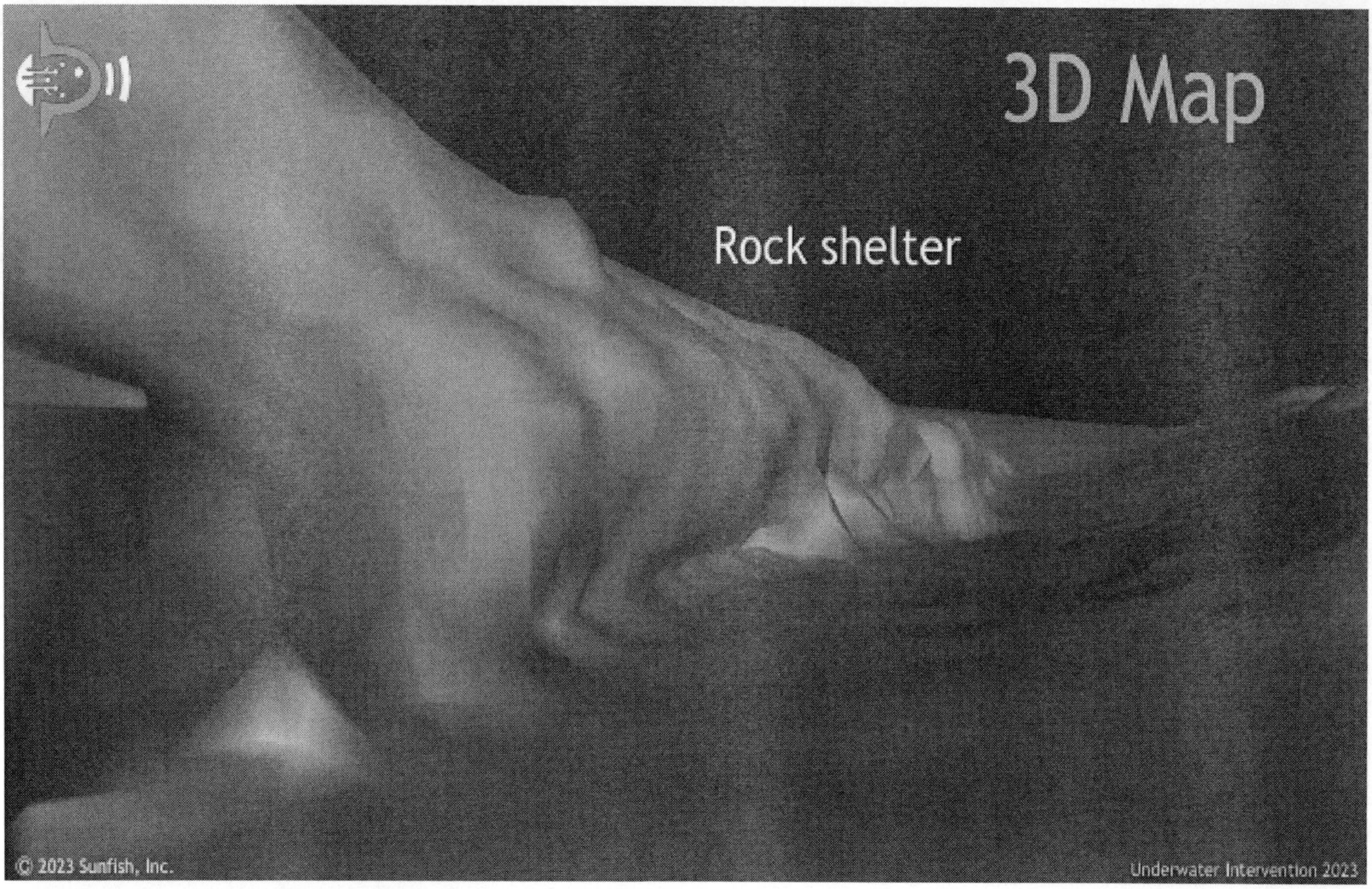
3D Map
Rock shelter
© 2023 Sunfish, Inc.
Underwater Intervention 2023

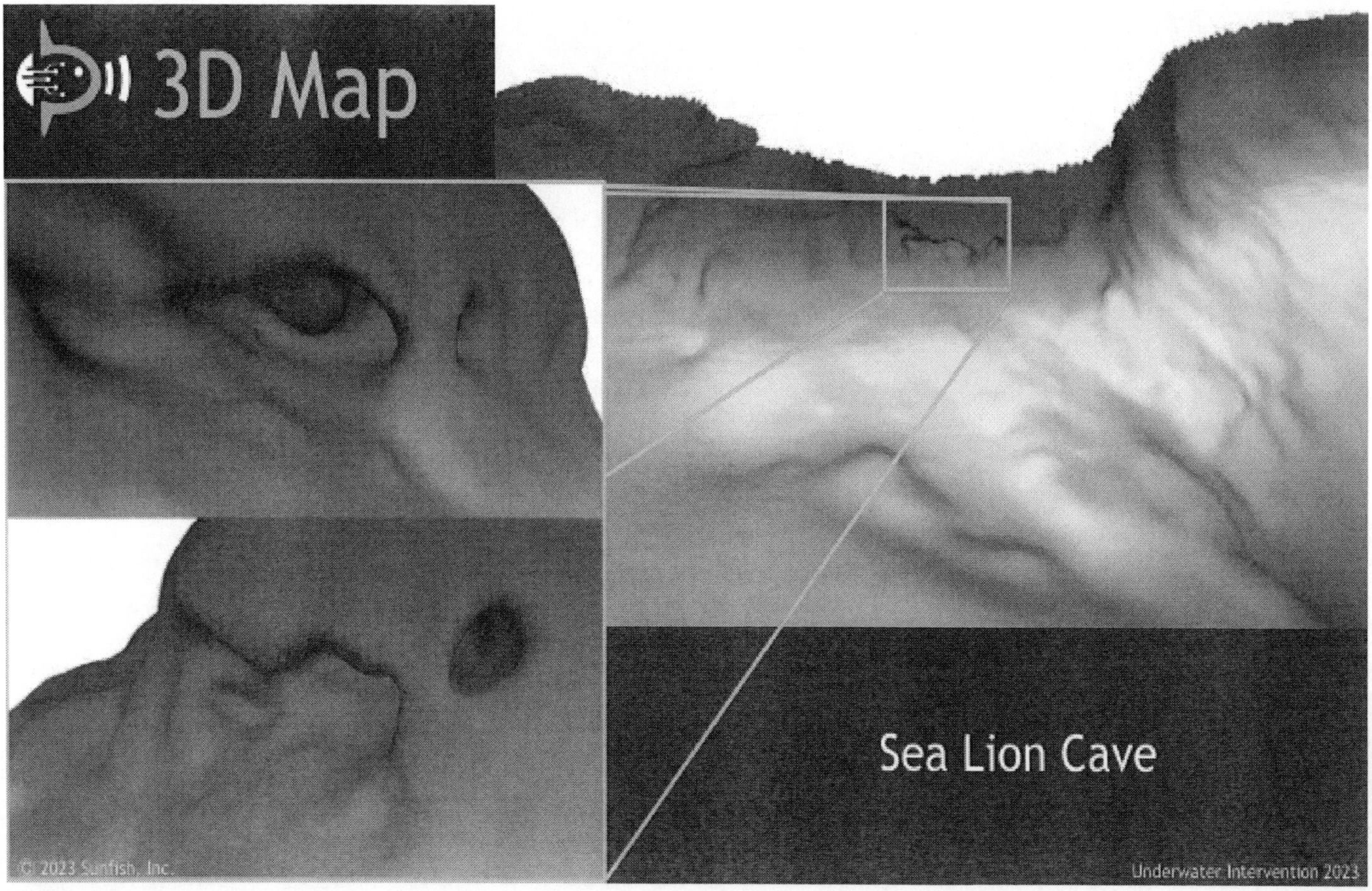
3D Map
Sea Lion Cave
© 2023 Sunfish, Inc.
Underwater Intervention 2023

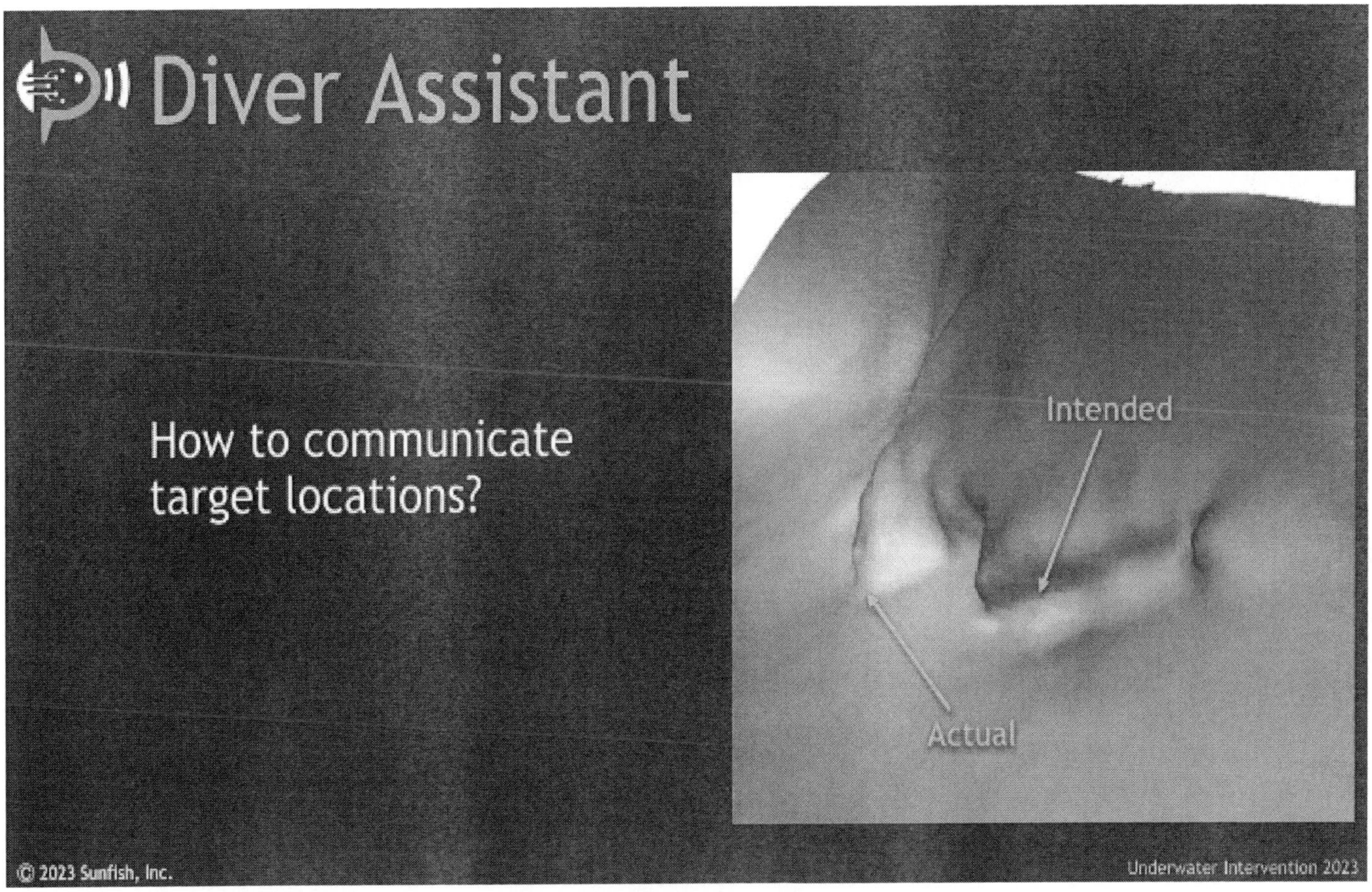
Diver Assistant
How to communicate target locations?
Intended
Actual
© 2023 Sunfish, Inc.
Underwater Intervention 2023

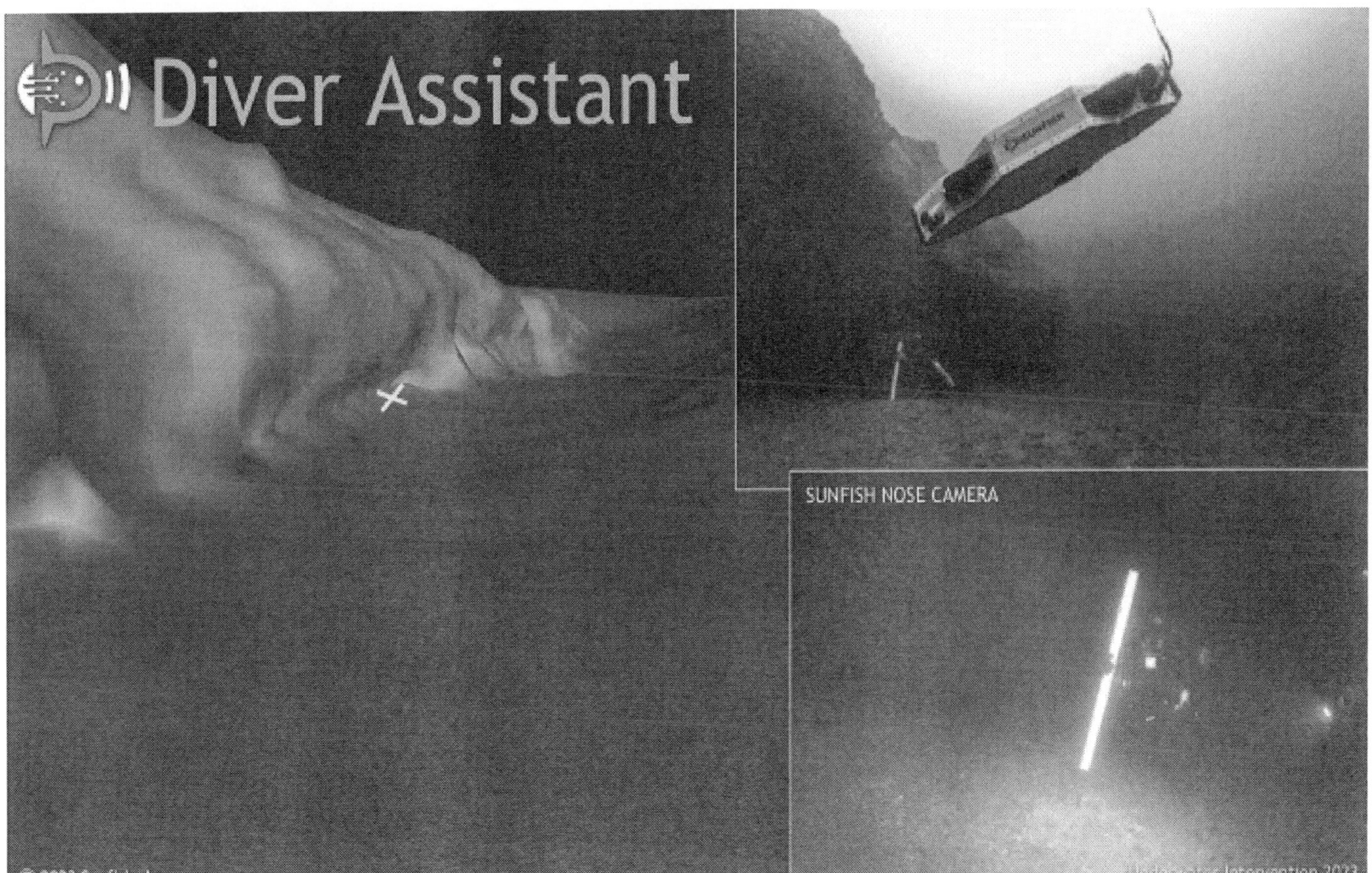
Diver Assistant
SUNFISH NOSE CAMERA
© 2023 Sunfish, Inc.
Underwater Intervention 2023

Acknowledgements

Team OSP

Tyler Haden, Vickie Siegel, Mimi Alexander, Jason Gulley, Tamara Adame, Taylor Heaton, Brant Baxter, Patrick Weston, Madison Stumpf, Roby Medina
Kristi Erickson, Russell King, Lee Kadinger, Alberto Lopez, Rosita Worl
Kelly Monteleone, Kristof Richmond

This project was supported by NOAA Ocean Exploration Grant **#NA21OAR0110198.**

Many thanks to the people of southeast Alaska for access to Hinyaa Aaní (land of the Hinyaa K̲wáan) and K'áyk'aanii (Haida Country in Alaska).

MTS
marine technology society

The Eagle Ray ROV:

AN INSTRUCTIONAL ROV KIT DESIGNED WITH PRACTICAL TECHNOLOGY TO ENHANCE STUDENT KNOWLEDGE AND SKILLS FOR THE MODERN WORKFORCE

Eagle Ray

mtsociety.org

SeaMATE Angelfish

SeaMATE Pufferfish

SeaMATE Triggerfish

SeaMATE Barracuda

PREVIOUS MATE ROV KITS

ANGELFISH: **How DPDT switches work, connecting wires**

PUFFERFISH: **Soldering (wires and components to printed circuit boards), printed circuit boards, LEDs, resistors**

TRIGGERFISH: **2-potentiometer joysticks, analog speed control – motor speed controllers, pulse width modulation, H-bridges**

BARRACUDA: **Analog control to digital speed control, programming, Arduino, relays, I2C**

NEW TECHNOLOGY COMPARED TO PREVIOUS MATE ROV KITS

The Eagle Ray has a number of features that replicate what students might find in industry.

- The Electronics are moved underwater onboard the ROV
- ROV uses a 48-volt power system
- 48 volts are sent down the tether and converted to lower voltages on the ROV
- The Eagle Ray uses an X-box controller and laptop to control the ROV
- The Eagle Ray uses Blue Robotics thrusters and other components.

Electronics Underwater

48 volts

Voltage converted on ROV

Joystick & Laptop Controls

Blue Robotics Thrusters

Electronics Underwater

Eagle Ray has the electronics moved underwater into a 6-inch Blue Robotics canister.

All previous MATE ROV kits had their electronics on the surface and sent control signals down the tether.

This teaches such skills as:

- Waterproofing a canister
- O-ring usage and maintenance
- Penetrators for wires
- Electronics space management

48 VOLTS

Upgrade from 12 volts to 48 volts

Conversion on ROV

The Eagle Ray *sends 48 volts down the tether and must be converted to lower voltages on the ROV. The control boards have converters from 48 volts to 12 volts and 12 volts to 5 volts.*

The MATE 48-volt power supply *instructions are available to teams.*

Joystick & Laptop

Companies provide their own Joystick *and* Laptop *for the Eagle Ray. MATE does not provide these to teams building the ROV.*

Designed to have an X-box controller *connected to a laptop.*

Control Program

MATE provides a basic control program in both Python *and* LabView*. Teams can program in any computer language they wish; they are not restricted to using the MATE provided control programs.*

GUI (graphic user interface) for the Eagle Ray *in* LabView .

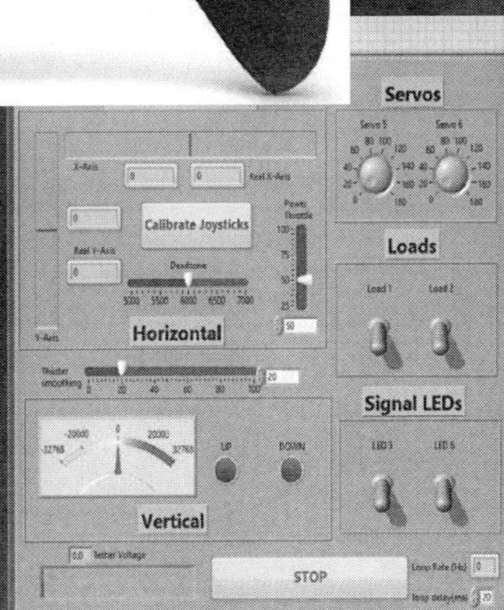

COMPONENTS

Blue Robotics thrusters *to drive the vehicle. Thruster wires used* penetrators *to pass through the wall of the 6-inch canister and connect to* ESCs *inside the canister.*

USB extender hub *inside the canister.* USB cameras *inside the canister, with the potential for more cameras exterior of the canister.*

Heat transfer *considerations. Converters create heat, USB hub and other electronic components are somewhat heat sensitive. Components are mounted to an* aluminum plate *which is connected (touching) the aluminum end caps of the canister.* Heat transfer *from inside to the outside environment.*

Some electronic components require surface-mount soldering.

In 2024 and beyond, creating/assembling the tether.

Additional Skills

3D printing *of components from a* Solidworks *files (.SLDPRT)*

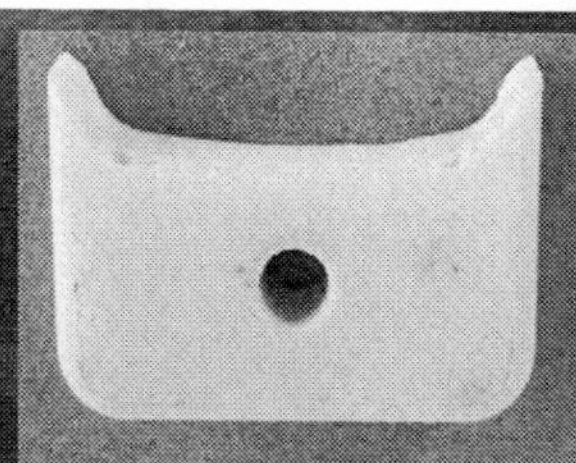

Metal and plastic fabrication. *Cutting, bending and shaping of metal (aluminum sheeting). Cutting and shaping of ½-inch HDPE (plastic) frame. Both using a CNC machine to cut/shape from* Solidworks *files*

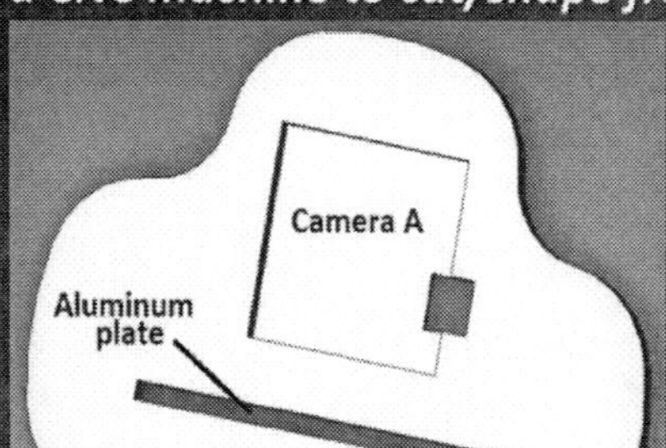

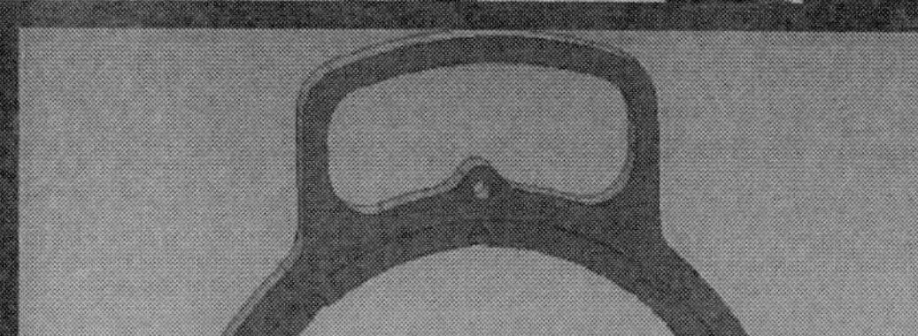

EAGLE RAY ROV KIT
Eagle Ray
MATE ROV has an NSF grant to supply community college and technical college teams with an Eagle Ray kit of parts. Community/technical college students assemble and program the kit and attend the MATE ROV Competition.

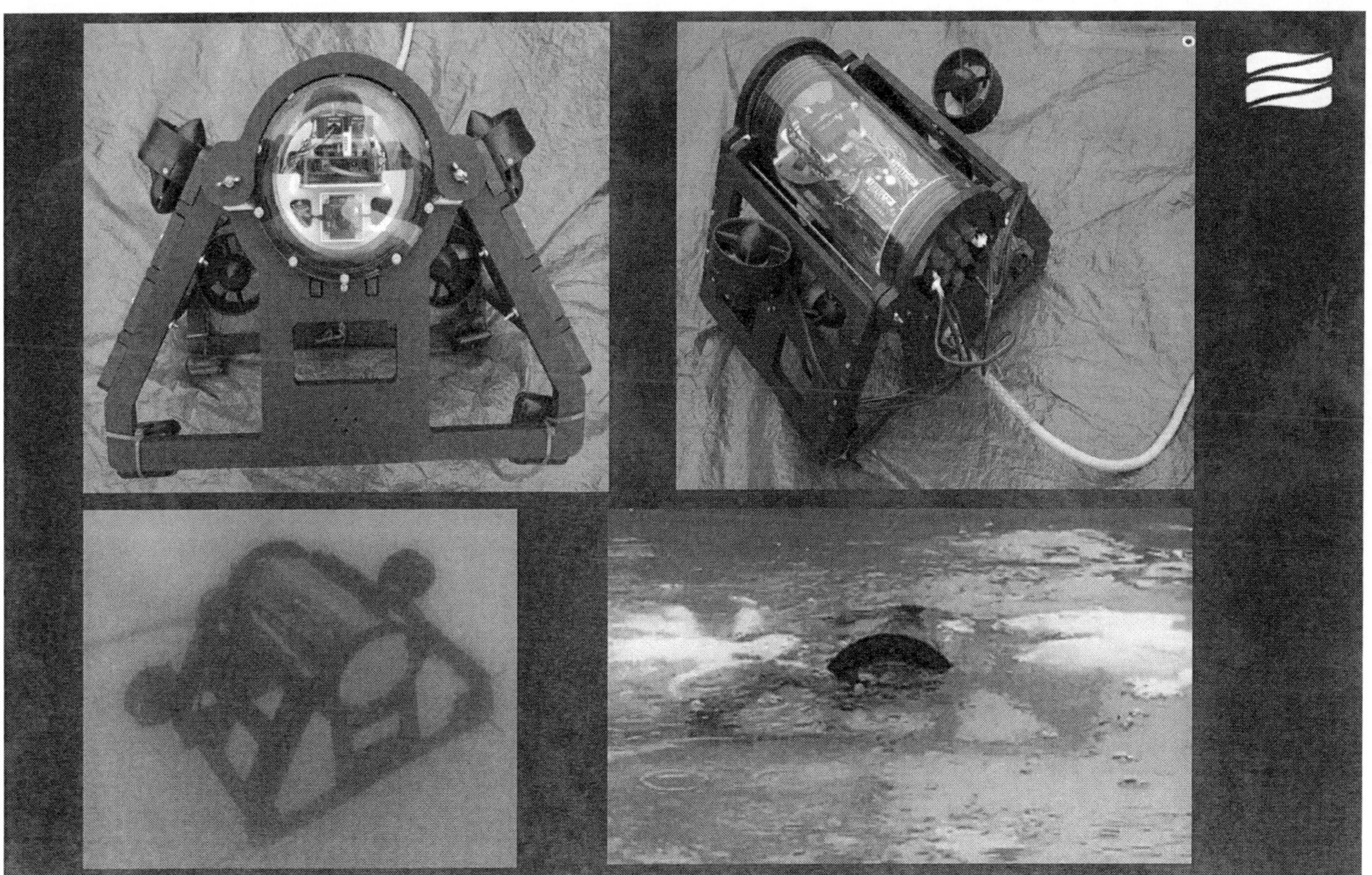

EAGLE RAY ROV KIT
TEAM MODIFICATIONS
Cabrillo College ROV SeaHawk
Southwest Virginia Community College Talon
Pasadena City College Hydra

MTS
marine technology society
Eagle Ray
QUESTIONS?
Matthew Gardner
Marine Technology Society
MATE ROV Competition – Competition Technical Manager
mgardner@marinetech.org
mtsociety.org

The impact of student competitions and other experiential learning on workforce development

Greg Mulder
Linn-Benton Community College
Albany, Oregon
mulderg@linnbenton.edu

LBCC Physical Science Outcomes:

Upon graduation, a physical science student from LBCC should have:

- A basic understanding of key topics in the physical sciences.
- The ability to utilize a variety of scientific equipment.
- The ability to function as a member of a team.
- The ability to communicate scientific understanding and discovery in a variety of manners to different audiences.
- An appreciation of how science interacts with society.

MATE ROV

The Marine Advanced Technology Education Center Remotely Operated Vehicle International Competition.

https://materovcompetition.org/

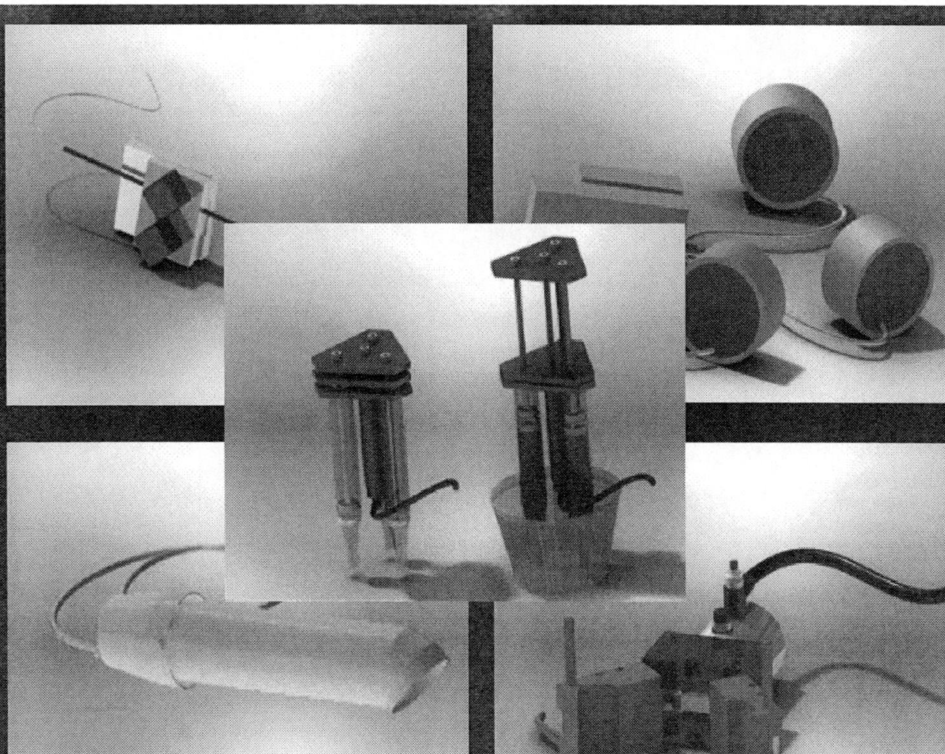

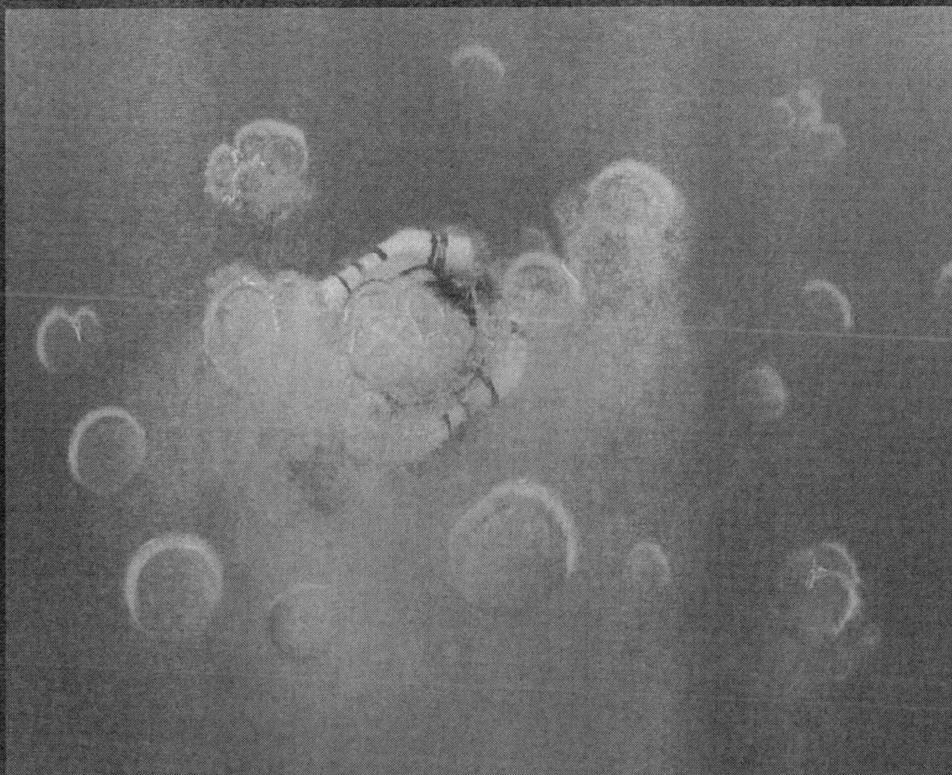

History of LBCC ROVs

“Allen Throop”

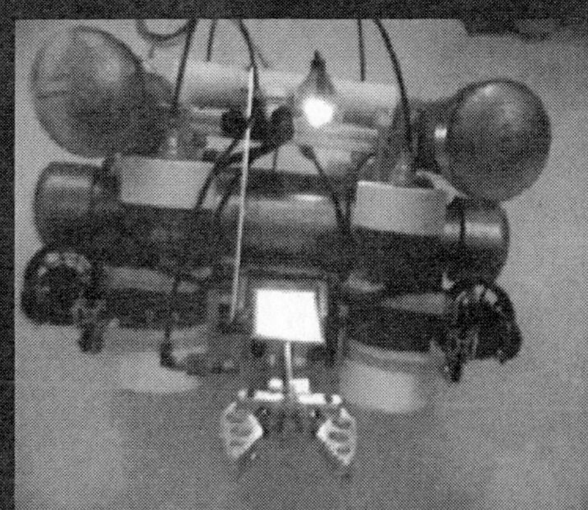

“$\varepsilon\delta_0$”

“NOVA”

2011 Houston Competition

The pool at NASA's testing facility

The Linn-Benton Community College "Phoenix".

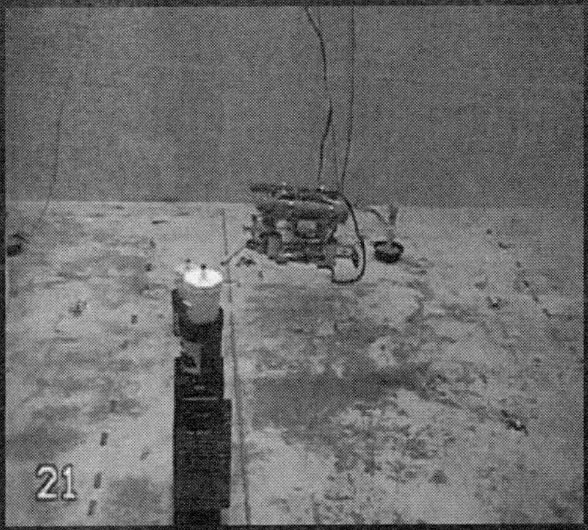

LBCC placed 3rd out
of roughly 450 teams
from across the world.

Paulina Lake

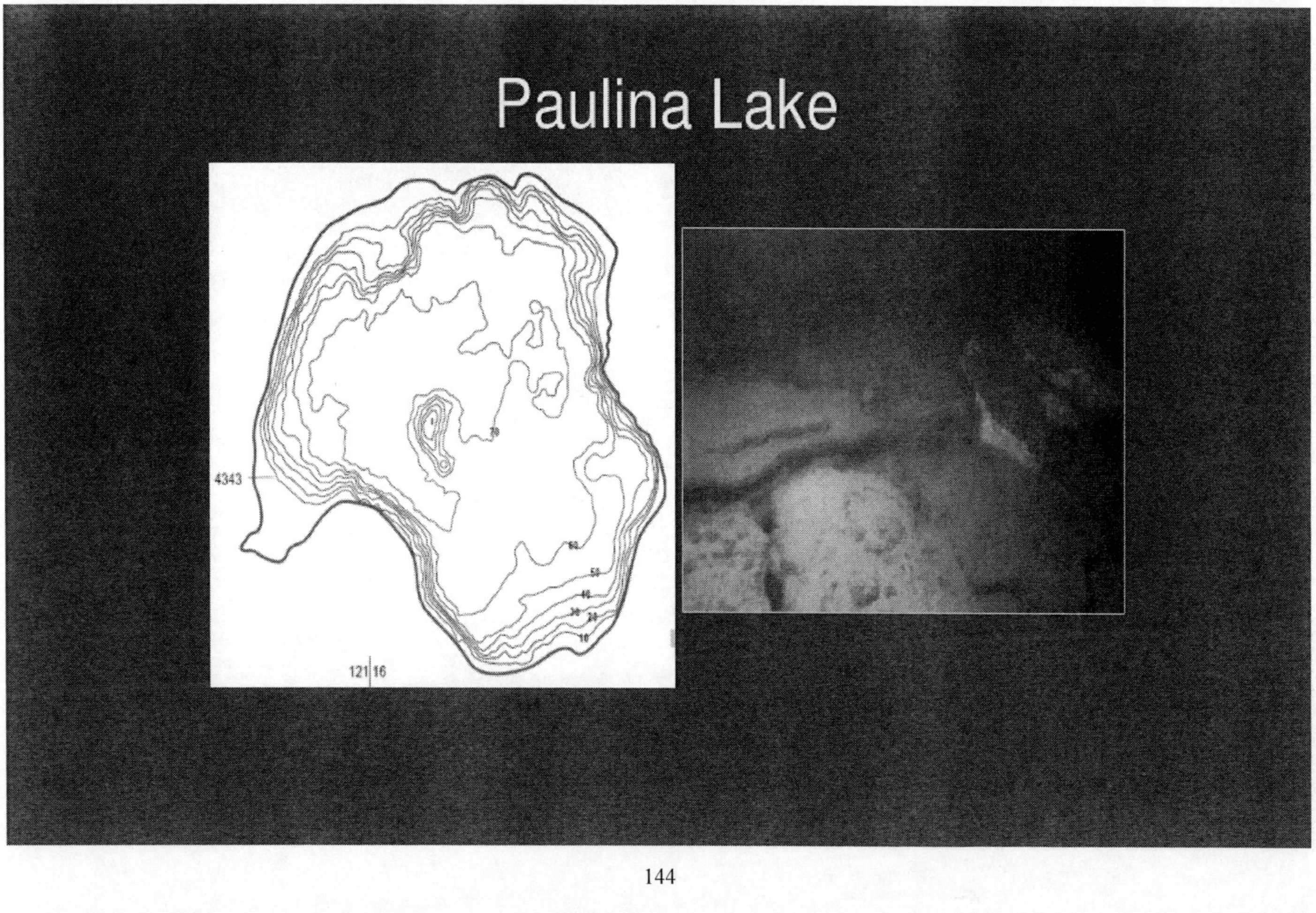
Paulina Lake
4343
121 16

OUVET

Thanks to funding from The Oregon Consortium and the Oregon Workforce Alliance (TOCOWA) LBCC students created the 2009 Oregon Underwater Volcanic Exploration Program

OUVET

ROV Youth camp

- A new group of research assistants (RAs) came out every week.
- They would design & create their own ROV.
- Help with research on the boat using the main ROV.

OUVET 2009

They would start their week with planning their ROV…

…and end it with a mini ROV to take home

OUVET 2009

Continued research of Paulina Lake's spire

- Everyday groups would rotate onto the boat.
- RAs would operate the ROV and capture footage of the spire.
- There was a successful dive all the way down to 75 meters

OUVET

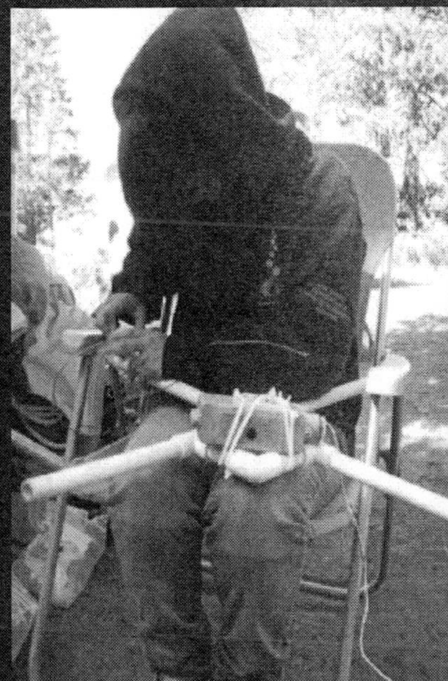

OUVET: Summer 2010

Continued research of Dante's Peak

- Cycled through shifts on the boat
- Collected rock samples and video footage of the spire
- Successful dive down to 75m

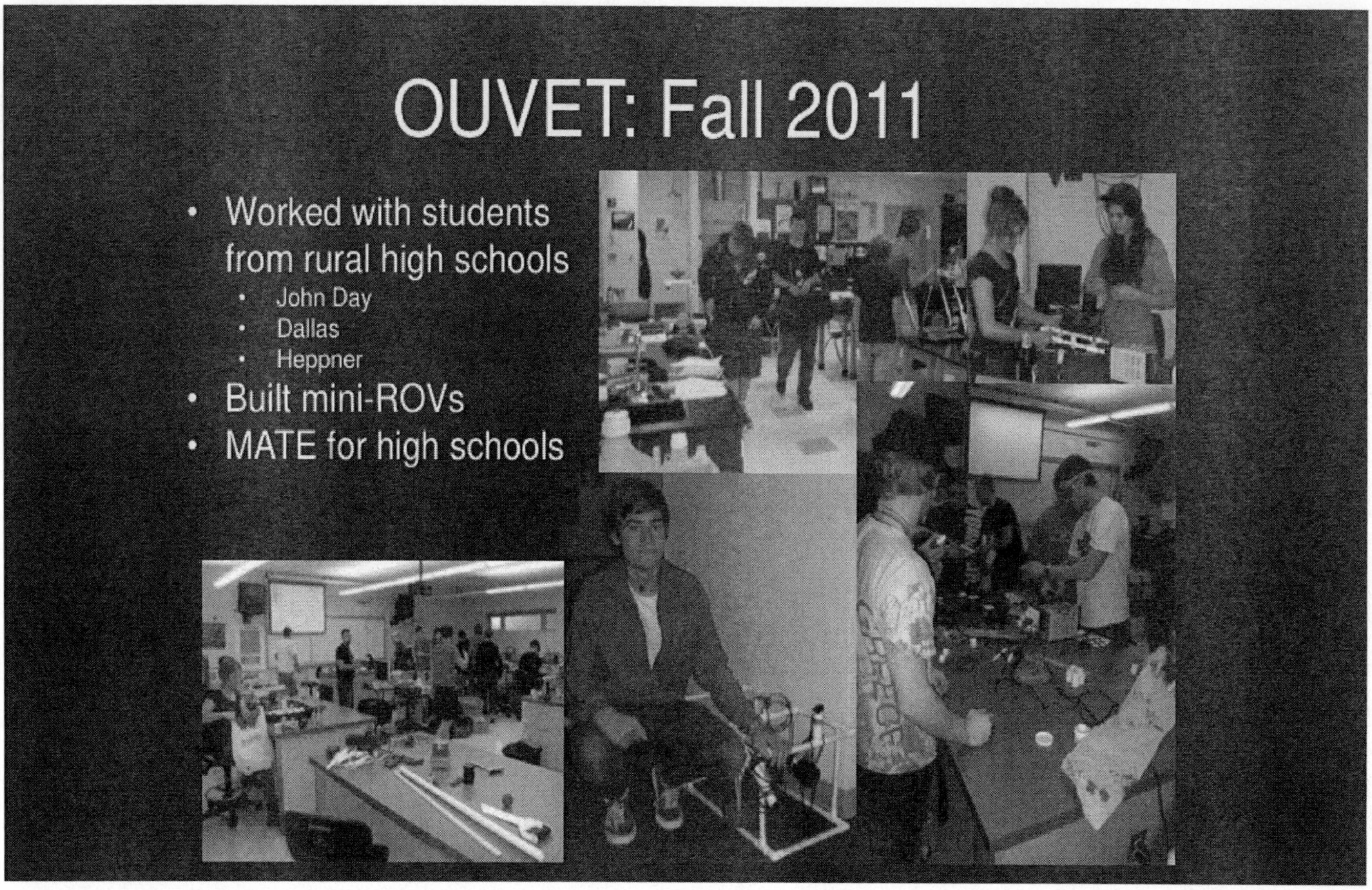
OUVET: Fall 2011
• Worked with students from rural high schools
• John Day
• Dallas
• Heppner
• Built mini-ROVs
• MATE for high schools

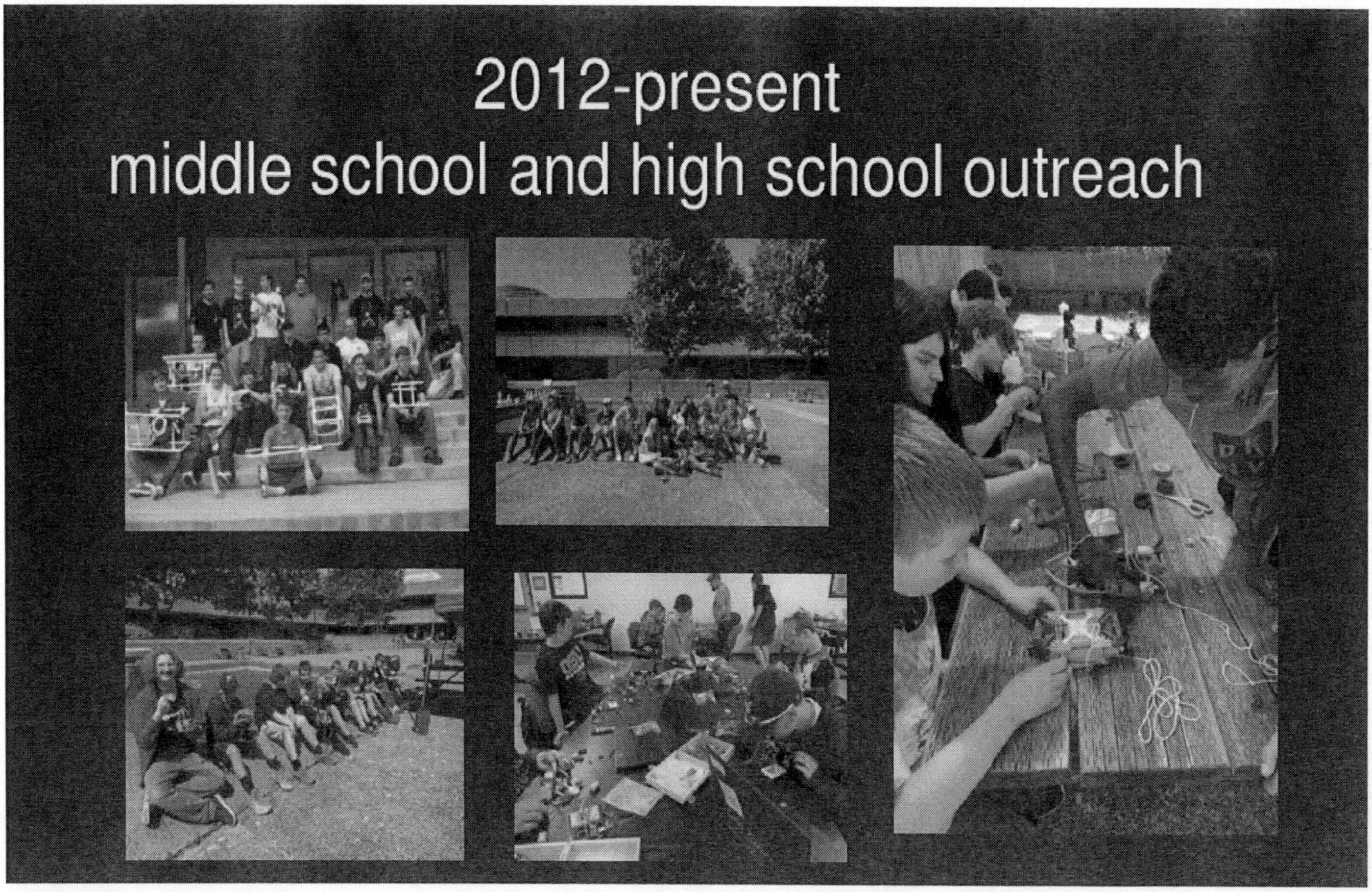
2012-present
middle school and high school outreach

AMERICAN MADE CHALLENGES
U.S. DEPARTMENT OF ENERGY

Marine Energy Collegiate Competition (MECC)

Marine energy—power generated from ocean waves, currents, tides, and temperature changes—is the world's largest untapped renewable energy resource. However, further work is needed to optimize marine energy device designs and reduce costs.

Follow the Competition

View Competition Rules

Marine Energy Collegiate Competition

Student teams identify a promising market within the blue economy and determine the best marine energy application with that marker to address.

Academic competitions, such as MATE, can promote inclusive behavior.

...have fun!!

...travel to exotic places...
ROAD
CLOSED

Departments at LBCC that have supported the team:

Physical Sciences

Engineering

Drafting

English

Speech

Computer Science

Mechatronics

Health and Human Performance

Sponsors
TOCOWA
white's
hp
Vernier
Measure. Analyze. Learn.
ESTERN POWER PRODUCTS INC.
UNITED PIPE & SUPPLY
Audio Files
541-753-0621
Salem Scuba
Best Western
NORGREN
Linn GEAR CO.
NORVAC ELECTRONICS, INC.
Tektronix
And the LBCC Foundation started it all

Sonardyne

Enabling a world of Multi-Robotic Ocean Exploration

Aidan Thorn
Marine Robotics Business Development Manager

https://www.sonardyne.com/marine-robotics/

 30 September 2020 sonardyne.com

Underpinning Autonomous Operations

Sonardyne

Covelya Group: Complete marine robotics payload and software solutions

Sonardyne

GROUP

Engineering Excellence and Innovation

Covelya Group companies operate in a wide range of sectors within the energy, science, defence and commercial markets. We develop, manufacture, sell and support instruments, systems and solutions worldwide.

Global representation across each operating company

Cohesive group of companies with synergistic technology offerings

Over 50 years of experience between group companies

Leading innovators and subject matter experts within the ocean sector

Sonardyne

Key Technologies

Comms is key - Acoustics

Sonardyne

Navigation: SPRINT-Nav Family

Sonardyne

Sonardyne

Use Case 1: Commercial

Sonardyne

2. Remote navigation and tracking

Remote navigation and tracking

Remote navigation and tracking

Remote navigation and tracking

Sonardyne

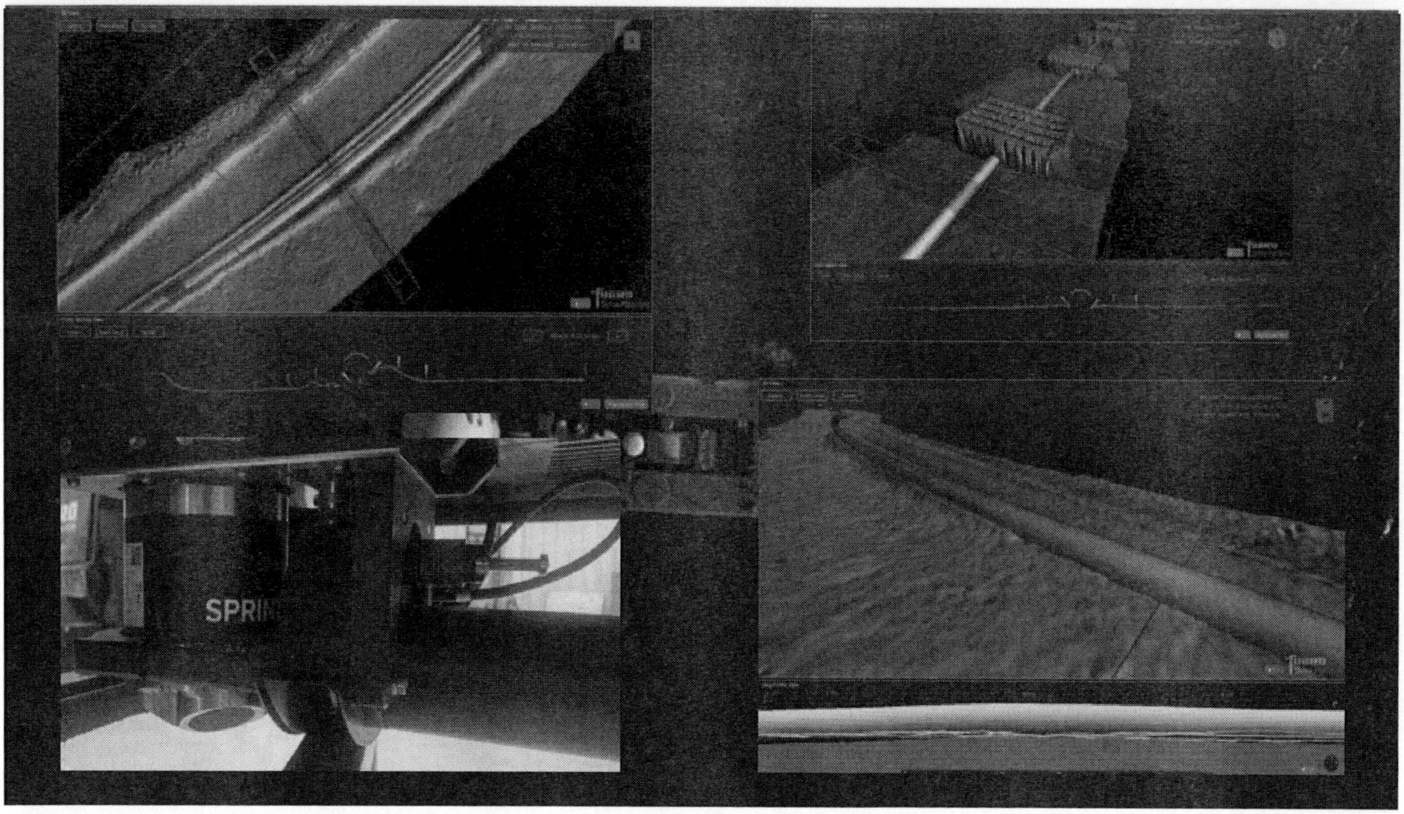

Sonardyne

Use Case 2: Ocean Science

All the robots... and a ship

Sonardyne

oeci

University of New Hampshire

CCOM JHC

Sonardyne

ROS Driver and message definitions

Sonardyne

ROS Sonardyne drivers (Developed by the University of New Hampshire)
https://github.com/CCOMJHC/sonardyne_usbl

modem_node

The modem_node.py node provides topics for sending and receiving SMS messages using `sonardyne_msgs/SMS` messages and for sending and receiving raw modem commands using `std_msgs/String` messages.

Supports serial, TCP or UDP connections.

ranger_node

The ranger_node.py node subscribes to asynchronous position updates from the Ranger software and publishes them as `geographic_msgs/GeoPointStamped` messages. Uses XML base UDP remote control protocol.

ROS

ROS Message Definitions
https://github.com/CCOMJHC/sonardyne_msgs

Position.msg

```
Header header
string UID
float32 age
string category
string name
float64 latitude
float64 longitude
float32 depth
string history
```

SMS.msg

```
time receive_time
string address
string message
```

OECI TECHNOLOGY **CHALLENGE** 06-22 May 2022

Sonardyne

During the 2022 NOAA Ocean Exploration Cooperative Institute Technology Integration Cruise (aka: OECI Tech Challenge), The surface robot DriX worked with the underwater robots Mesobot and NUI allowing Nautilus to freely map the seafloor nearby.

The Sonardyne ROS driver was a key component used with Project11 on DriX which provided situational awareness and command and control

Sonardyne

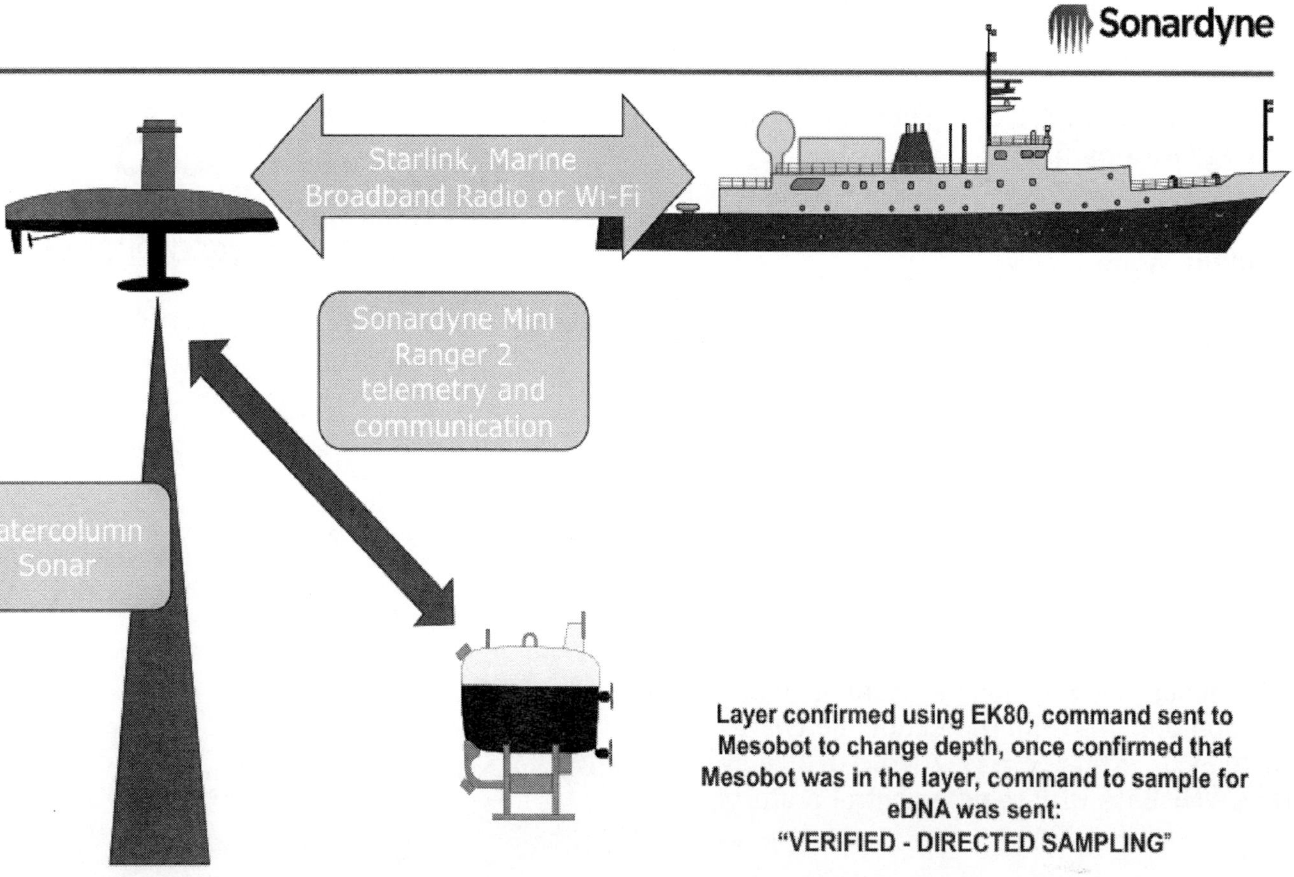

Layer confirmed using EK80, command sent to Mesobot to change depth, once confirmed that Mesobot was in the layer, command to sample for eDNA was sent:
"VERIFIED - DIRECTED SAMPLING"

Near Real-time Data for Operational Decision Making!

Sonardyne

- NUI "Co-Exploration"
- Both MBES mapping and camera data via DriX acoustic relay.

- Transmission of CTD measurements from Mesobot to ship via DriX for real-time input on water column properties

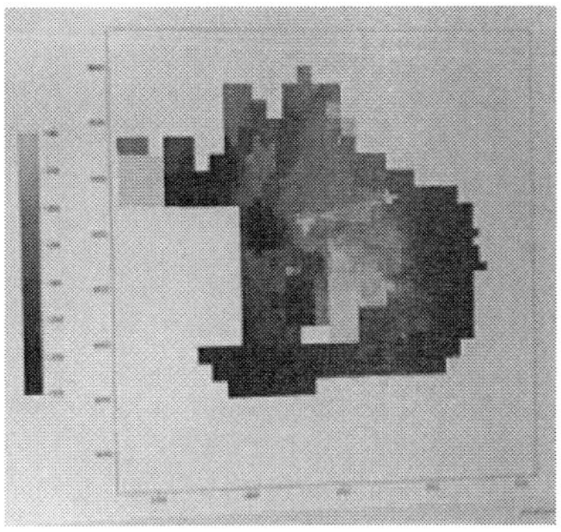

Realtime "CO-EX" Transmitted Data

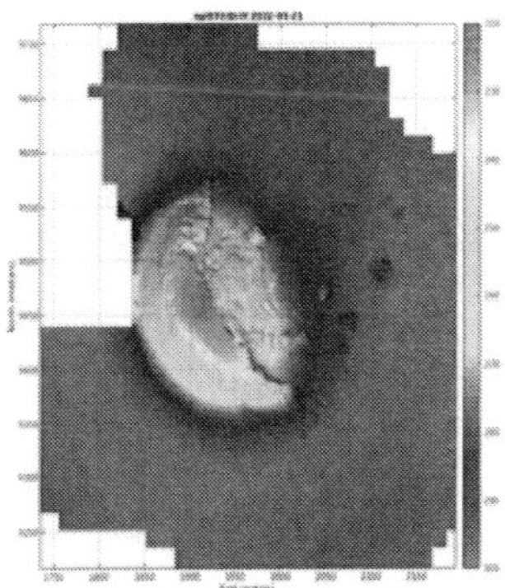

Post Deployment Data

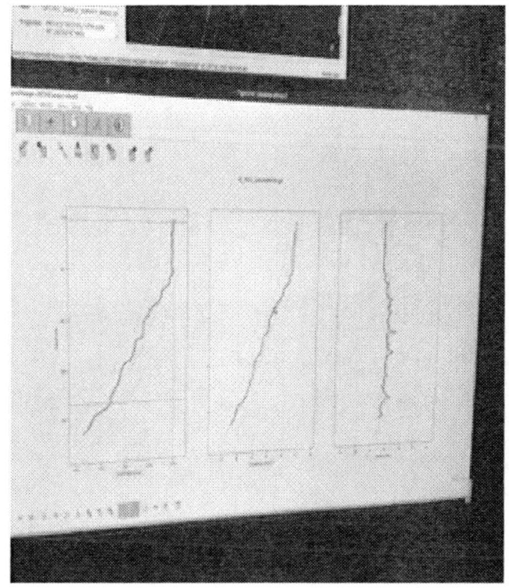

Conclusion... but not really, there's another slide and a video

Sonardyne

The combination of the Sonardyne Mini Ranger 2 system with a ROS driver for integration with Project11 was a key building block allowing the rapid development of technologies for marine robot cooperation

University of New Hampshire

Sonardyne

Sonardyne

Building on the success of the 2022 Tech Challenge cruise, Mesobot and DriX cooperation was improved.

- Verified Directed Sampling was refined using offset orbits
- Automatic layer tracking was implemented as well as layer following by Mesobot. The Sonardyne system using the ROS driver was a key component for making this happen.

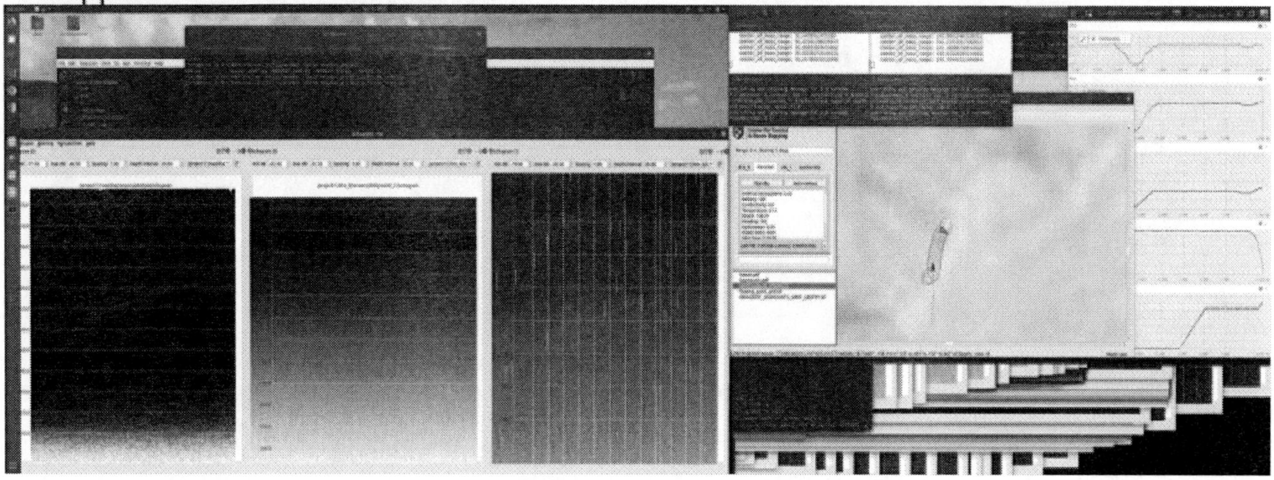

Sonardyne

Thank you for your time today - Visit us on Booth UI3737

Any Questions?

https://www.sonardyne.com/marine-robotics/

Aidan Thorn
Aidan.Thorn@sonardyne.com

sonardyne.com

US GoM
DSWG
Diving Safety
WORK GROUP

Agenda

- Overview of Diving Safety Work Group
- Underwater Ships Husbandry Discussion
- HBOT Therapy in Diving and Non-Diving Applications

2

US GoM DSWG Diving Safety WORK GROUP

Safety Starts With Me !

Working Safe ;

- When We do it for *Family* it's called "Love"
- When We do it for *Others* it's called "Integrity"

3

US GoM DSWG Diving Safety WORK GROUP

Steve Acton
Mark Begnaud
Chris Mouritensen
Patrick Ireland
Kevin Griffith
Chris Primeau
David Witherow
Chandon McGrath
Scott Mercer
Chris Hollifield
Matthew Koratko
Terrence Linston
John Chapman
Patrick Walker

Do you recognize any of the names on this list?

US GoM
DSWG
Diving Safety
WORK GROUP

Steve Acton	Explosion – Grinding
Mark Begnaud	Explosion- Oxy Arc Burning
Chris Mouritensen	Equipment Maintenance /Hose Management
Patrick Ireland	Embolism (Hose management - Manta Ray)
Kevin Griffith	Embolism (Hose management - Manta Ray)
Chris Primeau	Crushed – Vibratory Hammer dropped Piling
David Witherow	Crushed – Crane lowering spoolpiece
Chandon McGrath	Lost Gas / Drowned / Competancy
Scott Mercer	Explosion- Oxy Arc Burning
Chris Hollifield	Explosion- Oxy Arc Burning
Matthew Koratko	Crushed - Crane – Lack of Work Management
Terrence Linston	Strong current / Hat & Hose Management
John Chapman	LOTO- Propellers on Reverse design Jack-up
Patrick Walker	Working on Pressurized Pipeline Flange

These guys didn't go home

US GoM
DSWG
Diving Safety
WORK GROUP

US GoM Diving Safety Workgroup
Vision Statement

The US GoM Diving Safety Workgroup is a US GoM focused, non-competitive and non-commercial group of oil and gas operators, transmission companies, commercial diving companies, supporting sub-contractors, organizations and industry stake holders. The group will provide a unified voice to promote and improve diving safety, through the following:

- identification and sharing of best practices
- identify and seek solutions to industry challenges and issues
- review and comment of existing and proposed standards and guidelines
- provide input to the regulators and industry associations

Anti-Trust / Competition Law

US GoM DSWG Diving Safety WORK GROUP

- While the purpose of these meetings is improving safety which is a legitimate context for competitors to discuss, we must remember that we are competitors and that anti-trust laws apply.
- Therefore, we must adhere to the agenda and only discuss Diving Safety matters (of non-commercial, non-confidential nature).
- Do not discuss prices or costs of petroleum products, production and marketing decisions, sales and purchases, suppliers, customers, business plans, commercial strategies or other competitively sensitive information.
- If in doubt consult your own legal / intellectual property advisors. All in the meeting are to remain vigilant and if anyone believes the rules are transgressed then they are expected to intervene and alert the meeting of your concern.

US GoM Diving Safety Work Group

2023 Executive Board / Roles

US GoM DSWG Diving Safety WORK GROUP

- Chairman – David Gilbert – Operator- ExxonMobil
- Vice Chairman – Ted Roche- Contractor- Aqueos
- Presentations – Kevin Lord- Operator- Shell
- Website - Billy Bratkowski- Contractor- Helix/Triton
- Incident Shares and LFI- Operator-John Hocutt- Quarter North
- Membership, Incident Shares - Troy Turner- Contractor- Phoenix
- Secretary/Scribe – Gary Kane- Affiliate- Kane Company

US GoM Diving Safety Work Group 8

About the DSWG

History-

- **Concept discussed in 2011**
- **First DSWG General meeting held in February 2012**
- **By-Laws adopted in May 2012**
- **DSWG website rolled out in January 2013**
  - http://usgomdswg.com
- **LinkedIn Page-** ***US GOM DSWG*** created in August 2017

- **2012 Membership – 33 General Members**
  24 Affiliate Members

- **2023 Membership – 148 Member Companies**
  - **60 General Members**
  - **88 Affiliate Members**

There are no dues or fees for DSWG membership, all we ask for is a passion for safety and active participation- (Membership Application is on the website)

US GoM Diving Safety Work Group

DSWG General Members

American Inshore Divers
American Marine Corporation
Aqueos Corporation
Ardent Global
Arena
Anchor Diving
Arendal
Axis Subsea Construction
BHP Billiton Petroleum
Blackwater Diving LLC.
BP
Bosarge Diving Inc.
Blue Marine
Brown Integrity Group
Brennan Marine
Central States Underwater
C-Dive, LLC.
Chevron North American
Chevron Environmental Mgmt Grp
Commerce Construuction
CIDCO
Couvillion Group
Cross Diving
Dean Equipment & Dredge
Dive Ops
DOF Subsea
Enbridge Energy Partners, LLC.
Energy Transfer
EXND Diving and Marine
ExxonMobil Upstream Oil and Gas
Global Diving and Salvage
Helix ESG
Hydra Offshore
Kinetica Energy Express, LLC.
Leviathan
MADCON Corporation
Mainsteam Diving
Marlin Oilfield Divers
MSI Marine Solutions
Morrison Energy
Neptune Diving
Oceaneering International
Phoenix International
Quarter North Energy
Sapura
Sea-Force Diving
Shell E&P Company
Shell Pipeline Company
Specialty Offshore
Spectra Energy
Subsea 7
SubSea Global Solutions
Technip FMC
T&T Subsea, LLC.
Titan Salvage
TransCanada Offshore
Triton Diving
U.S. Underwater Services,
Williams Pipeline
W&T Offshore

US GoM Diving Safety Work Group

US GoM
DSWG
Diving Safety
WORK GROUP

About the DSWG

Continual Interaction with Regulatory and Industry Groups

- USCG (U.S. Coast Guard)
- ADCI (Association of Diving Contractors International)
- IMCA (International Marine Contractors Association)
- IOGP (International Oil and Gas Producers)
- OOC (Offshore Operators Committee)
- API (American Petroleum Institute)
- CBP JADE (Customs and Border Patrol - Jones Act Division of Enforcement)
- National Board of Hyperbaric Physicians

  - ✓ Recognized by these groups
  - ✓ Requested to comment on proposed rules, Recommended Practices, etc.
  - ✓ Representatives from some of these groups regularly attend our meetings

US GoM Diving Safety Work Group

US GoM
DSWG
Diving Safety
WORK GROUP

About the DSWG

Accomplishments

- ✓ Through group comments and vetting, provided input into future USCG Commercial Diving Regulations
- ✓ Provided comments to IMCA on document revisions
- ✓ Provided comments and influence related to Diver Certification acknowledgment through face to face meetings in Washington, DC
- ✓ Developed Guidance documents
- ✓ Provide an Industry Forum for Incident, Safety, and Information Shares (over 600 Incident Shares to date)
- ✓ Our Incident shares are provided as shares in IOGP (DOSC / IDIF)
- ✓ Hyperbaric physicians participation and presentations regarding diver medical concerns. Serio / Sanders/ Allemand / Bourgeous/ Dr B /VanMeter

US GoM Diving Safety Work Group

DSWG Incident Shares Spreadsheet

US GoM DSWG Diving Safety WORK GROUP

DSWG Industry Incident Near Miss Report

DSWG Industry Incident & Near Miss Record

DSWG Meeting Date	Location	Occurrence Type	Injury or Damage	Actual Incident Rating	Primary Cause of Incident	Topside/ Method of Diving/Other Location	Depth of Dive - Feet	Description of Incident
11/15/2023		Non-Diving	None		Personnel			Subsea beacon holder fell off crane wire and fell 3 meters to deck. All stop was called, site was secured and then lifting operations resumed. Subsea beacon is normally attached by two clamps - one for the holder and one for the beacon itself - this one was just held by one. When worker went to remove what he thought was just the first clamp from the beacon, the beacon holder was released and fell to deck. No injuries, no damage. Stand down was immediately called and proper installation and removal of beacon revisited. Crane operator will now verify, and second strap was also added to ensure proper attachment in the future.
11/15/2023		Diving	None		Personnel	Air Diving		DP diving ops, recovering diver on a No-D dive, miscommunication occurred. Diving supervisor told the LARS operator that "diver was No-D, come up on him". LARS misunderstood and thought he said to start moving up on the stage. Diver called All Stop and clipped hose into stage and off of crossover. Moving forward the supervisor is going to confirm the stage is ready to travel, and will then tell LARS operator to "recover the diver to the surface."
11/15/2023			None			Air Diving	35	Two suspected DCS initially but ruled out. 1 - Diver experienced left foot pain 2 days after completing a 32 foot dive. Ruled out due to length of time before symptoms showed up, and also because pain only occurred when weight was put on the foot. 2 - Another diver experienced shoulder pain 4 hours after completion of a 35 foot dive. Also ruled out through investigation, partially due to the discovery that diver had pre-existing condition, likely spinal stenosis. When suspected DCS hits occur at the jobsite, personnel are immediately put in the chamber and run on a Table 6.
11/15/2023		Non-Diving	None	Near Miss	Public/Others	Other		ROV boat on location with ROV on bottom and seismic boat entered the cone of safety. Neither vessel was aware of the other initially. Seismic boat told dive boat to leave, dive boat couldn't due to ROV being on bottom. Near miss with a potential for entanglement.
11/15/2023	GoM	Non-Diving	Head Injury	First Aid	Other Equipment	Saturation		Dropped object incident; frozen gallon jugs were being hung in the bell due to high temperature in the bell. Tied with a rope through the handles. One handle broke and the jug fell and hit a diver in the head, causing a cut. Neurological testing and first aid done, diver was all right and made a full recovery. Will install cooler to better regulate temperature in the bell, and secure hanging objects much more carefully.
9/20/2023	Inland	Diving	Illness	First Aid	Environment			Diver in S. LA finishing dive, brushed himself off. Started acting strange and unresponsive, but vitals were ok. EMS was called, and by the time they came he was coming back to normal, but tired. Doctors determined he had a severe allergy to fire ants, previously unknown.
9/20/2023		Diving	Illness	Recordable Incident		Air Diving		Commercial diver who had taken a recreational diving course before going to dive school. He felt he'd possibly gotten bent in the recreational course, but nobody believed him. Worked for quite a while professionally, but kept having complaints. MRI revealed he does have brain infarcs, likely from the original recreational course. Diver also had a PFO, and can no longer dive.

US GoM Diving Safety Work Group

Barrier Model for Incident Prevention

US GoM DSWG Diving Safety WORK GROUP

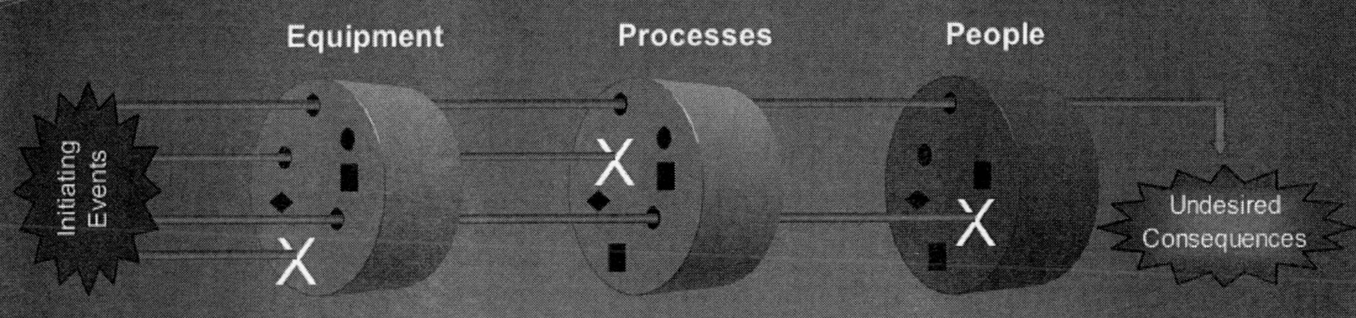

Design Standards
Equipment Strategies

- Equipment is certified & available for use
- Safety Equipment
- Beacons/sonar/ Survey Systems
- Vessel DP Systems
- Mooring Systems
- Fire fighting / detection
- Dive Equipment Audits
- Crane Inspections
- Rescue Equipment / Emergency Response Equipment

Operating Procedures
Work Management

- Regulations and Standards
- Industry Guidance's
- Life Saving Actions
- Risk Assessments
- JSAs/Work Permits
- Procedures/Work Aids
- SIMOPS Matrix
- Pre start check lists
- Integrity Test/Inspections
- Weather forecast
- Emergency/Rescue Plans

Leadership & Communication
Accountability / Training

- Required personnel in place
- Training and Competency verified
- Sufficient controls in place.
- Experienced Supervisor/Lead
- Stop, Think, & Act in use
- SAFE Choices- Our last line of defense
- STOP Work Authority
- Supervisor/Lead attended Risk Assessment
- Pre-Job Crew Meetings /Level 2 RA
- **Emergency Drills completed**

About the DSWG

2022-2023 Presentations

Monthly ADCI and IMCA Updates
Industry Safety Shares
Incident Shares Spreadsheet
Presentations from Hperbaric Physicians
IOGP Lifesaving Rules
Caviblaster / Waterblaster Noise
Marine Assurance / Marine Crew Competancy
Behavior Based Safety
Gates Hose presentation
Ladder Safety
Diving School status
Subsea Excavations (Comments on IMCA document)
Management of Change
Stop Work Authority
Linde Gas Management
Compact Chest Compression Device
Understanding the Risks in Mooring Operations
International Code of Safety for Diving Operations - IMO Subcommittee (Newsum)
HM Recompression Services
811 / Camo Marine Safety and operations
GOM Shelf Decommissioning Review
ICSDO / IMO Implementation – (Jessop)
Dr. Brian Bourgeois – DCS /PFOs in Diving
Dr. Keith Van Meter – Hyperbaric Oxygen therapy in Diving and Non-Diving Applications-

US GoM Diving Safety Work Group

About the DSWG

2024 Planned Presentations

- JSEA Audit Process by managers
- PVHO/ASME Presentation -- FMS Engineering
- Scott Angelle – LAGCOE - State of the Oil and Gas
- Significant Sediment Resource Areas (SSRA) – Need a government agency to present
- Seismic Presentations – Wait for IMCA Seismic survey impact on diving operations
- Launch and Recovery system presentation – LARs inspections/maintenance
- Dive Schools –(Ocean Corp, Young Memorial or DIT) overview of diving curriculum
- Doctors- (DMAC) follow up tests after COVID – Lung O2 pulmonary level test
- Kirby Morgan – New Reclaim hat – KMB presentation – Hat Maintenance
- Real time breathing air quality
- Risk Assessment process

US GoM Diving Safety Work Group

DSWG Faces Industry Challenges

US GoM DSWG Diving Safety WORK GROUP

Challenges we face moving forward:

- Regulations – or Lack of..
- The Big Crew Change
- Changes in Work Group Membership
- **Changes in Shelf Operators with Diving Work and Exit of the Majors**
- Economics -How to sustain profitability
- The Potential for Erosion in Safety Standards to cut costs
- Changes in Contractors and Strategies
- Aging Fleet - Vessels and Equipment

US GoM Diving Safety Work Group

2024 DSWG Meeting Schedule

US GoM DSWG Diving Safety WORK GROUP

Month	Day	Meeting Type	Location	Hosts
January	17-Jan	General Membership	QNE Houston	TBD
February	21-Feb	Executive Board	Virtual	TBD
March	20-Mar	General Membership	Virtual	TBD
April	17-Apr	Executive Board	Virtual	TBD
May	15-May	General Membership	Virtual	TBD
June	19-Jun	Executive Board	Virtual	TBD
July	17-Jul	General Membership	Virtual	TBD
August	21-Aug	Executive Board	Virtual	TBD
September	18-Sep	General Membership	Virtual	TBD
October	16-Oct	Executive Board	Virtual	TBD
November	20-Nov	General Membership	Virtual	TBD
December	18-Dec	Executive Board	Virtual	TBD

US GoM Diving Safety Work Group 21

US GoM
DSWG
Diving Safety
WORK GROUP

Underwater Ships Husbandry Discussion

Statistics/Dangers/Potential Solutions

Errol Gritten – Phoenix International

Commercial Underwater Ship Husbandry (CUSH)
Technical Operations Manager

Chris Williams – Phoenix International

Vice President Manned Operations

US GoM Diving Safety Work Group

Underwater Ships Husbandry Discussion

Statistics/Dangers/Potential Solutions

NO SCUBA

UWSH SCUBA Fatalities 2013 – 2023

- Panama – 6 fatalities (SCUBA)
- Fujairah, UAE – 7 fatalities (SCUBA)
- Singapore – 8 fatalities (SCUBA)
- Greece – 4 fatalities (SCUBA)
- Taiwan – 3 fatalities (SCUBA)

Total SCUBA Fatalities: 28

Many dive companies can secure projects due to the low overhead cost of maintaining substandard equipment and training protocols. This translates to lower cost to the customer, but they are typically unaware that these companies are not operating within any diving industry safety standards.

25

LOTO (Lockout/Tagout)

- New Zealand – 1 near miss with injuries (SSA) Prop activated and umbilical entanglement
- Mauritius – 1 Fatality (SSA) Umbilical entangled in brush cart/band mask torn off
- China – 3 fatalities (SSA) limited details available

TOTAL SSA LOTO Fatalities: 5

LOTO Items

Lockout-Tagout (LOTO) is a safety procedure which is used in the commercial diving and maritime industry to ensure that potentially hazardous machines (thrusters, intakes, rudders, propellers, ICCS systems) are properly shut off and not able to be started up again prior to the completion of maintenance or repair work. It requires that hazardous energy sources be "isolated and rendered inoperative" before work is started on the equipment in question. The isolated power sources are then locked, and a tag is placed on the lock identifying the designated personnel who placed it. The diving supervisor then holds the key(s) for the lock, ensuring that only the designated personnel can remove the lock(s) and reenergize the machinery.

NOTE: The ADCI mandates that Lockout-Tagout methods be employed when ship husbandry operations are performed.

•

LOTO Items

US GoM DSWG Diving Safety WORK GROUP

• Before diving operations begins on a ship's hull, the diving supervisor should first meet with the master (or mate) and chief engineer and lock out all machinery with intakes, including the main engines. Signed paperwork and possession of keys (by the diving supervisor) for the locks are the only confirmation that lockout procedures have been followed. Depending upon the vessel type, other items that may need to be locked out include (but may not be limited to):

- **Thrusters** - Completely Rack Out breakers
- **Steering Pumps**
- **Sea Chests** -Advise which remain Active
- **Z Drives** - Rack Out breakers
- **Pod propulsion systems**
  - -Apply brake in high current
  - -Rack out breakers from propulsion and steering
- **Impressed current cathodic protection systems** - Switch off, Fwd and Aft
- **Transducers and Sonar.**
- **Main Propulsion**
  - -CCP/CRP isolated
  - -Turning gear engaged
  - -Air start valves off

*** All alarms should be visible in ECR and should be confirmed by diving supervisor**

*** It is important to make certain that no auxiliary equipment onboard the vessel is set on automatic start if that equipment has intakes outside of the hull.**

LOTO Continued

US GoM DSWG Diving Safety WORK GROUP

- Propellers on ITB / tugboats. Air start valves are secured and bled down.
- Bow / stern thrusters, rack out breakers, etc. must be engaged - not free spinning.
- SC operational, port or starboard high or low, reduce flow, mark frame number referenced
- Steering gear isolated
- Anchor windless locked down and secure
- Bunkering schedule
- Azipod, brakes engaged, steering secured
- Stabilizer fins, pitch and extend hydraulics isolated
- Sounding devices switched off on Bridge
- Tags placed on controls in ECR and Bridge
- Bear in mind that international vessels can have language barriers within the vessel crew - consider a universal drawing/depiction of LOTO procedures as well as posted written procedures in the language of the vessel's flag.

Typical Underwater Maintenance Items

US GoM DSWG Diving Safety WORK GROUP

- Underwater hull cleaning/propeller polishing to remove fouling organisms - Hull cleaning may be done by divers using hand-held tools or self-propelled mechanical brushing equipment, water jets or scrapers.
- Hull inspection - Several methods may be used, including visual inspection, video recording, magnetic particle testing and ultrasonic thickness testing.
- Electrical component replacement, to include sacrificial and/or Impressed Current Cathodic Protection (ICCP) systems and transducers.
- Underwater welding is either done in a submerged dry or hyperbaric habitat, or in water. The AWS D3.6M:2010 Underwater Welding Code defines important variables associated with underwater welding (e.g. metal transfer characteristics, solidification behavior, weld appearance, mechanical properties, etc.) and to describe welding and inspection procedures so that work of a known quality level can conveniently be specified. The AWS D3.6 Underwater Welding Standard is currently the only standard available for qualifying wet or hyperbaric underwater welding. Prior to any wet or hyperbaric welding, diving contractors should qualify welder/divers to existing procedures or qualify their own procedures to the class weld that fits their applications.
- Removing obstructions from thrusters, propellers or rudders of the vessel.
- Blanking of discharges and intakes for internal works

NOTE: Several of these operations will release some quantity of harmful material into the water, particularly hull cleaning operations, which will release antifouling toxins. Underwater ship husbandry may cause an adverse environmental effect as significant amounts of copper and zinc are released by underwater hull scrubbing. Alien biofouling organisms may also be released during this process. Environmental regulations regarding the release of these materials vary by location and must be considered as part of the project plan.

Safety Considerations

US GoM DSWG Diving Safety WORK GROUP

- It is critical that divers understand how ships are constructed and understand the terminology used to describe the various areas and parts of vessels.
- An assessment should be made on the suitability of the vessel from which the ship husbandry diving is taking place. Vessel size, available working deck space, adequate space for equipment and supplies, as well as adequate space to address any diver related emergencies should be considered. Reaction from working next to a large ship in swell and heave, as well as mooring arrangements must also be considered.
- A thorough JHA and RA shall be conducted prior to dive operations.
- All divers' must be established as fit-to-dive before beginning operations.
- A means for the safe recovery of an injured/unconscious diver must be present at the dive site.
- All machinery identified in the JHA/RA that poses a risk to the diver should be adequately controlled using appropriate LOTO and hierarchy of control.
- Diver umbilical management must be outlined to consider all identified hazards.
- There should be adequate umbilical management procedures to restrain divers from accessing live water intakes.
- LOTO warning signs must be posted at the bridge and control room console.
- Ensure LOTO criteria is applied to any diver deployment vessel if utilized.
- All anchor stoppers/chain locks must be engaged.

Safety Considerations

US GoM DSWG Diving Safety WORK GROUP

- Periodic announcements should be made from the vessel's bridge that divers are in the water, and to not take suction from or discharge into the sea, operate propulsion or steering equipment or pay out on moorings or anchors.
- Diver umbilical management must be outlined to consider all identified hazards.
- Divers should be restricted from transiting further than the keel of the vessel from the side of entry i.e., the side the diver is being tended.
- Divers' excursion distance should be relative to the diver-worn emergency gas supply and duration at working depth.
- The bridging of the diving contractor's emergency response plan with that of the vessel's emergency response plan should be performed.
- No work may take place above the diver's worksite (no scaffolding, lifting operations, repairs, etc.).
- Assessments should be performed when working on a quayside for other land-based risks such as intakes for seawater, outlets from drains, etc.
- Consideration of differential pressure (Delta P) situations shall be considered where the hull of the vessel is ruptured.
- The Alpha Flag and all other dive-operations warning signs must be displayed as required.
- Periodic updates on weather conditions should be conducted.
- The ship's draft, depth below the hull, and time of tide must be checked. Ensure a safe depth is under the vessel to prevent a diver from being trapped or crushed under the hull during low tides or if the river level drops. Passing ship traffic can turn a "safe" area into a crush hazard area. This can also include the loading and offloading of supplies and materials (or bunkering) while the diver is in the water.

Delta P (Differential Pressure)

US GoM DSWG Diving Safety WORK GROUP

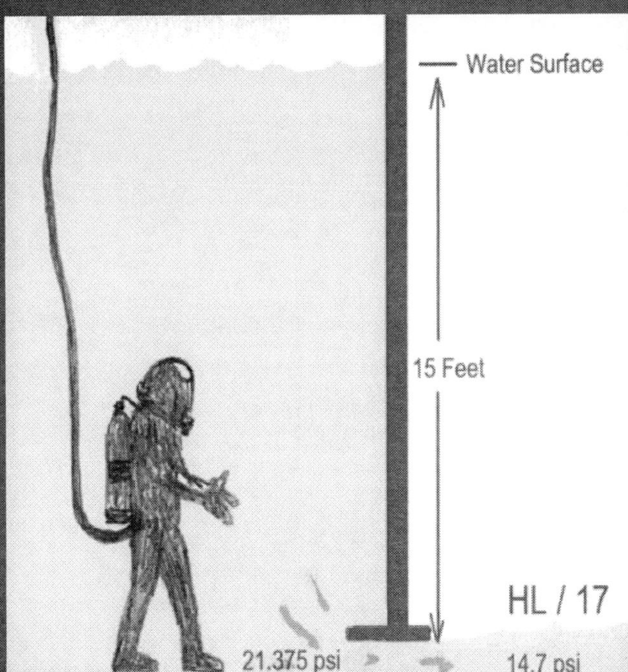

- China – (SSA) limited details available/2 fatalities are known to be delta P related

US GoM DSWG Diving Safety WORK GROUP

Delta P Continued

- When using DC plugs you should never use a plug with a diameter that does not allow 50% or more of the plug to be outside the overboard opening @ the shell plate.
- Once the internal work is completed the pipe should be flooded internally to equalize any pressure on the system for easy removal of the plug / cofferdam.

Damage Control Plugs - DC Plugs

These conical tapered softwood plugs are used for damage control on Naval vessels and are manufactured according to DOD specifications. DC Plugs are manufactured from various softwoods such as Hem Fir, White Pine, Ponderosa Pine, Sugar Pine, and Spruce. They are designed to be used in the patching and plugging of hulls and pipes. These damage control plugs are also used by many Diving and Salvage companies.

US GoM DSWG Diving Safety WORK GROUP

DC (Damage Control) Plug - proper vs improper usage

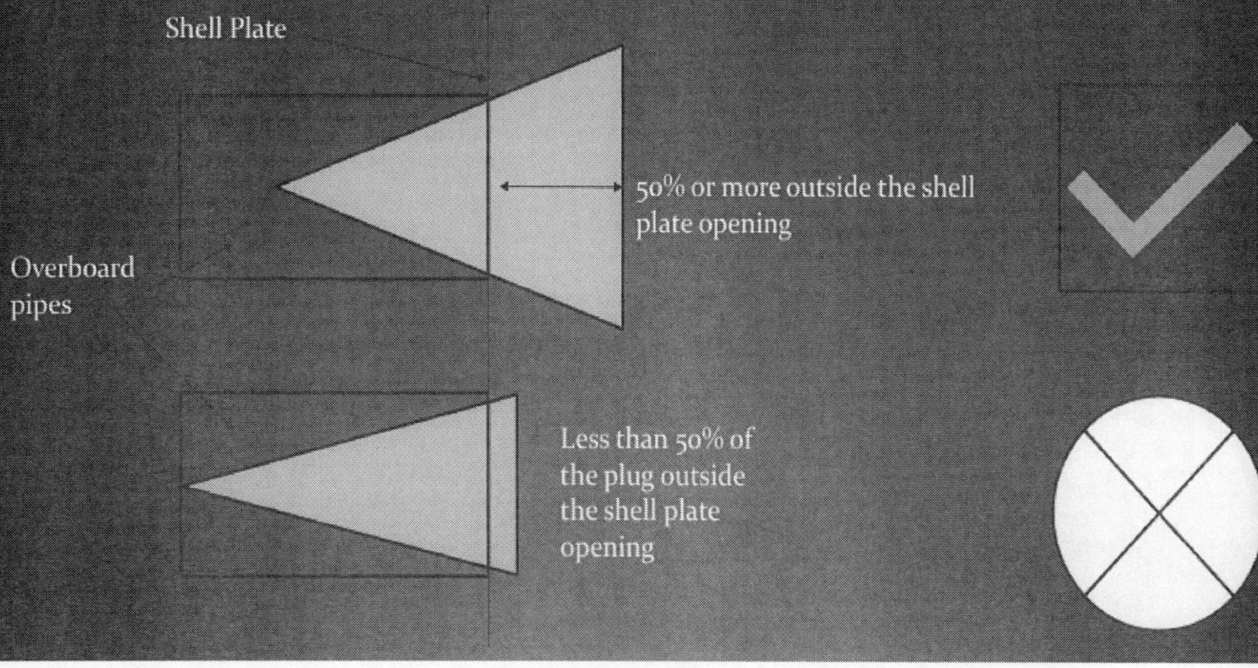

US GoM DSWG Diving Safety WORK GROUP

Flat patch and cofferdam

- Preferred method of isolating valves
- Installed with valve to allow for equalization of differential pressure and additionally allows for testing of watertight integrity without completely removing patch

US GoM DSWG Diving Safety WORK GROUP

Stay away from operating SC grates

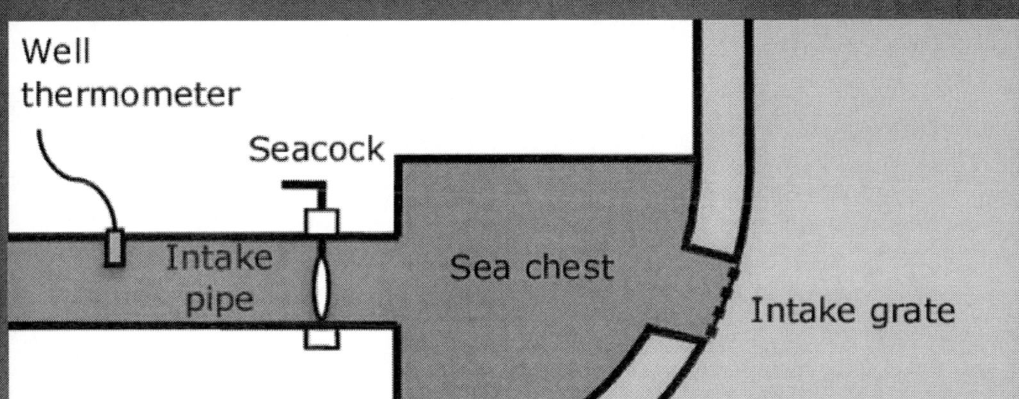

Proper deployment / position of boat fendering

- Dockside - ensure enough fendering is available dockside
- Offshore – ensure there are properly sized fenders to keep diver from being pinched between DSV and vessel / unit
- Diver should never surface if it can be avoided between DSV and vessel / unit unless mooring is secure and sea state is calm

Operational Guidelines

The use of SCUBA is NOT authorized for the performance of Underwater Ship Husbandry Operations.

1. The maximum depth of each dive shall be determined prior to the start of operations.

2. The breathing mixture supplied to the diver must be composed of a mixture of gasses that is appropriate for the depth of the dive. All mixed breathing gasses must be analyzed before they go on-line for O2 content and for proper mixture necessary to support the maximum depth of the planned dive.

3. A separate dive team member shall continuously tend each diver while the diver is in the water.

4. Diver-worn (or carried) emergency gas supply (EGS) shall be utilized and calculations for the gas supply should be performed based on distance, depth, ingress, and egress of the diver.

5. If no decompression chamber is on site, the nearest manned operational chamber (capable of providing treatment for dive-related illnesses) should be known, and an evacuation plan should be in place. A thorough risk assessment should be conducted to determine if a chamber is needed at the dive site. Dives with planned decompression and deeper than 100 fsw [30 msw]) are required to have at least one double-lock decompression chamber and adequate air source to recompress the chamber to 165 fsw. An adequate supply of gases for the planned dive profile and a potential treatment plan are required.

Operational Guidelines

US GoM DSWG Diving Safety WORK GROUP

6. Both the diver's umbilical and the hull-cleaning umbilical must be actively tended to ensure the hull cleaning machine does not cut or entangle the diver's umbilical. Risk assessments must be completed for this type of operation. The diver's umbilical and hull cleaning machine umbilical are not to be mated and must always remain separate from each other.

7. Hull cleaning machines that recover all debris and have large bore recovery pipes pose a greater challenge for umbilical management. A separate machine operator is to be provided at the power pack control or the dive supervisor must have immediate ability to independently stop the hull cleaning machine's rotating brushes.

8. All machines should have a dead-man handle to allow the diver to immediately stop the rotation of the brushes. The standby diver must be able to respond with the ability to remove the hull cleaning rotating brushes should the diver's umbilical become entangled. Note: Some machines cannot be moved unless the brushes are moving

9. All elements of potential differential pressure should be considered.

Equipment Requirements

US GoM DSWG Diving Safety WORK GROUP

- One air source to independently support two divers (working diver and standby diver).

- Topside secondary air source.

- Adequate supply of gasses for the planned dive profile.

- Two hose groups consisting of:
  - Air hose.
  - Strength member/strain relief. (The strength member may be the entire hose assembly, if so designed.)
  - Communications cable.
  - Pneumofathometer hose.

- One set of air decompression and treatment tables.

- One control station consisting of:
  - Communication systems.
  - Depth gauges and gas distribution system with the capability to supply and control two divers at the maximum work depth. The rack box/manifold must be equipped with a non-return valve (NRV).

Equipment Requirements

US GoM DSWG Diving Safety WORK GROUP

• One basic first aid kit with ADCI required contents. Local regulatory authorities may require additional equipment and training.

• Emergency O2 administration kit, with sufficient O2 supply for transit to the nearest hyperbaric facility, capable of treating diving-related illnesses.

• Emergency rescue equipment for the recovery of an unconscious/injured diver.
• Two sets of divers' personal diving equipment consisting of:
- Helmet or mask.
- Diver-worn EGS.
- Weight belt if needed.
- Protective clothing.
- Tools as required.
- Safety harness.
- Knife(s).

• Spare parts, tools, and manuals as required, for the preventive maintenance of equipment.

• Logbooks, dive sheets, safe practices manual, first aid handbook and written JHA applicable for the project.

Vessel Owners/Operators Safety Input

US GoM DSWG Diving Safety WORK GROUP

- Ensure that the vessel agent/Captain/Operator (or whoever is requesting the work) is aware of minimum safety requirements for acceptable diving operations, and specify that when requesting the work, they will verify that the provider will meet all safety requirements as prescribed by ADCI/IOGP/IMCA or other governing body.
- ADCI Section 5.39: Underwater Ships Husbandry (UWSH) for Cargo Ships/Freighters
- IOGP 411 Rev 1 Appendix D: Underwater Ships Husbandry
- IMCA: Guiding Operations in Support of Underwater Operations Ships Husbandry (new document – still under review)

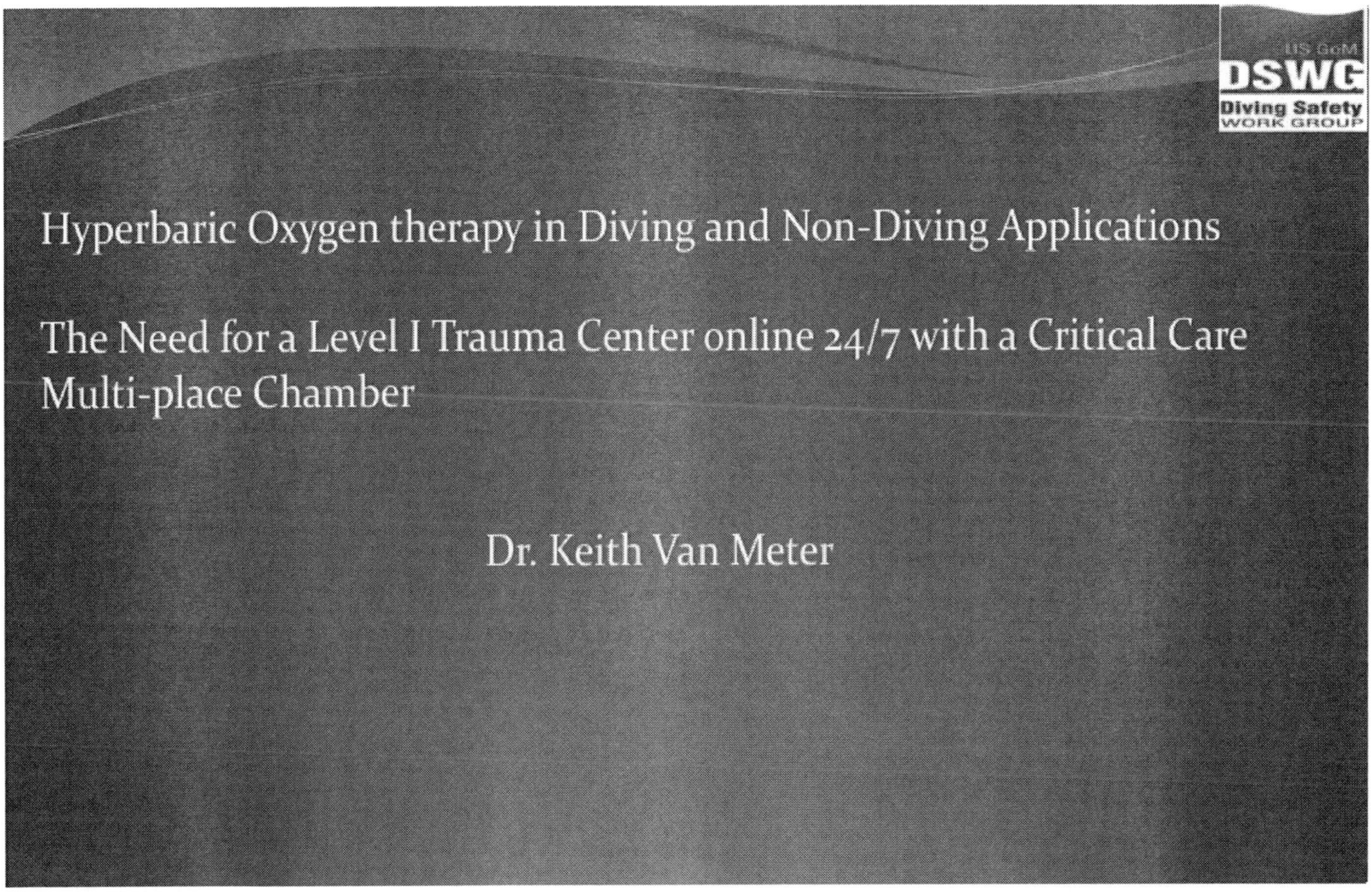
US GoM
DSWG
Diving Safety
WORK GROUP
Hyperbaric Oxygen therapy in Diving and Non-Diving Applications
The Need for a Level I Trauma Center online 24/7 with a Critical Care Multi-place Chamber
Dr. Keith Van Meter

Disclosures:

1. I have had neither a contract nor have I received compensation from manufacturers or drug companies for clinical presentations, patient care, or research
2. I am in process of donating three clinical multiplace hyperbaric chambers and monoplace chambers that I own to hospitals and universities.

Presentation for the GoM Diving Safety Work Group at UI

THE NEED FOR A LEVEL I TRAUMA CENTER
TO BE ON LINE 24/7
WITH A CRITICAL CARE MULTIPLACE CHAMBER
STAFFED BY SURGEONS/FPS/INTERNISTS/EM PHYSICIANS
SUBSPECIALTY BOARDED IN HYPERBARIC MEDICINE

11/30/2023

GoM Diving Safety Work Group

Why?

FDA/CMS Approved Hyperbaric Oxygen Treatable Conditions

(with emphasis on emergent conditions needing 24/7 availability of treatment)

1. CRUSH INJUREY
2. COMPARTMENT SYNDROME
3. NECROTIZING FASCIITIS
4. CRAO/CRVO
5. ACUTE PERIPHERAL LIMB ISCHEMIA
6. GAS GANGRENE
7. FAILING FLAPS AND GRAFTS
8. LIMB THREATING DFUs (WG 4-5)
9. CO/HCN/H_2S TOXICITY
10. BURNS
11. IATROGENIC ARTERIAL GAS EMBOLI
12. DECOMPRESSION ILLNESS

(soon to be approved)

13. TBI
14. ACUTE HEARING LOSS

(non-emergent approved conditions)

15. DFU(WG3)/STRN/ORN/CROM

GoM Diving Safety Work Group

24/7 Critical Care Multiplace HBOT Discontinued at WJMC on 1/25/2019

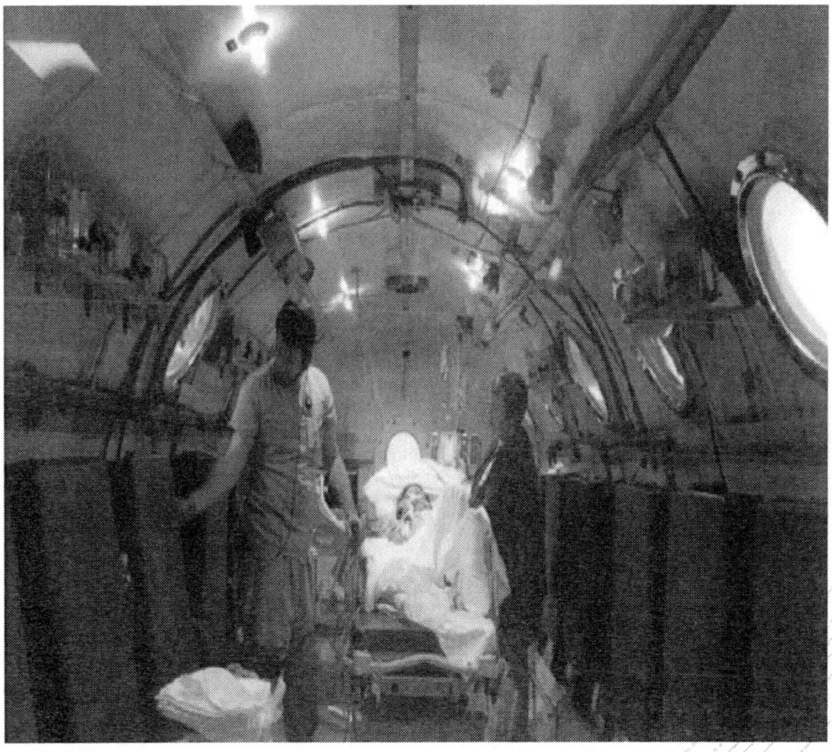

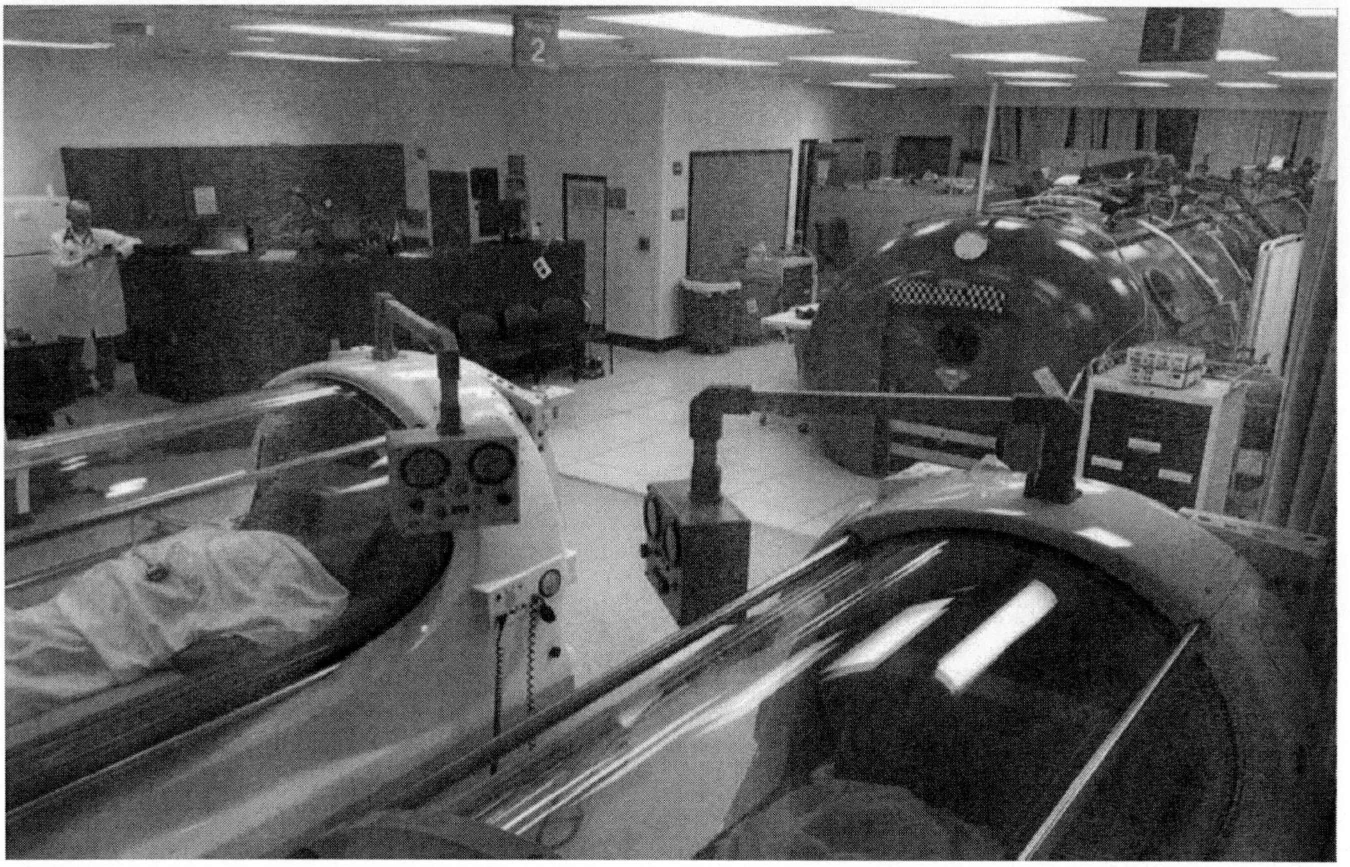

Deep Sixed! Jan/2019

GoM Diving Safety Work Group

Critical Care
HBOT Multiplace
Chamber
Capability
Needed 24/7!

Yesteryear

(where the concept began)

1979

Strange commercial diving operation in the press for Sabine Pass 1979

Chamber is prepared for shipment from Sabine Pass

A Longshot at a Shortshot: The potential use of hyperbaric oxygen in ACLS/ATLS suggested by commercial diving medicine experiences

Keith Van Meter, M.D.

2018 UHMS Annual Scientific Assembly

Section of Emergency Medicine
Dept of Medicine, LSU Health Sciences Center
Dept of Surgery, Tulane University School of Medicine
University Medical Center of New Orleans
New Orleans, Louisiana
2018

D.I.
45 y.o. male
Commercial diver tasked with a surface-supplied SurO_2D air dive (190/75)

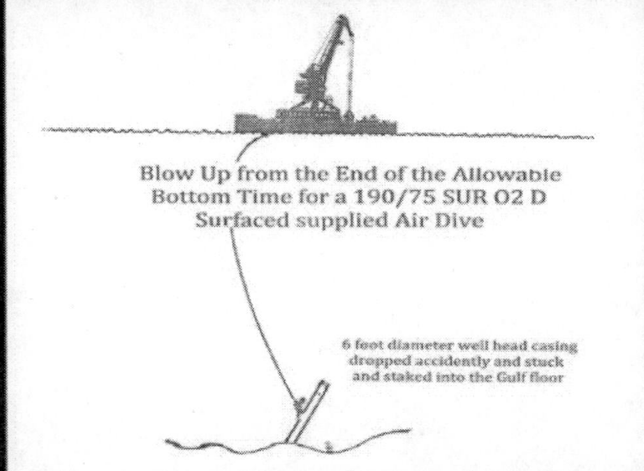

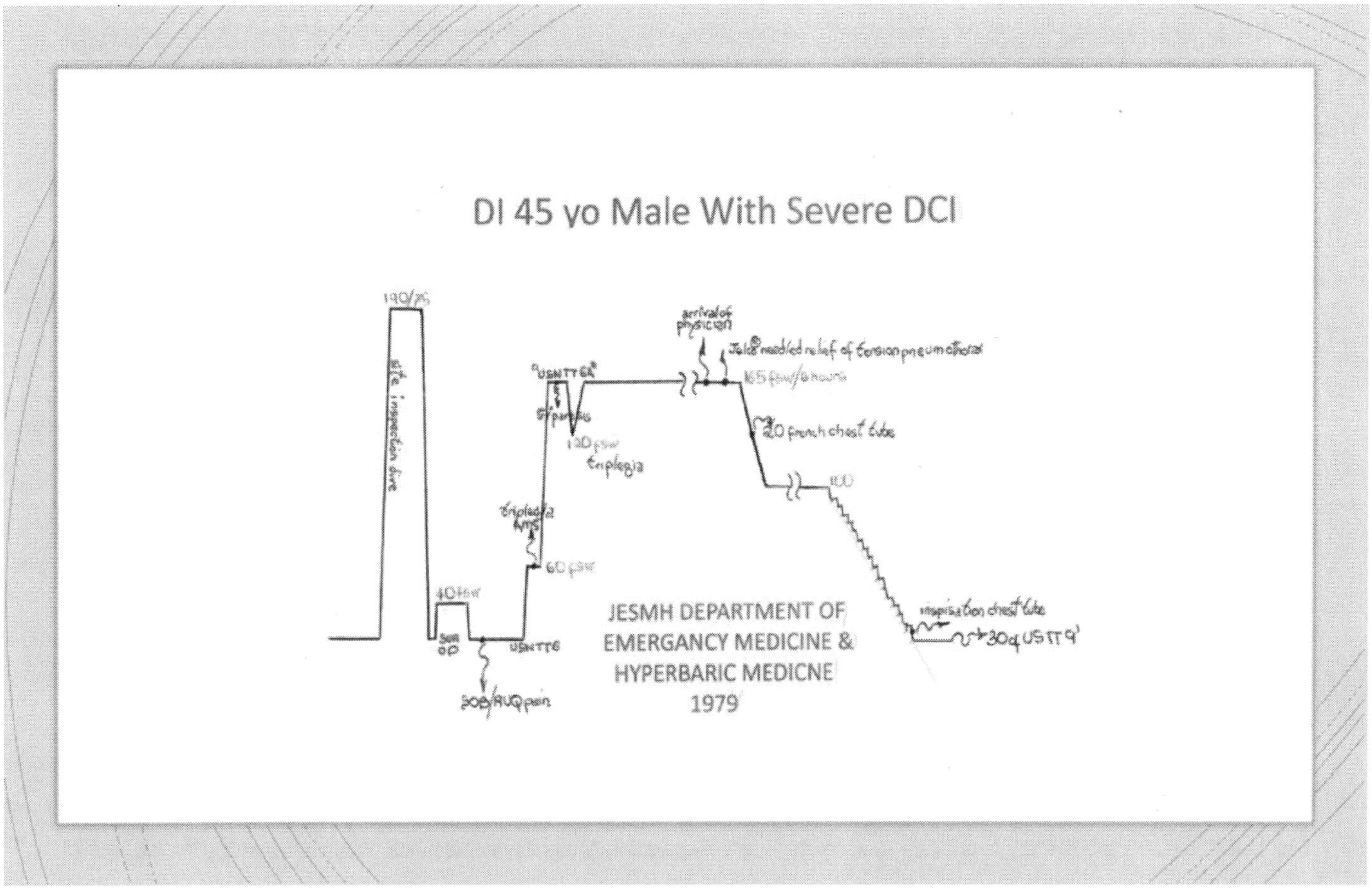
DI 45 yo Male With Severe DCI
190/75
site inspection dive
40 fsw
60 fsw
120 fsw
triplegia
arrival of physician
165 fsw/6 hours
20 french chest tube
100
inspiration chest tube
JESMH DEPARTMENT OF
EMERGANCY MEDICINE &
HYPERBARIC MEDICNE
1979

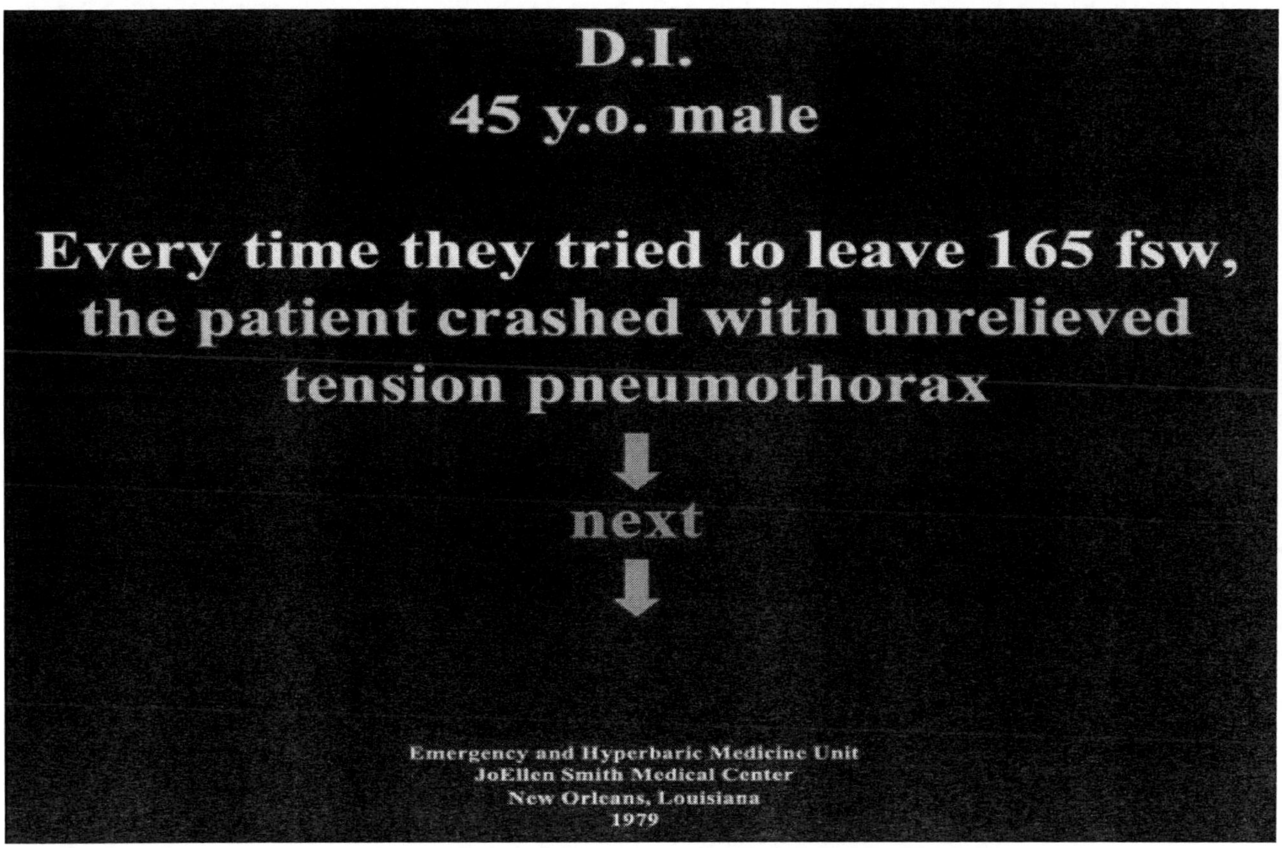
D.I.
45 y.o. male
Every time they tried to leave 165 fsw,
the patient crashed with unrelieved
tension pneumothorax
next
Emergency and Hyperbaric Medicine Unit
JoEllen Smith Medical Center
New Orleans, Louisiana
1979

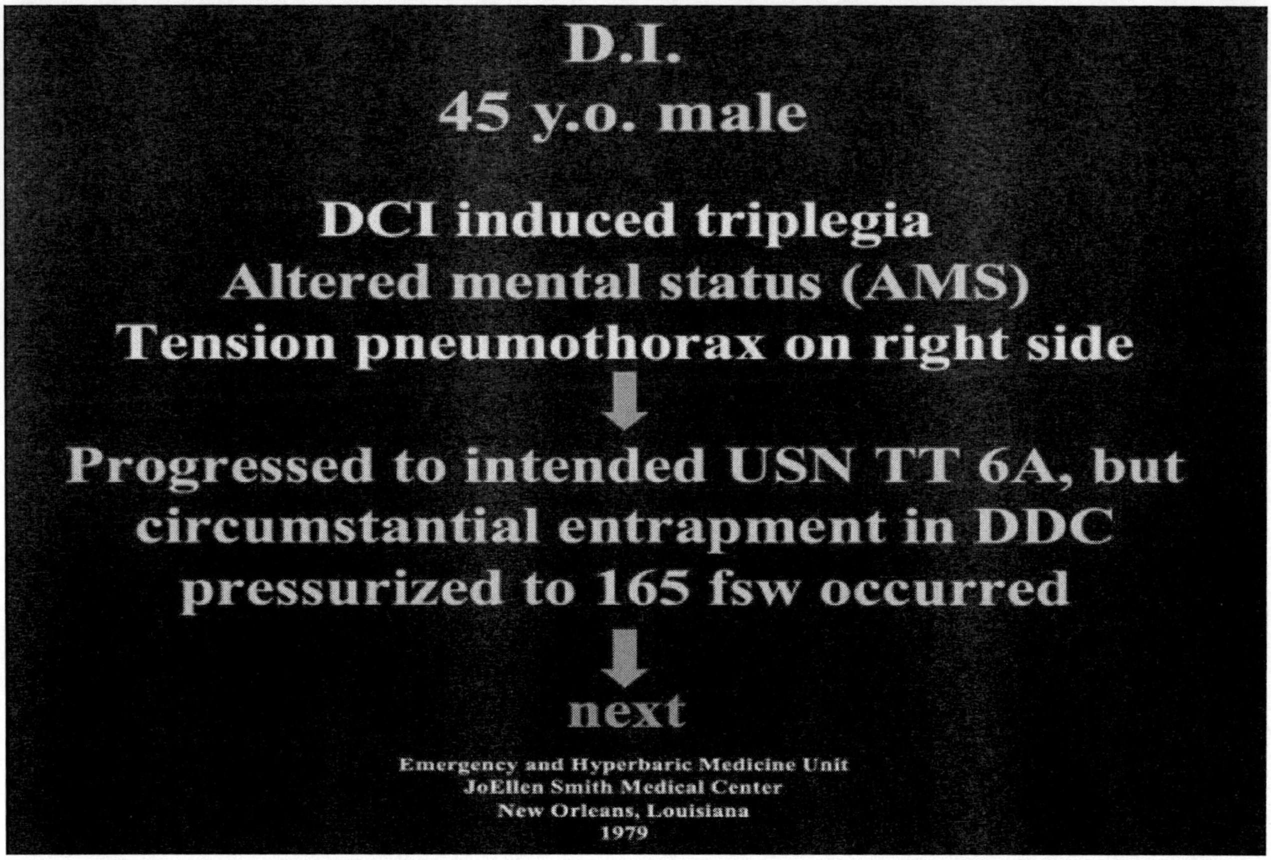
D.I.
45 y.o. male
DCI induced triplegia
Altered mental status (AMS)
Tension pneumothorax on right side
Progressed to intended USN TT 6A, but
circumstantial entrapment in DDC
pressurized to 165 fsw occurred
next
Emergency and Hyperbaric Medicine Unit
JoEllen Smith Medical Center
New Orleans, Louisiana
1979

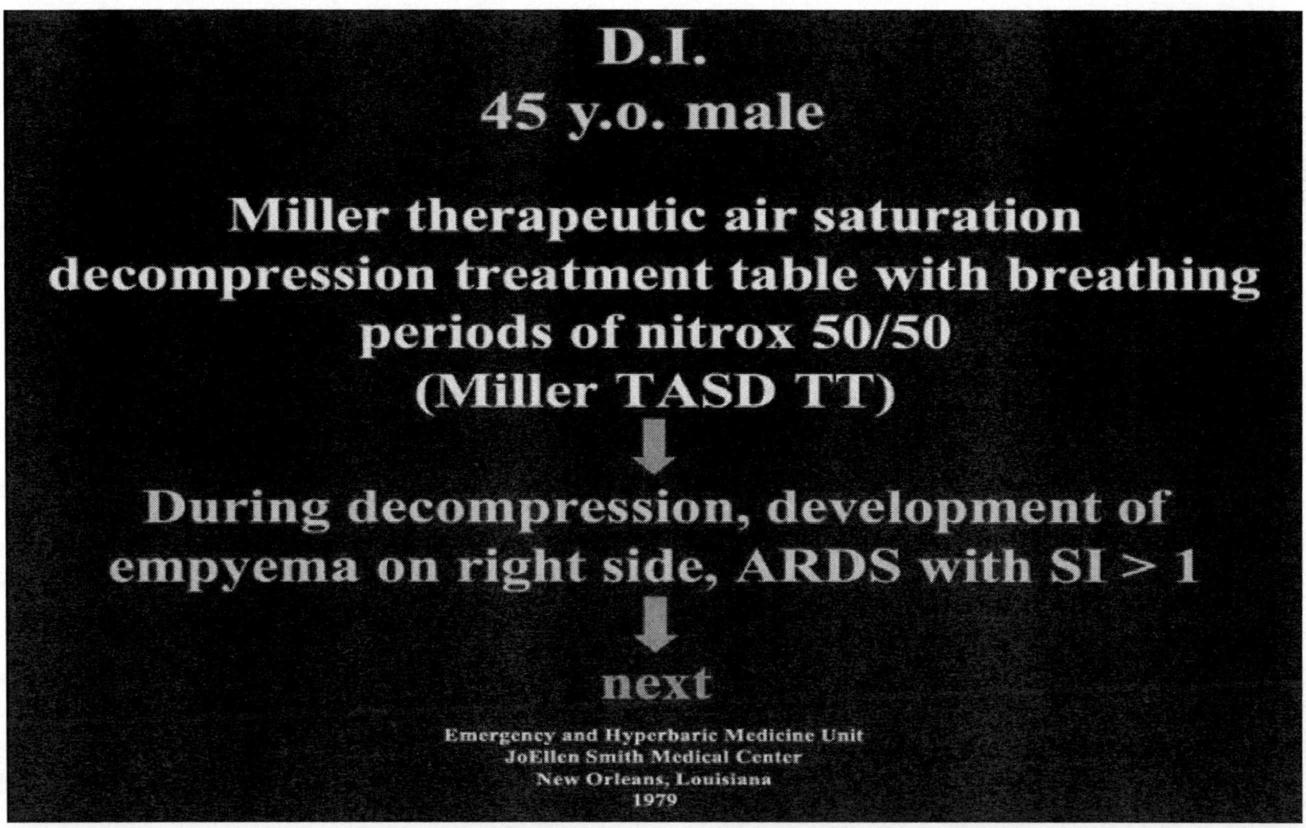
D.I.
45 y.o. male
Miller therapeutic air saturation
decompression treatment table with breathing
periods of nitrox 50/50
(Miller TASD TT)
During decompression, development of
empyema on right side, ARDS with SI > 1
next
Emergency and Hyperbaric Medicine Unit
JoEllen Smith Medical Center
New Orleans, Louisiana
1979

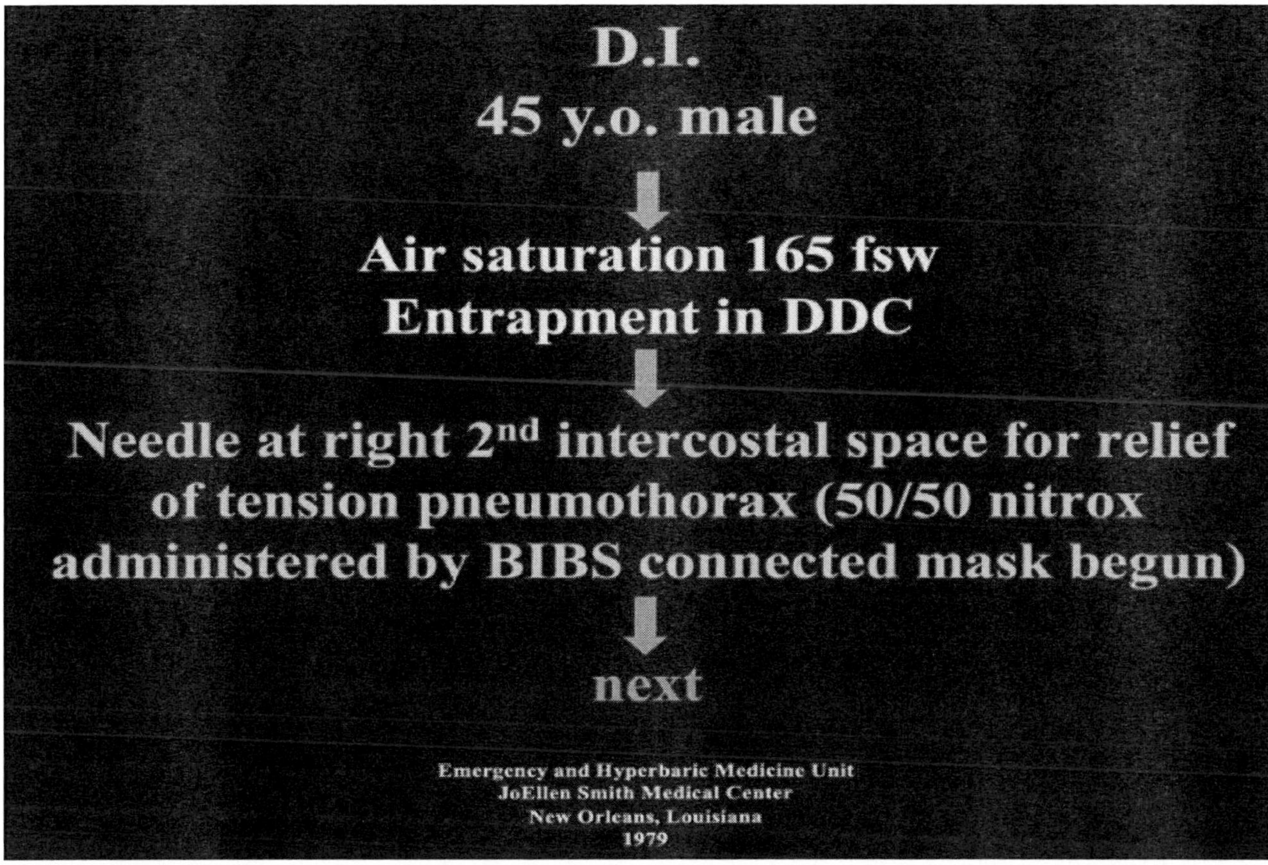
D.I.
45 y.o. male
Air saturation 165 fsw
Entrapment in DDC
Needle at right 2nd intercostal space for relief of tension pneumothorax (50/50 nitrox administered by BIBS connected mask begun)
next
Emergency and Hyperbaric Medicine Unit
JoEllen Smith Medical Center
New Orleans, Louisiana
1979

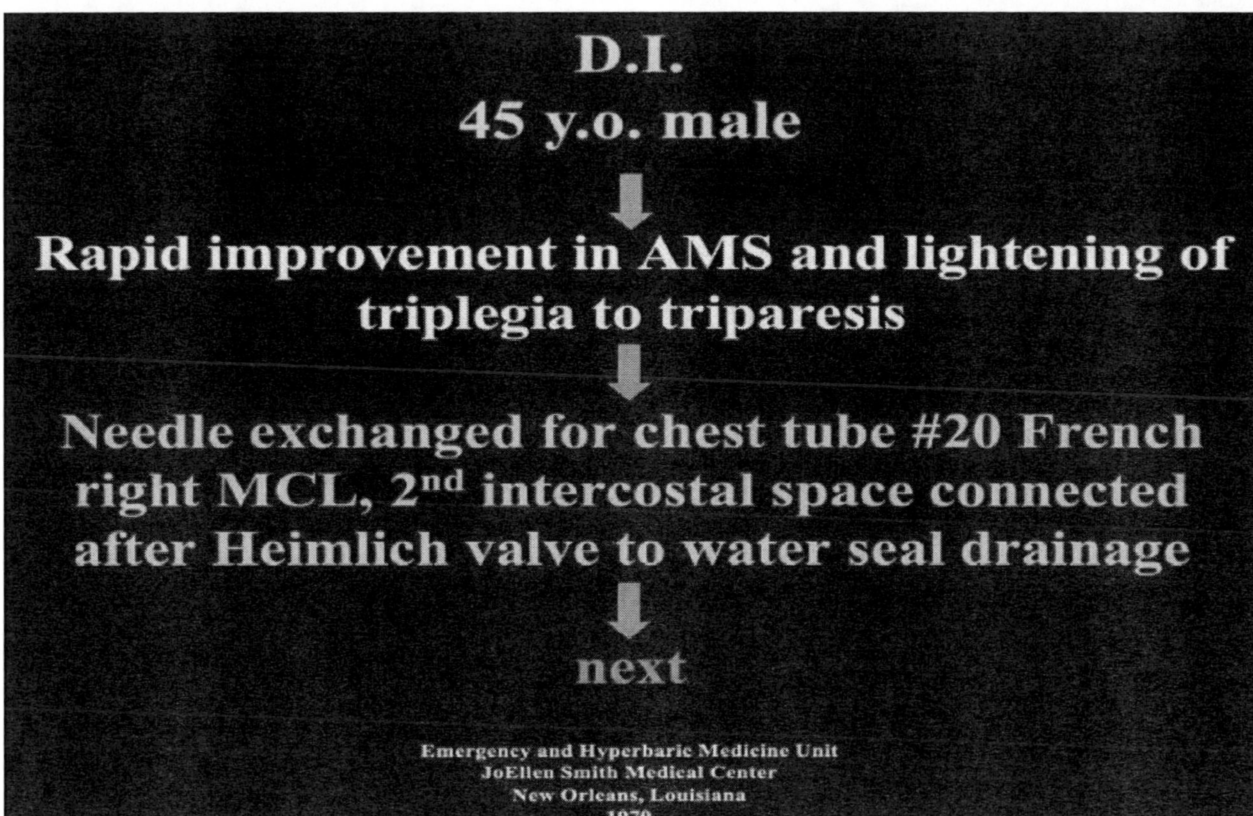
D.I.
45 y.o. male
Rapid improvement in AMS and lightening of triplegia to triparesis
Needle exchanged for chest tube #20 French right MCL, 2nd intercostal space connected after Heimlich valve to water seal drainage
next
Emergency and Hyperbaric Medicine Unit
JoEllen Smith Medical Center
New Orleans, Louisiana
1979

D.I.
45 y.o. male

⬇

IV antibiotics and IV fluid resuscitation
SI > 1!

⬇

Normalization of vital signs with continued progress in resolution of triplegia

Emergency and Hyperbaric Medicine Unit
JoEllen Smith Medical Center
New Orleans, Louisiana
1979

45 yo male, 1979, first DCI case in literature with long series of tailing treatments to accelerate resolution of lingering neurologic injury

D.I.

45 y.o. male

⬇ **Total time in DDC – 60 hours**

Dried inspissation developed in patient's chest tube shortly before the DDC reached surface

⬇

Retensioning of right pneumothorax

Emergency and Hyperbaric Medicine Unit
JoEllen Smith Medical Center
New Orleans, Louisiana
1979

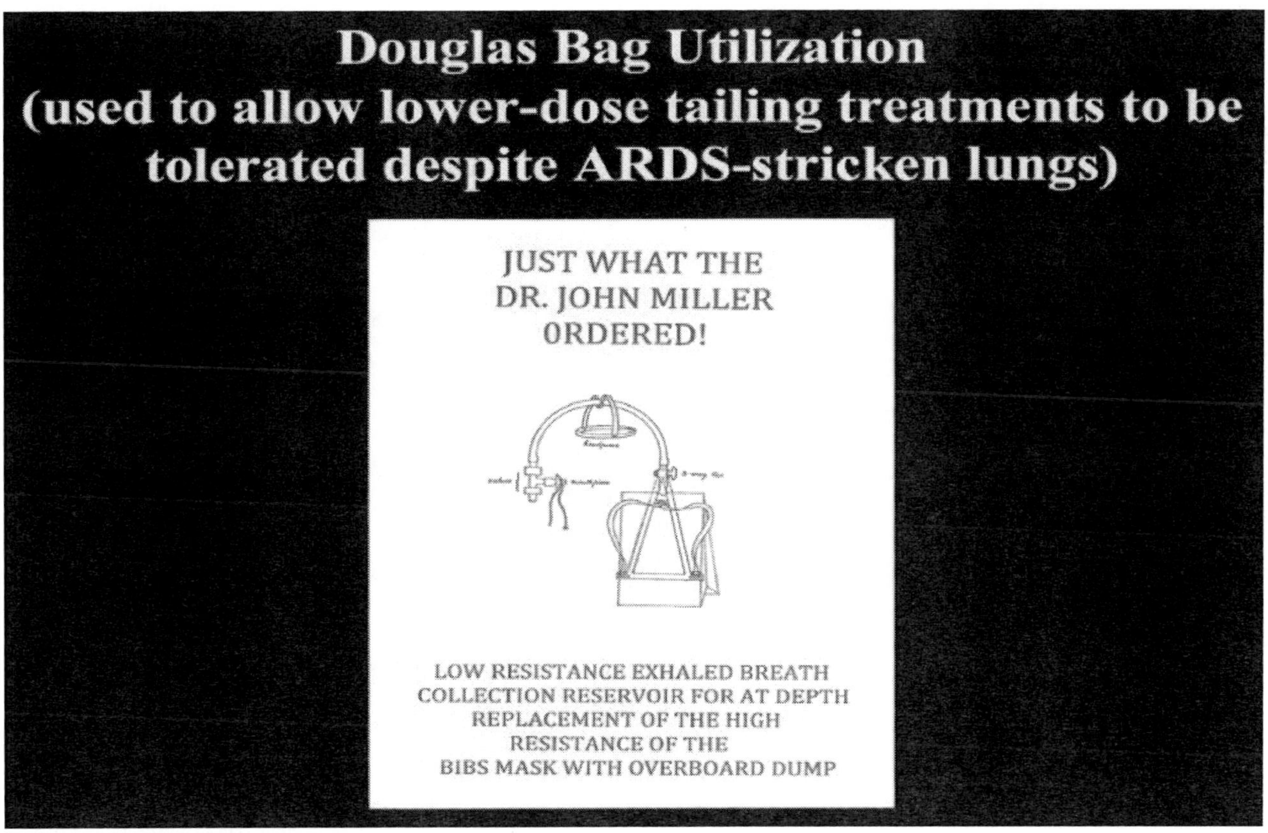
Douglas Bag Utilization
(used to allow lower-dose tailing treatments to be tolerated despite ARDS-stricken lungs)
JUST WHAT THE
DR. JOHN MILLER
ORDERED!
LOW RESISTANCE EXHALED BREATH
COLLECTION RESERVOIR FOR AT DEPTH
REPLACEMENT OF THE HIGH
RESISTANCE OF THE
BIBS MASK WITH OVERBOARD DUMP

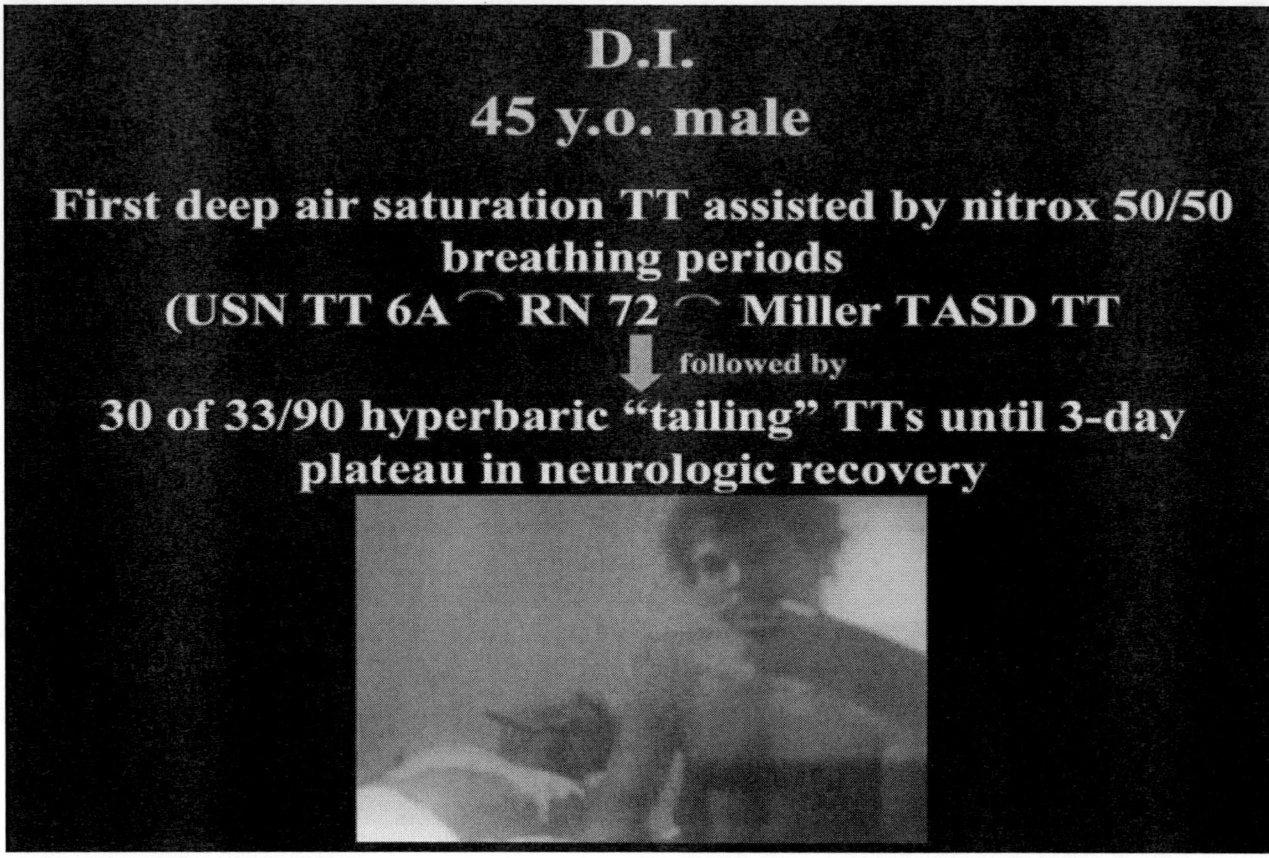
D.I.
45 y.o. male
First deep air saturation TT assisted by nitrox 50/50 breathing periods
(USN TT 6A ⌒ RN 72 ⌒ Miller TASD TT
followed by
30 of 33/90 hyperbaric “tailing” TTs until 3-day plateau in neurologic recovery

GoM DIVING SAFETY WORK GROUP
UNTARNISHED!

GoM DIVING SAFETY WORK GROUP

A relook at the emergent conditions treated by HBOT

FDA/CMS Approved Hyperbaric Oxygen Treatable Conditions

(with emphasis on emergent conditions needing 24/7 availability of treatment)

1. CRUSH INJUREY
2. COMPARTMENT SYNDROME
3. NECROTIZING FASCIITIS (NNT for limb salvage = 3 with [illegible]/antibiotics/HBOT)
4. CRAO/CRVO
5. ACUTE PERIPHERAL LIMB ISCHEMIA
6. GAS GANGRENE
7. FAILING FLAPS AND GRAFTS
8. LIMB THREATING DFUs (WG 4-5) (NNT for limb salvage=4 with antibiotics/HBOT)
9. CO/HCN/H_2S TOXICITY
10. BURNS
11. IATROGENIC ARTERIAL GAS EMBOLI
12. DECOMPRESSION ILLNESS

(soon to be approved)

13. TBI
14. ACUTE HEARING LOSS

(non-emergent approved conditions)

15. DFU(WG3)/STRN/ORN/CROM

R.P.
35 y.o. male

23 minute, non-intervened-upon cardiopulmonary arrest

(Morel-Levallée head/neck suction injury before blow-up ascent during commercial dive)

Mississippi River
Destrehan, Louisiana
1983

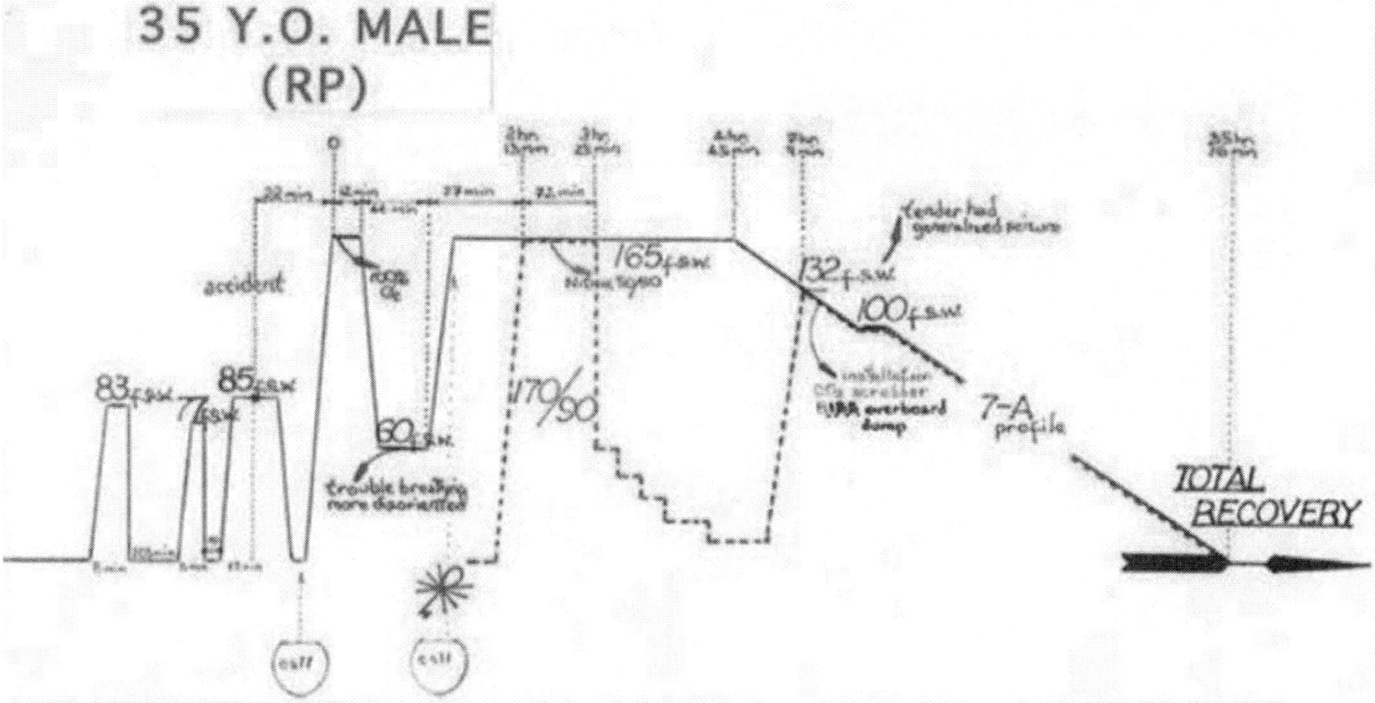

J.E.S.M.H. DEPARTMENT OF EMERGENCY & HYPERBARIC MEDICINE, NEW ORLEANS, LOUISIANA, 1983

GoM Diving Safety Work Group

Adjusting Gear Before The Fateful Dive

(Construction of the Desterhan Cable Stay Bridge support towers)

Mississippi River 1983

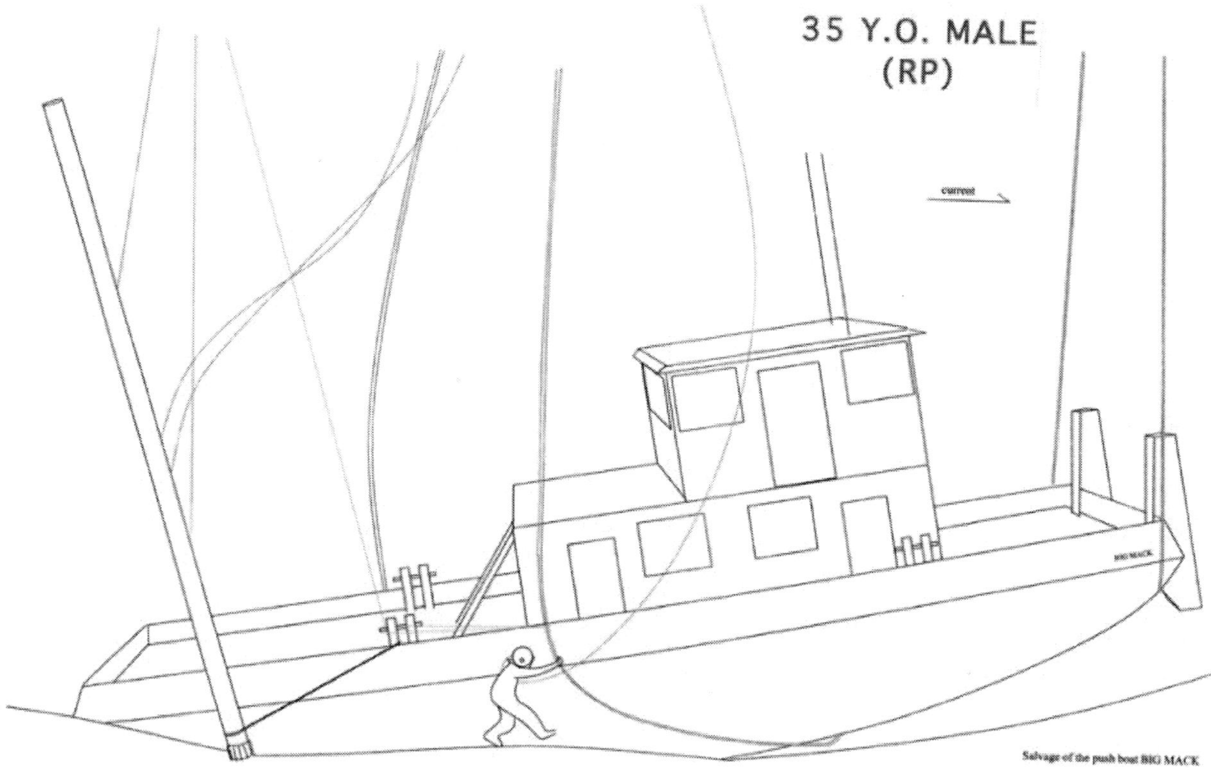

Salvage of the push boat BIG MACK

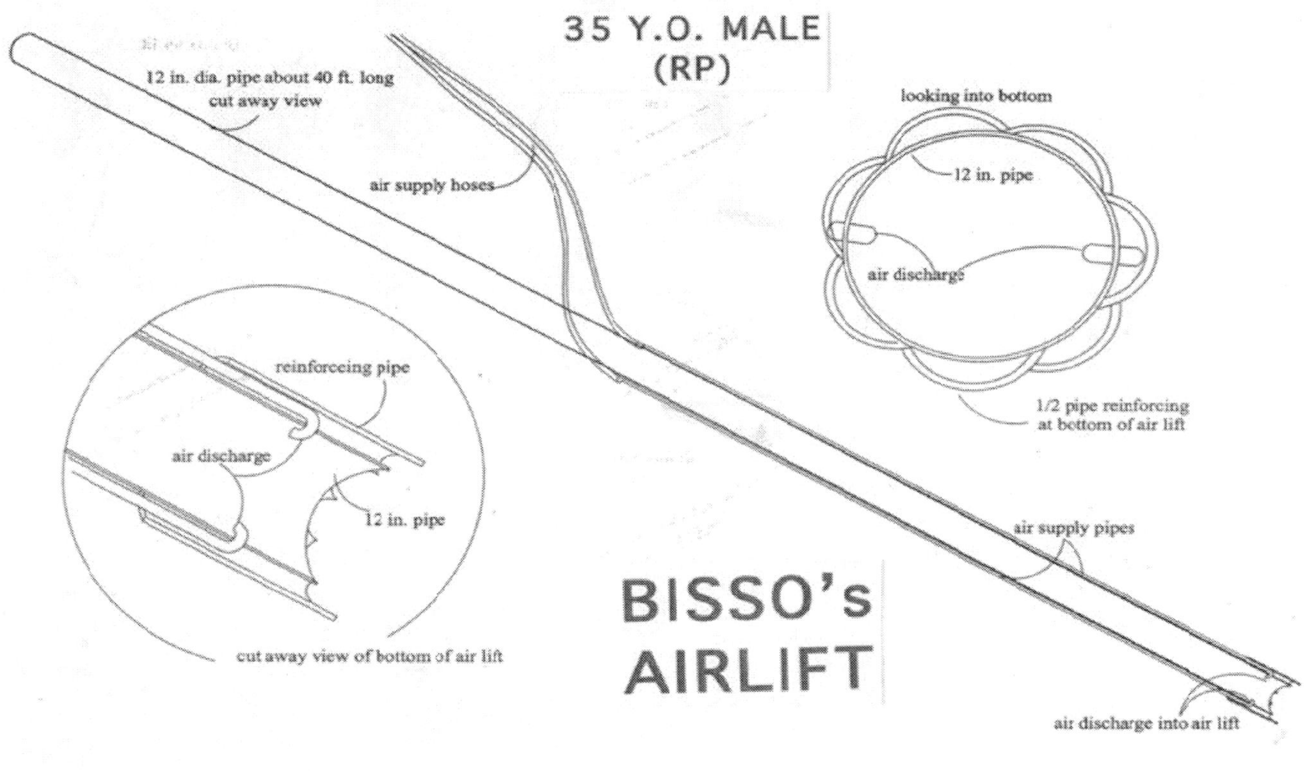

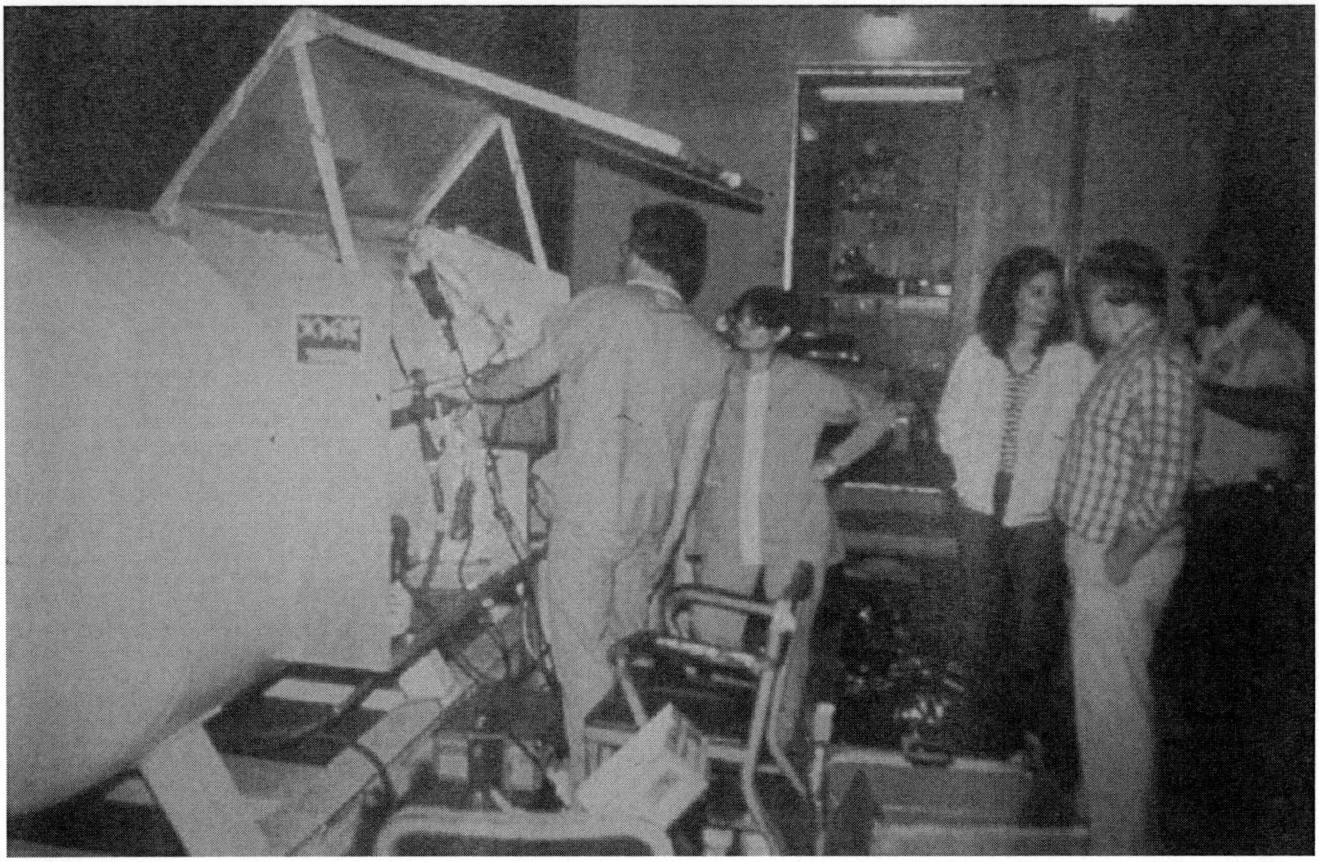

GoM Diving Safety Work Group

EXITED CHAMBER AFTER 3 ½ DAYS OF THERAPEUTIC DECOMPRESSION

MOREL LEVALLEE

SCALP/FACE DEGLOVING INJURY WITH ASPHYXIATION/CARDIOPULMONNARY ARREST

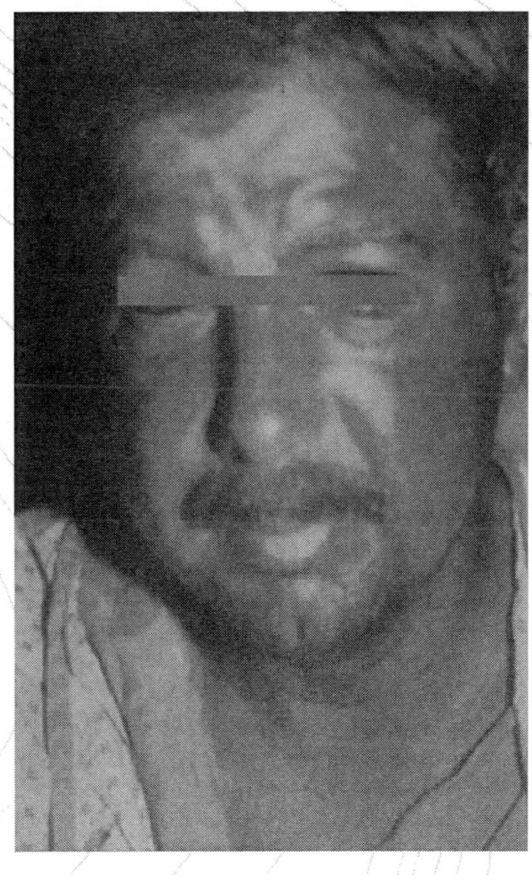

GoM DIVING SAFETY WORK GROUP

ONE MONTH LATER

Day long psychometric testing at the Tulane School of Medicine Department of Neurology

120 IQ Fully Intact!!

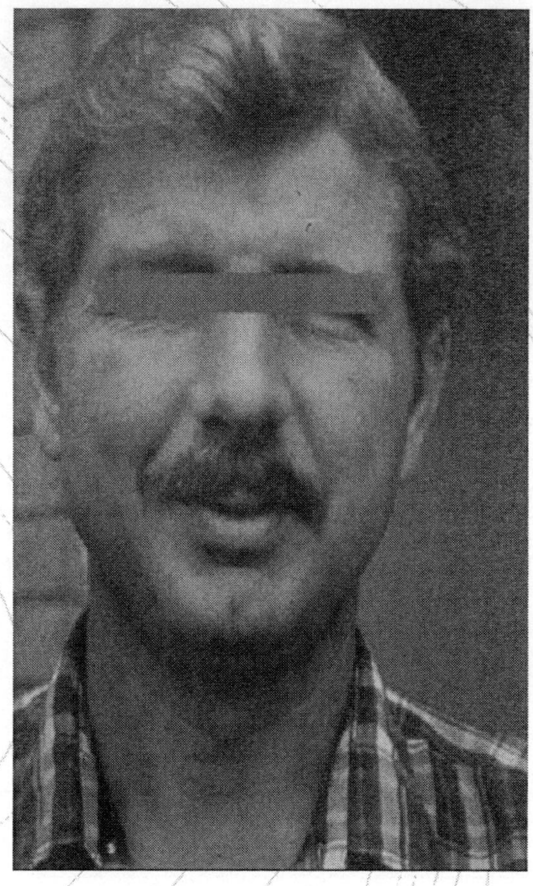

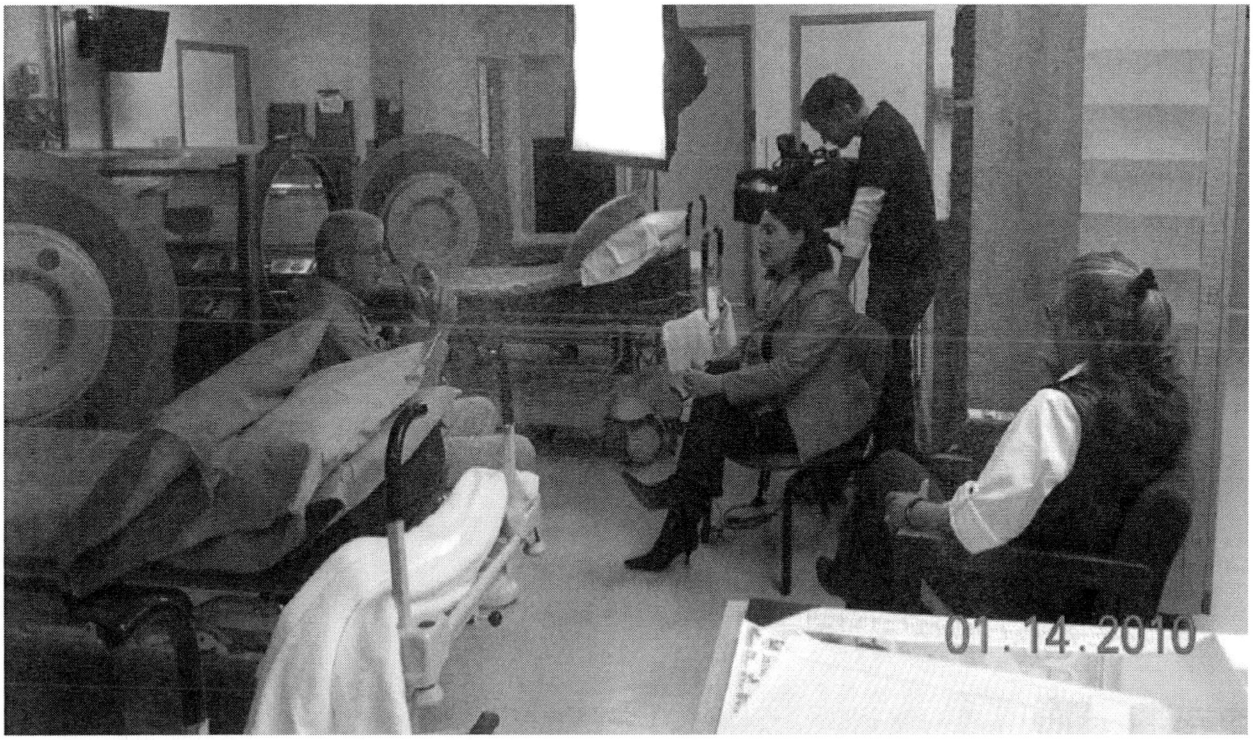

GoM DIVING SAFETY WORK GROUP

TO THE RESEARCH LAB TO REPRODUCE THE INSULT IN A CONTROLLED PORCINE STUDY

23 MINUTE CARDIOPULMONARY ARREST

WITHOUT CPR TO BEGIN FOR 23 MINUTES

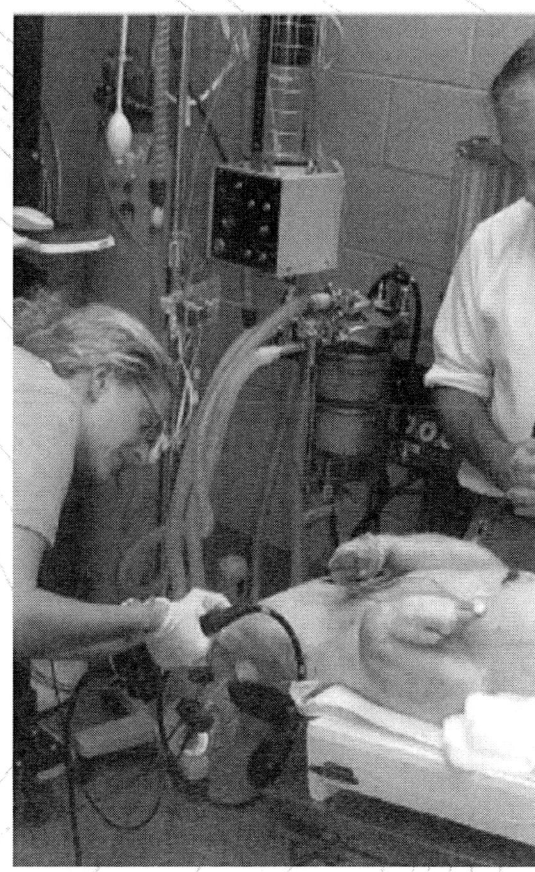

BRI (NOLA) Experimental
Hyperbaric Resuscitation Chamber

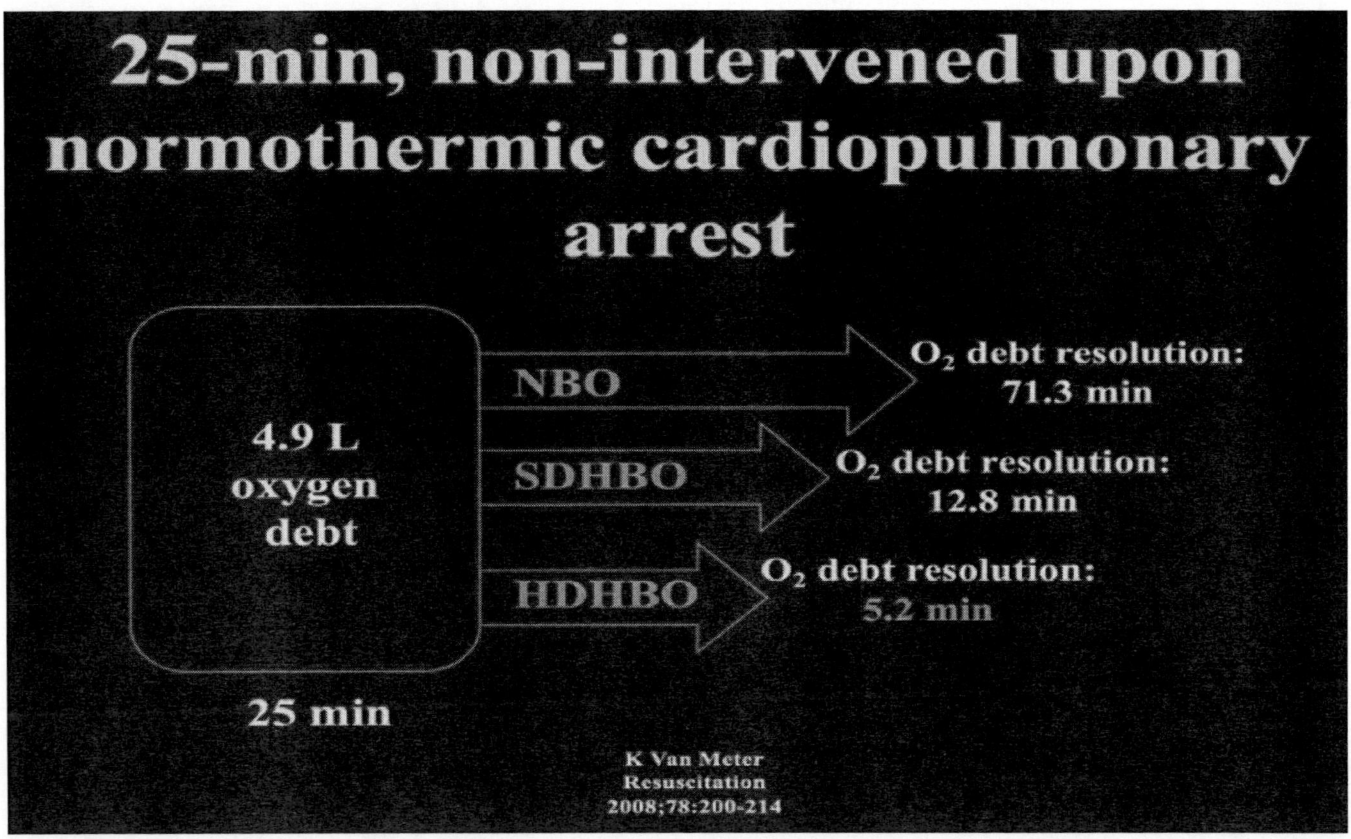
25-min, non-intervened upon normothermic cardiopulmonary arrest
4.9 L oxygen debt
25 min
NBO
O2 debt resolution: 71.3 min
SDHBO
O2 debt resolution: 12.8 min
HDHBO
O2 debt resolution: 5.2 min
K Van Meter
Resuscitation
2008;78:200-214

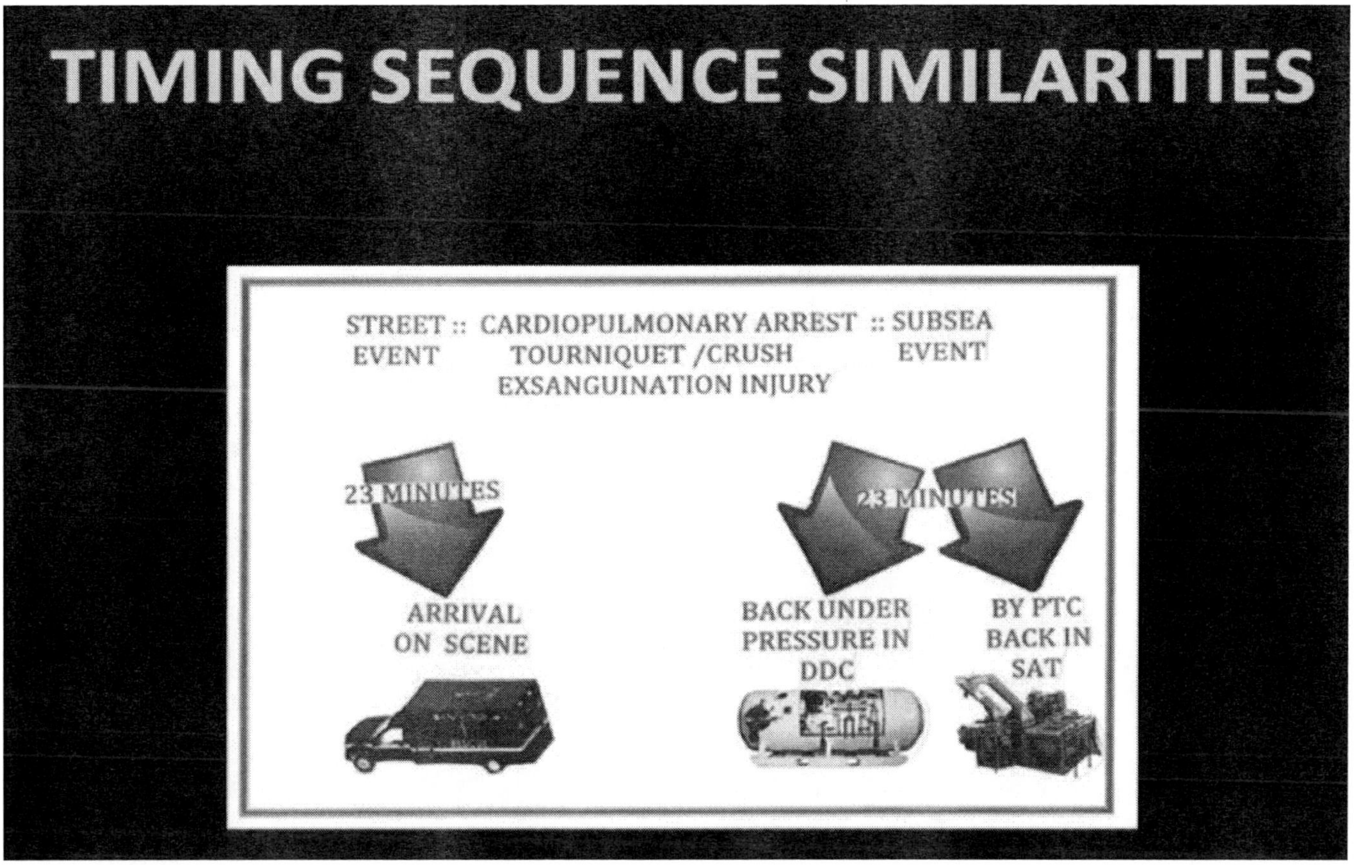
TIMING SEQUENCE SIMILARITIES
STREET :: CARDIOPULMONARY ARREST :: SUBSEA
EVENT TOURNIQUET /CRUSH EVENT
EXSANGUINATION INJURY
23 MINUTES
23 MINUTES
ARRIVAL ON SCENE
BACK UNDER PRESSURE IN DDC
BY PTC BACK IN SAT

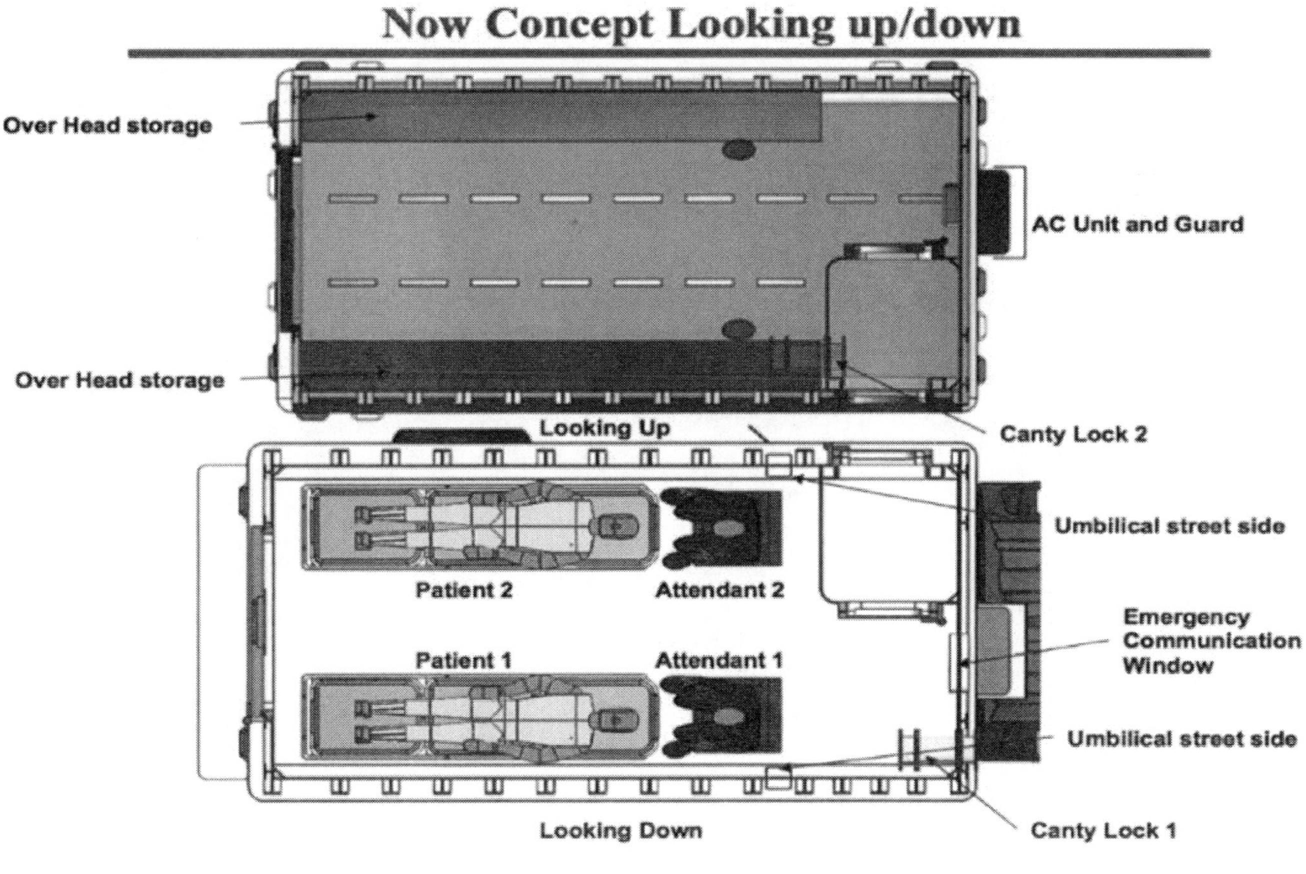
Now Concept Looking up/down
Over Head storage
AC Unit and Guard
Over Head storage
Looking Up
Canty Lock 2
Umbilical street side
Patient 2
Attendant 2
Emergency Communication Window
Patient 1
Attendant 1
Umbilical street side
Looking Down
Canty Lock 1

Key point to be made:

The commercial diving field site provides a hyperbaric chamber on top of the accident awaiting immediate intake of the sick and injured
(often inadvertently to provide HBOT for septic shock/crush injury.cardiopulmonary arrest/exsanguination)

Today non-hyperbaric ambulances provide on site immediate intake of the sick and injured
(septic shock/crush injury/cardiopulmonary arrest/exsanguination)

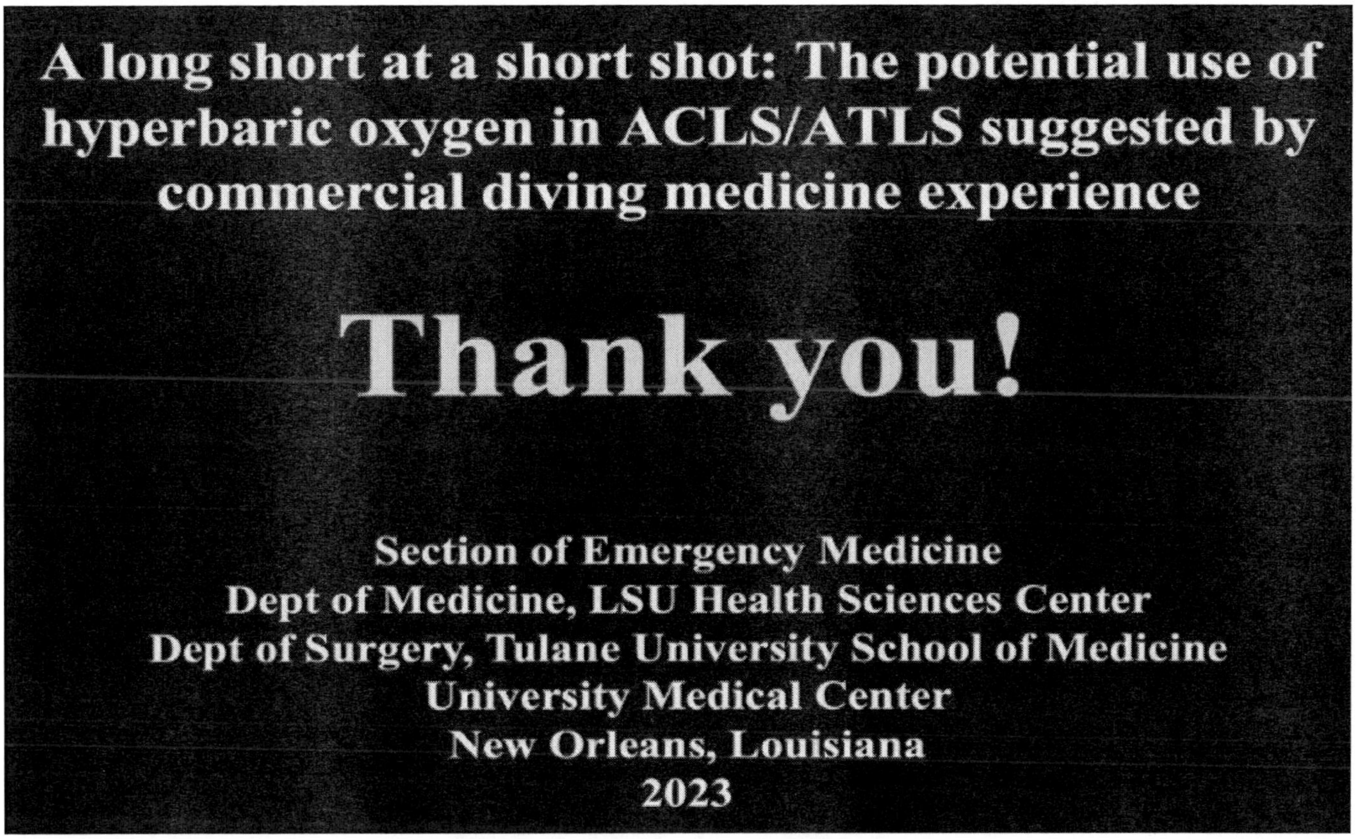
A long short at a short shot: The potential use of hyperbaric oxygen in ACLS/ATLS suggested by commercial diving medicine experience
Thank you!
Section of Emergency Medicine
Dept of Medicine, LSU Health Sciences Center
Dept of Surgery, Tulane University School of Medicine
University Medical Center
New Orleans, Louisiana
2023

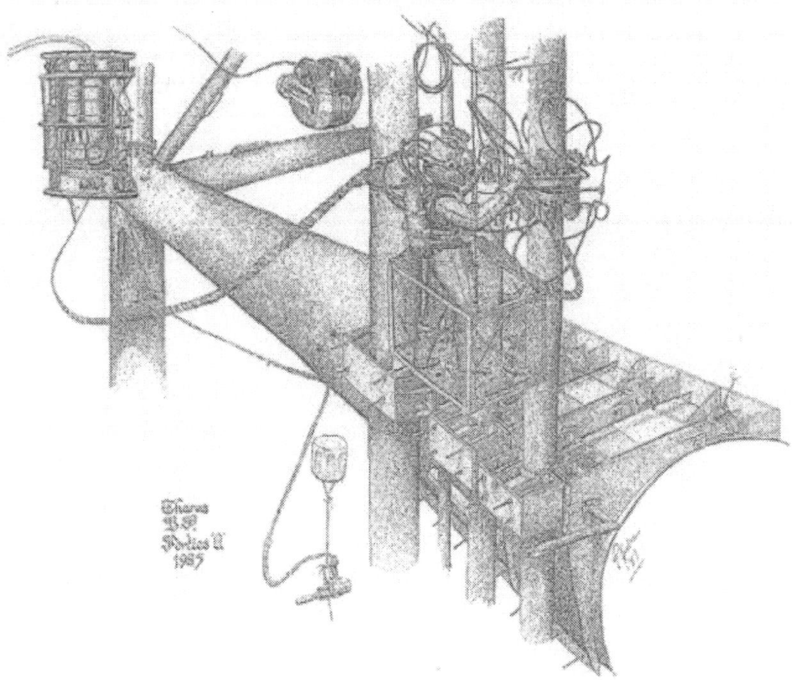

"ECMOID" ROLL OF HYPERBARIC OXYGEN THERAPY AS AN ADJUNCTIVE ADDITION TO ACLS/ATLS

QUESTIONS?

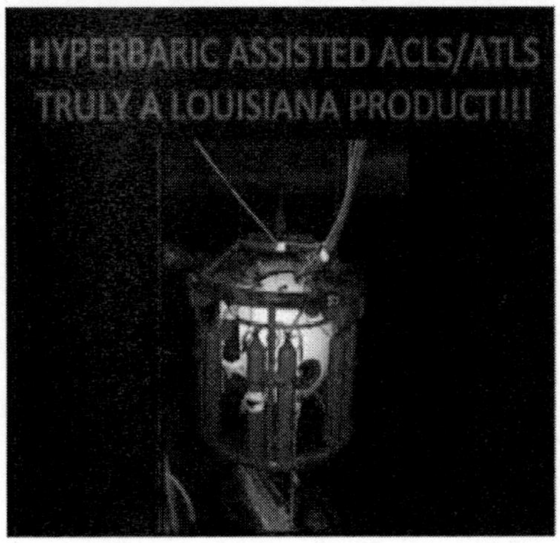

DSWG AT UI 2023

What the LCMC-UMC LEVEL 1 Trauma Center Is Temporarily Left With!

We Ask For Help To Advocate For More

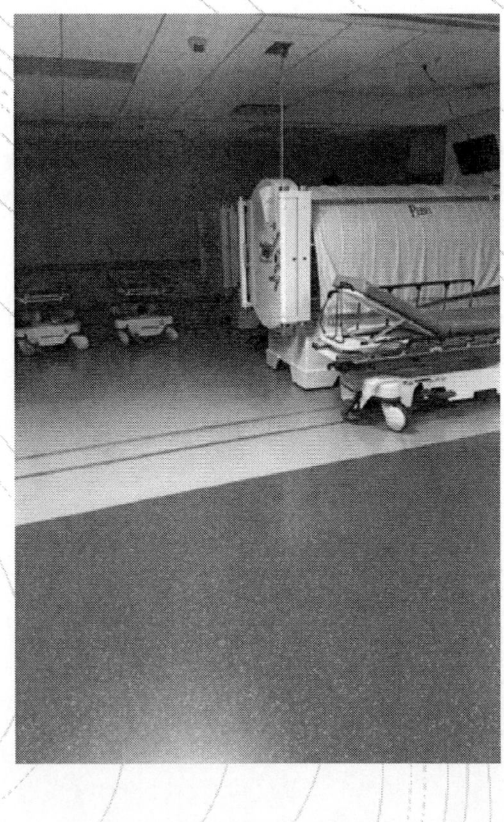

GoM Diving Safety Work Group

Please on Letterhead
Send Letters of Support
for a Critical Care Multiplace
Hyperbaric Chamber at
a Level 1 Trauma Center

Email Letter of Support
For a Level 1
Trauma Center Critical Care
Multiplace HBOT Chamber
Available 24/7 to:
Chancellor Steve Nelson MD
SNelso1@lsuhsc.edu

Or Snail-Mail to :
Chancellor Steve Nelson MD
LSUHSC
433 Bolivar Street
New Orleans,
Louisiana, 70112

GRi
SIMULATIONS
Net-Zero Digital Twin Approaches to Marine Autonomy R&D Presentation
INTERNATIONAL
WORKBOAT

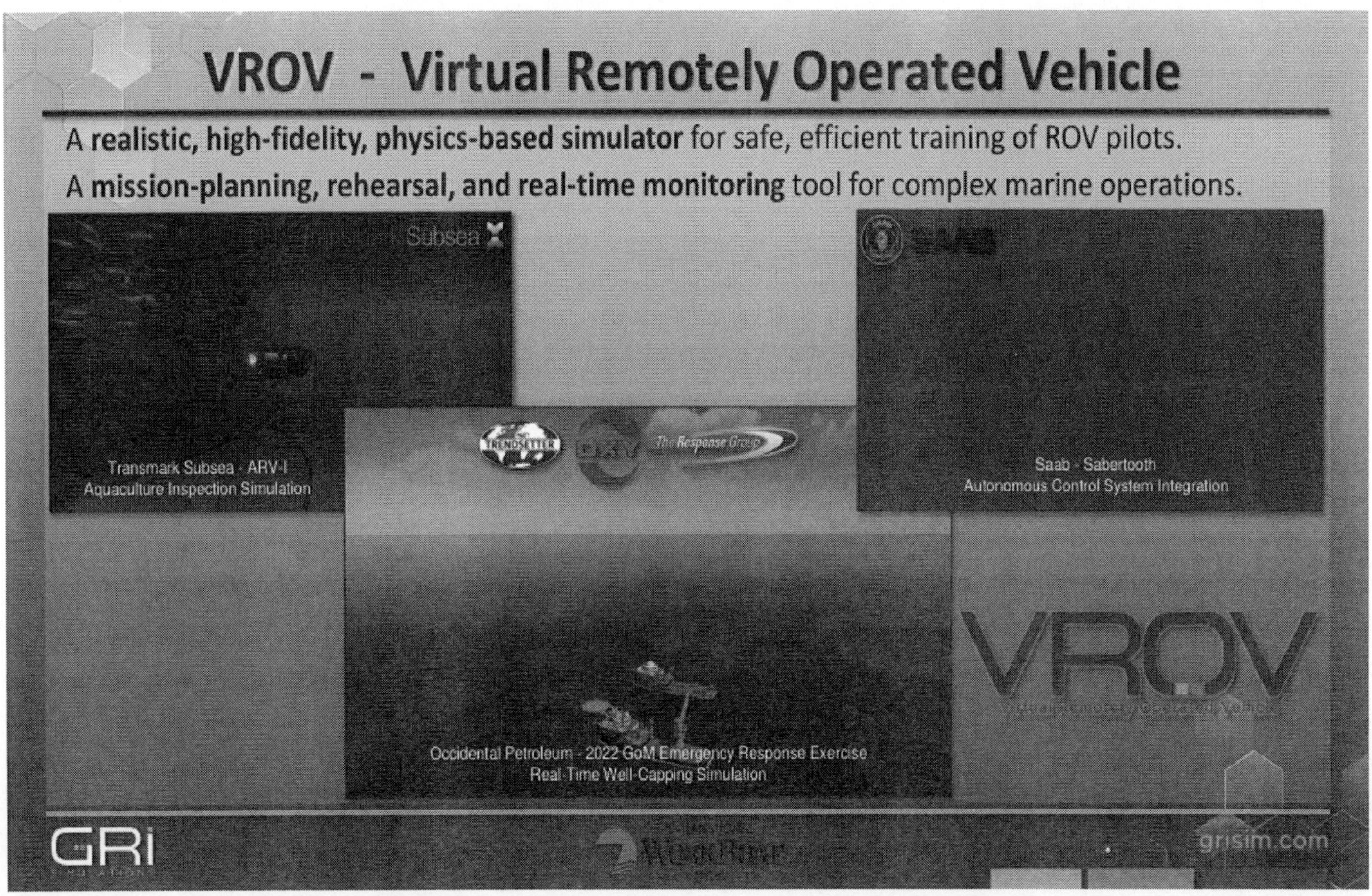
VROV - Virtual Remotely Operated Vehicle
A realistic, high-fidelity, physics-based simulator for safe, efficient training of ROV pilots.
A mission-planning, rehearsal, and real-time monitoring tool for complex marine operations.
Subsea
SAAB
Transmark Subsea - ARV-I
Aquaculture Inspection Simulation
TRENDSETTER
The Response Group
Saab - Sabertooth
Autonomous Control System Integration
VROV
Virtual Remotely Operated Vehicle
Occidental Petroleum - 2022 GoM Emergency Response Exercise
Real-Time Well-Capping Simulation
GRi
SIMULATIONS
grisim.com

Simulation for Robotics Operator Training (Canadian Navy)

Simulation serves as **an invaluable asset in the training of robotics operators**, enhancing the learning experience and ensuring that operators are well-equipped with the necessary skills and knowledge to handle real-world robotic systems efficiently and safely. Here are several reasons why simulation is so beneficial in this context:

- **Safety** - risk reduction, emergency scenarios
- **Cost-Efficiency** - reduced wear and tear, resource saving
- **Standardized Training** - consistency, customized scenarios
- **Hands-on Experience** - practical skills, familiarity
- **Variety of Scenarios** - diverse situations, adaptability
- **Performance Analysis** - immediate feedback, improvement tracking
- **Customizability and Flexibility** - adjustable difficulty, personalized learning paths
- **Access and Availability** - remote access, 24/7 availability
- **Environmentally Friendly** - reduced carbon footprint, sustainable
- **Realistic Environment** - high-fidelity simulations, immersive experience
- **Continual Learning and Updating** - up-to-date content

Royal Canadian Navy - Build in Canada Innovation Program (BCIP)

GRi grisim.com

Simulation for Robotics Operator Training (MI)

Marine Institute's ROV Operator program is the only program of its kind in Canada, with GRi's VROV Simulator setup providing an integral part of the training experience.

BP Atlantis - Equipment Installation and Operation Simulations (donated to Marine Institute following project completion)

Marine Institute - Underwater Exploration Laboratory Ridge Road Campus, NL

GRi grisim.com

Simulation for Robotics Development (AIV)

Simulation is **a critically valuable tool for robotics development** for several reasons, including:

- **Cost-Efficiency** - virtual prototyping, cost reduction of maintenance and repair
- **Safety** - risk-free testing in virtual harsh environments
- **Flexibility and Scalability** - variety of environments, scenarios, and robotics systems
- **Repeatability** - consistent conditions, automated testing
- **Algorithm Development and Testing** - machine learning, real-time testing
- **Debugging and Troubleshooting** - problem identification, code testing
- **Human-Robot Interaction** - user experience, behaviour testing
- **Community and Collaboration** - open source/open architecture, shared environments and tasks
- **Accelerated Development** - quick iterations, parallel testing
- **Customizability** - task specificity, sensor and actuator configuration

Simulation for Robotics Development (Loggerhead)

VROV Comms integration with ROS (Robot Operating System) to enhance development of autonomous systems.

VROV Comms allowed Honuworx to control all aspects of the prototype Loggerhead vehicle, including the remote control of:

ROS HonuWorx

- Unlatching and relatching of the Loggerhead Comms buoy
- Control of the Comms buoy buoyancy
- Operation of the Comms buoy winch, including cable length feedback
- Sensor feedback for water detects and buoy redocking with the Loggerhead sub
- Control of the Loggerhead buoyancy modules

Honuworx also took advantage of the VROV Comms waypoint transit and hover modes, allowing for rapid control of the Loggerhead vehicle to help test its remote communications system.

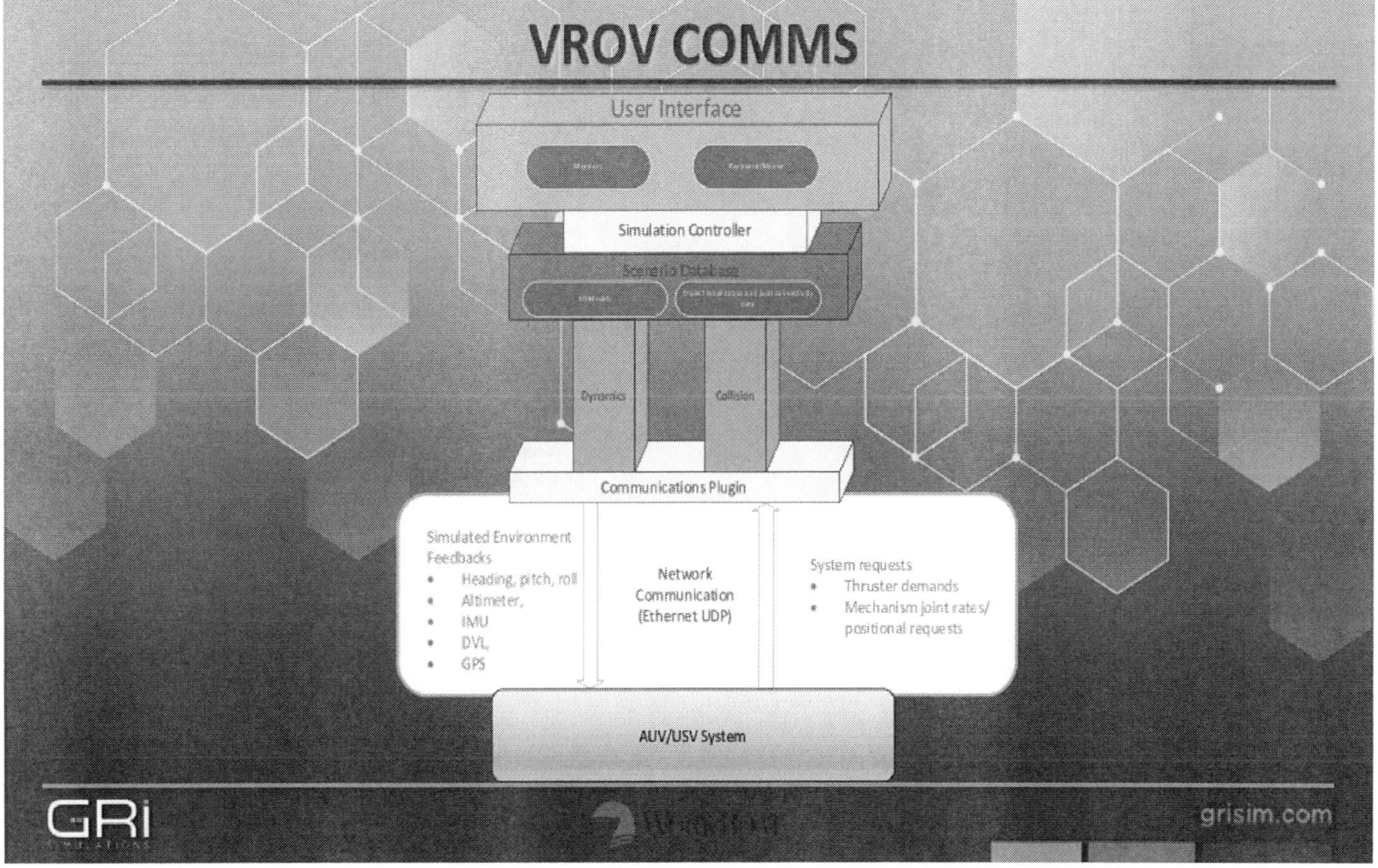
VROV COMMS
User Interface
Simulation Controller
Scenario Database
Dynamics
Collision
Communications Plugin
Simulated Environment Feedbacks
• Heading, pitch, roll
• Altimeter,
• IMU
• DVL,
• GPS
Network Communication (Ethernet UDP)
System requests
• Thruster demands
• Mechanism joint rates/ positional requests
AUV/USV System
GRi
grisim.com

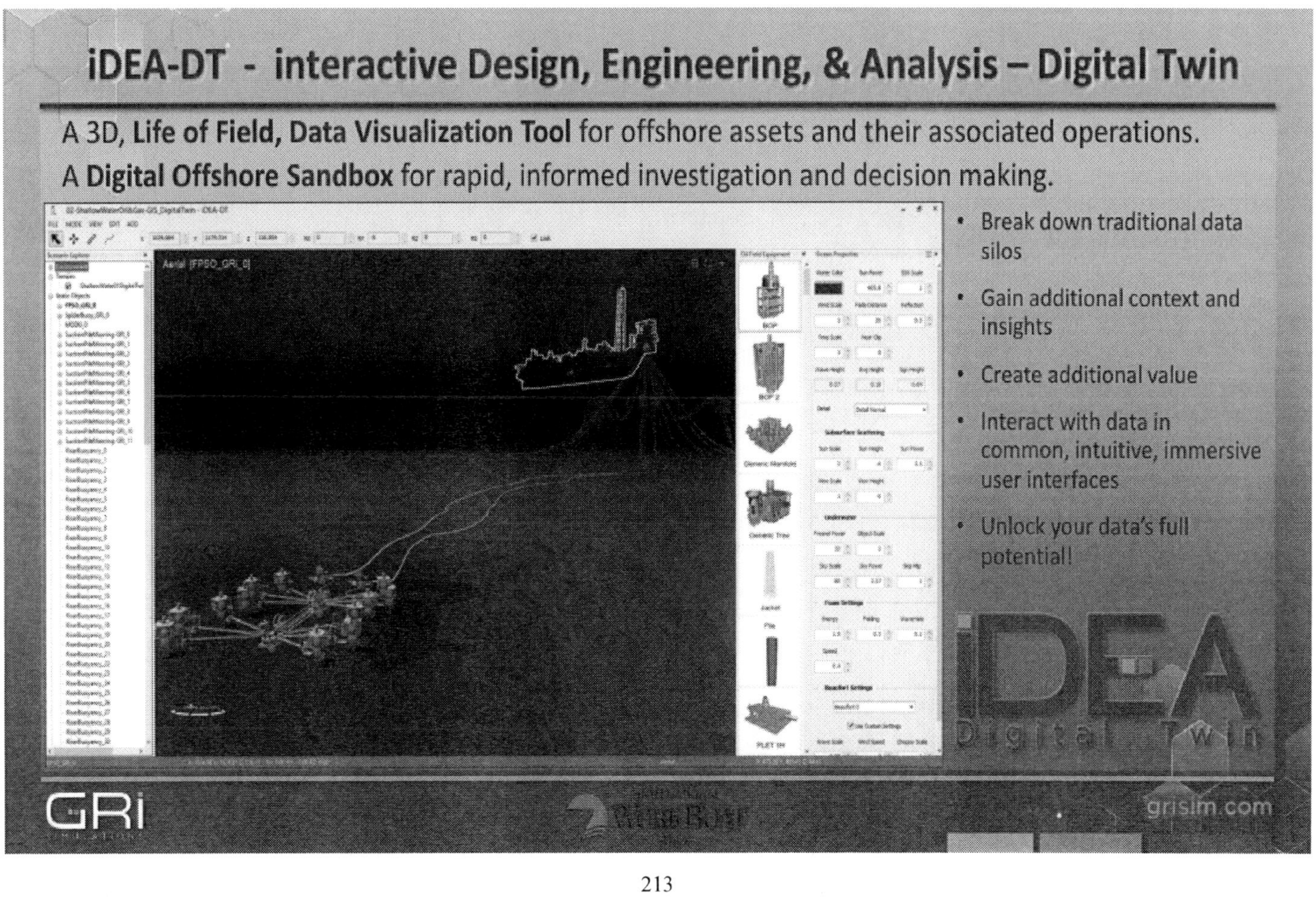
iDEA-DT - interactive Design, Engineering, & Analysis – Digital Twin
A 3D, Life of Field, Data Visualization Tool for offshore assets and their associated operations.
A Digital Offshore Sandbox for rapid, informed investigation and decision making.
• Break down traditional data silos
• Gain additional context and insights
• Create additional value
• Interact with data in common, intuitive, immersive user interfaces
• Unlock your data's full potential!
iDEA
Digital Twin
GRi
grisim.com

Digital Twin Data to Enhance Simulation Environments

Digital twin data is highly beneficial for simulations due to several key reasons:

- **High Fidelity Replication** - more accurate predictions and analyses
- **Real-Time Data Integration** - immediate and relevant decision-making
- **Predictive Analysis and Forecasting** - understanding potential outcomes, identifying risks, and planning preventive maintenance
- **Iterative Testing and Optimization** - optimizing designs, processes, and systems
- **Cost and Time Efficiency** - especially valuable in complex systems like manufacturing processes, urban planning, and aerospace
- **Risk Reduction** - crucial in high-stakes environments like offshore energy, and mining
- **Customization and Flexibility** - allows more focused and relevant simulations
- **Lifecycle Management** - enable monitoring and analysis throughout the entire lifecycle of an asset
- **Integration with Other Technologies** - enhanced predictive analytics and automated decision-making

GRi grisim.com

Digital Oceans Canada (DOC) Platform Overview

The DOC Platform **drives innovation, enhances value creation, and reduces time to market** by providing **research, development and digital twin technologies and data** to **Canada's Ocean Industry** through a uniquely centralized and **secure open collaboration cloud environment**

- Industry-standard interoperability using OSDU
- Data model definition and transformation
- Structured and unstructured data storage and ingestion
- Software hosting and licensing
- Sandbox environment with industry-wide data and applications for research and development
- Accelerated development in a pre-built, secure cloud environment

DIGITAL OCEANS CANADA

- Openly available, limited access or controlled-access environments to manage data and IP
- Digital Twin simulations
- Process Modelling and Optimization
- AI and Machine Learning Capability
- Hybrid Connectivity and Integration to on-prem technology
- Identity control and access management

GRi grisim.com

DOC Use Case 2 - Intelligent Subsea Field Models

Pilot Project Goal - Build Intelligent Subsea Field Models (Digital Twins) to support development of advanced autonomy in offshore remote intervention systems (ROVs, AUVs, UIDs), and enhanced Asset Integrity Management (AIM) services for offshore assets.

Challenge - Sea-bed resident UIDs are on the cusp of becoming commercially viable for IMR tasks for subsea installations. Their usage in the harsh environment of Canada's East Coast is unproven, and subsea inspection challenges are compounded by unique subsea architectures (e.g. placement in excavated drill centers).

Core R&D - Build models of subsea field architecture and other marine robotics operating environments and assess feasibility/limits of a seabed-resident UIDs to support a variety of IMR tasks. (Starting TRL: 4/Final TRL: 8)

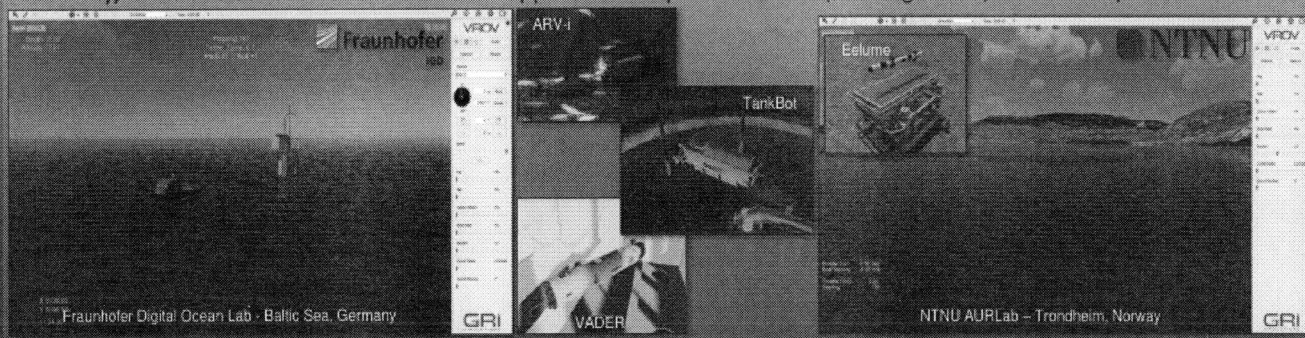

GRi grisim.com

MI - The Launch (Newfoundland, Canada)

- Topology pulled from Google Maps
- Bathymetry data supplied by MI
- Subsea Production Tree donated by Suncor
- Environmental data integrated from Smart Atlantic buoys.
- Lidar scans of facility captured by Geomatics students from College of the North Atlantic

GRi grisim.com

Q&A
GRi
grisim.com

Contact Us!
Des Greene
Senior Project Manager,
Simulation Design & Development
des.greene@grisim.com
1-709-747-5599 (ext. 204)
Mitch Winter
Training & Technical Sales
Representative
mitch.winter@grisim.com
1-709-747-5599 (ext. 209)
Steve Dodd
Vice President, Operations
& Business Development
steve.dodd@grisim.com
1-709-747-5599 (ext. 202)
GRi
SIMULATIONS

The Oilfield Divers Monument is a project of the International Petroleum Museum & Exposition

HOW TO BUILD A MONUMENT WITHOUT EVEN TRYING.

Rusty Wright
Bryce Merrill

The Oilfield Divers Monument is a project of the International Petroleum Museum & Exposition

Oilfield Divers Monument Project

- A project of the International Petroleum Museum & Exposition in Morgan City, Louisiana.
  - A 501 3c corporation
  - Qualified to accept tax deductible donations

The Oilfield Divers Monument is a project of the International Petroleum Museum & Exposition

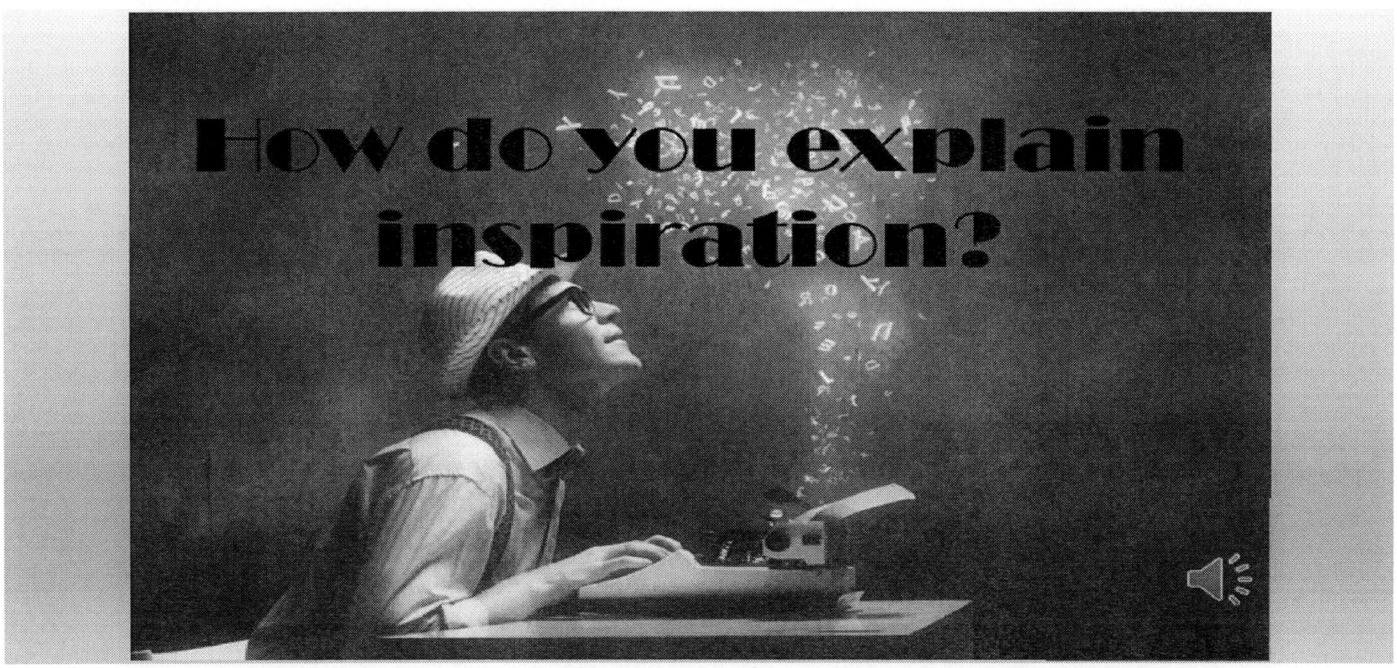

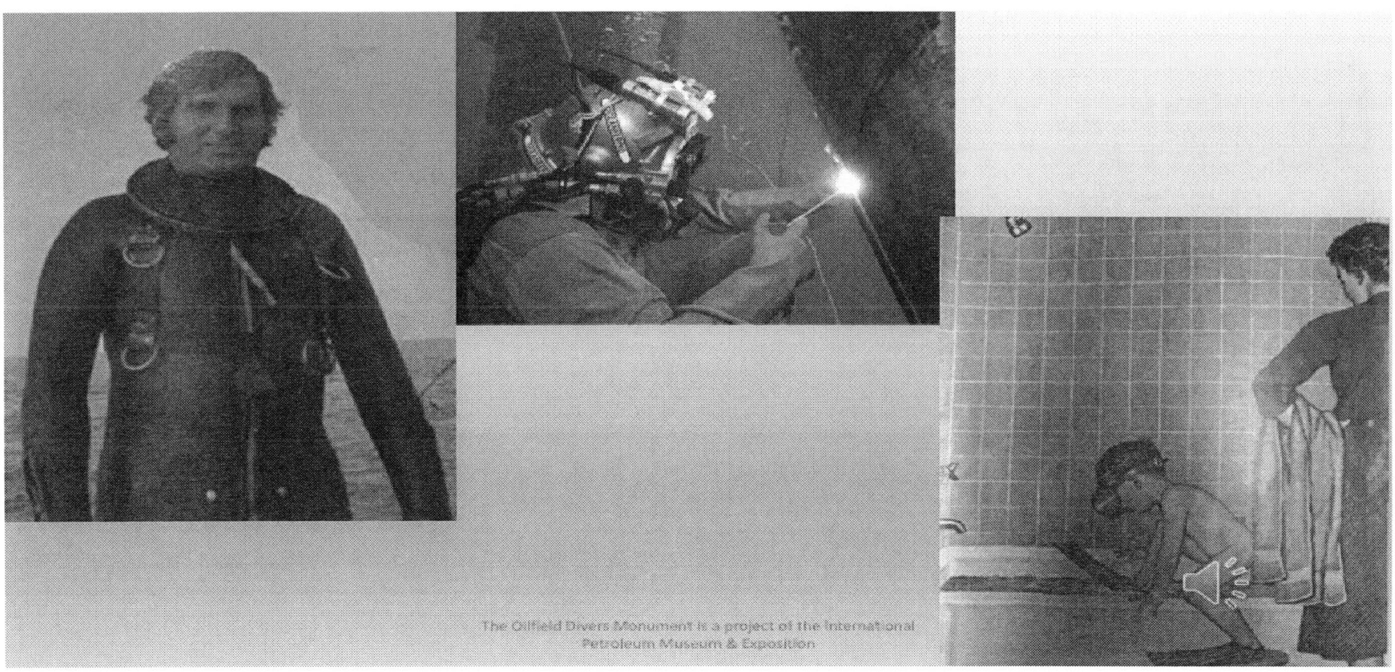

The Oilfield Divers Monument is a project of the International
Petroleum Museum & Exposition

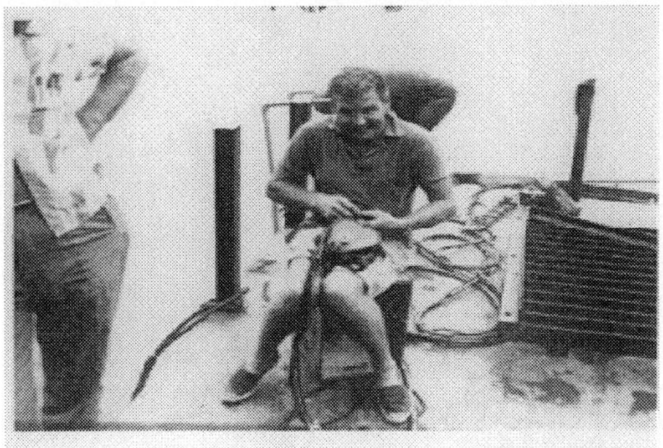
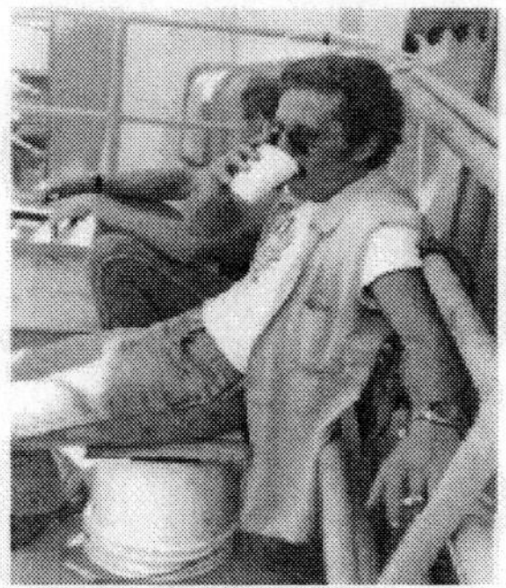
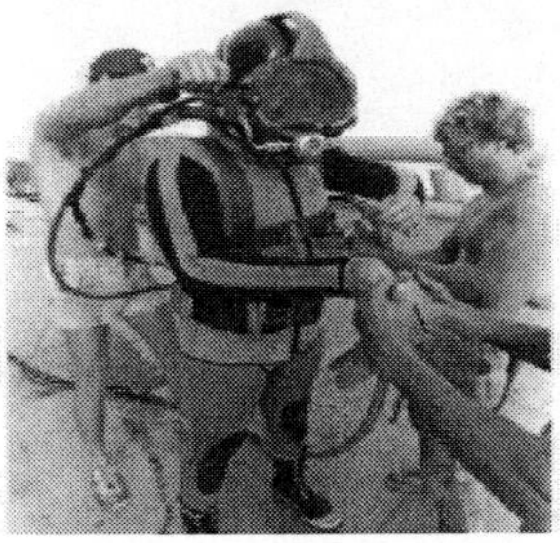

AUG 20 79
Louisiana

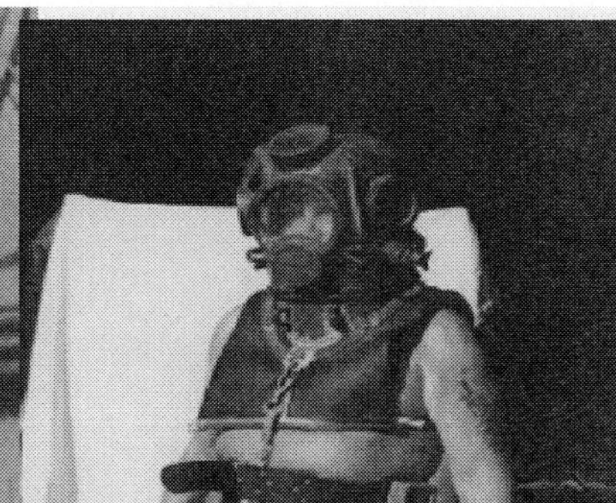

74-182A
8 MAY 1974
) David L Bull 2021
(c) David L Bull 2021
OILFIELD DIVERS MEMORIAL
MORGAN CITY LA.
The Oilfield Divers Monument is a project of the International
Petroleum Museum & Exposition

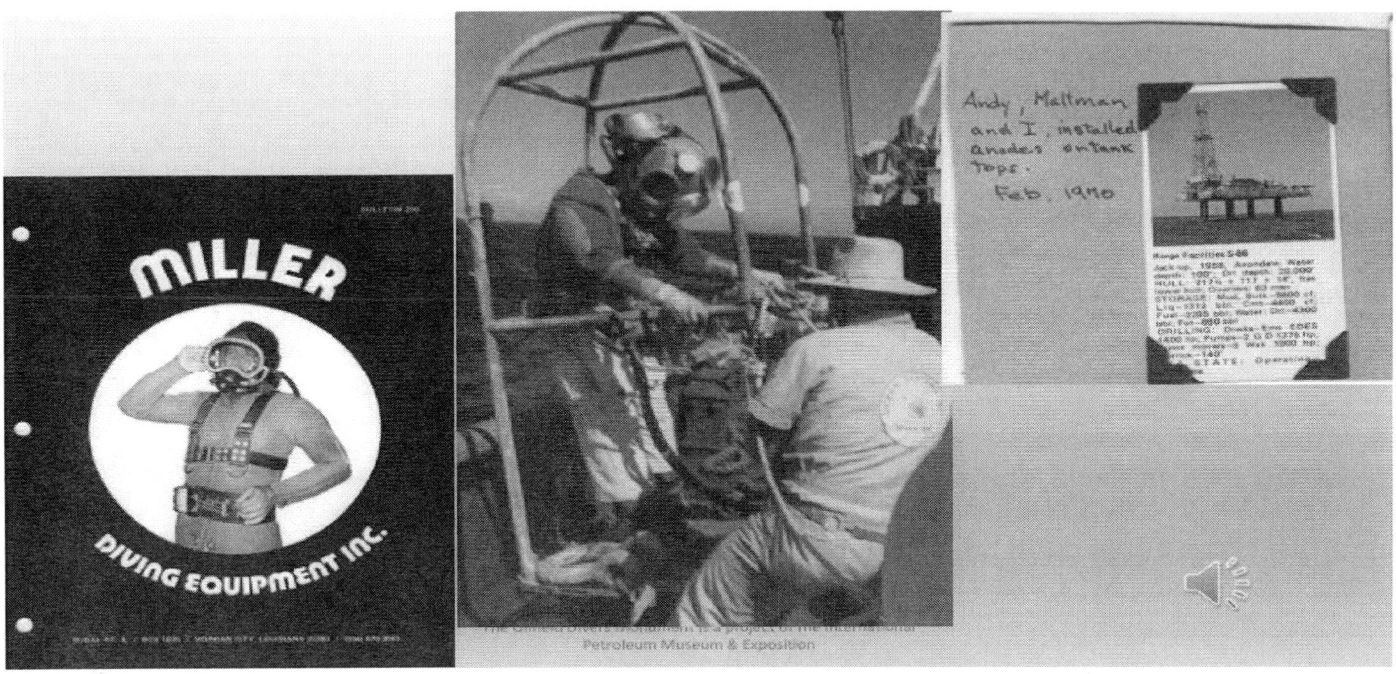
MILLER
DIVING EQUIPMENT INC.
Andy, Mattman
and I, installed
anodes on tank
tops.
Feb, 1990
Petroleum Museum & Exposition

The Oilfield Divers Monument is a project of the International
Petroleum Museum & Exposition

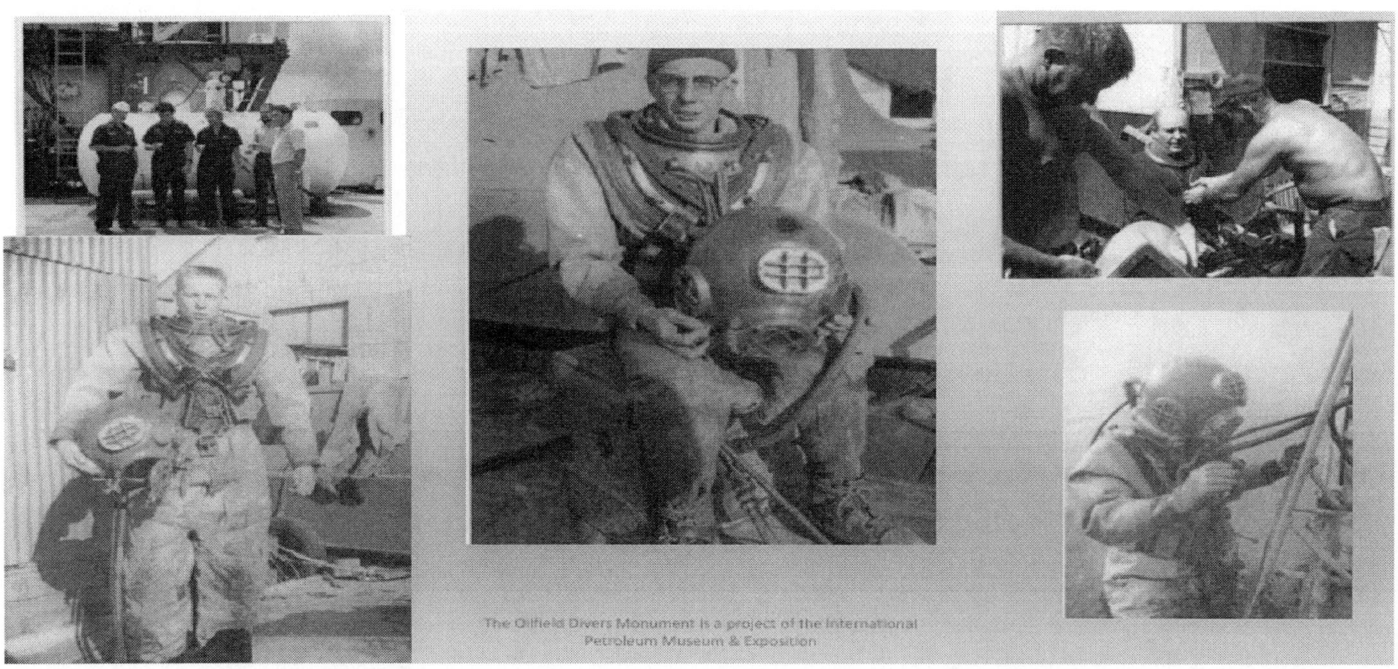
The Oilfield Divers Monument is a project of the International
Petroleum Museum & Exposition

(c) David L Bull 2021
The Oilfield Divers Monument is a project of the International Petroleum Museum & Exposition

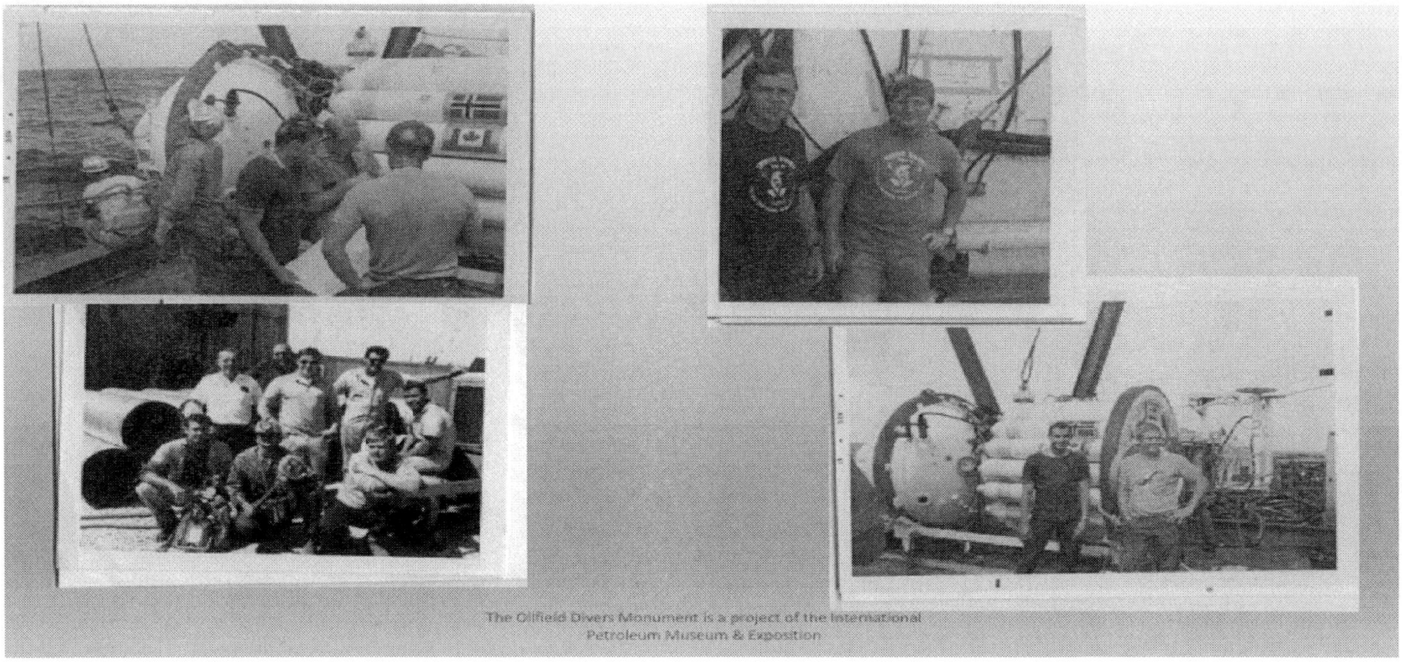
The Oilfield Divers Monument is a project of the International
Petroleum Museum & Exposition

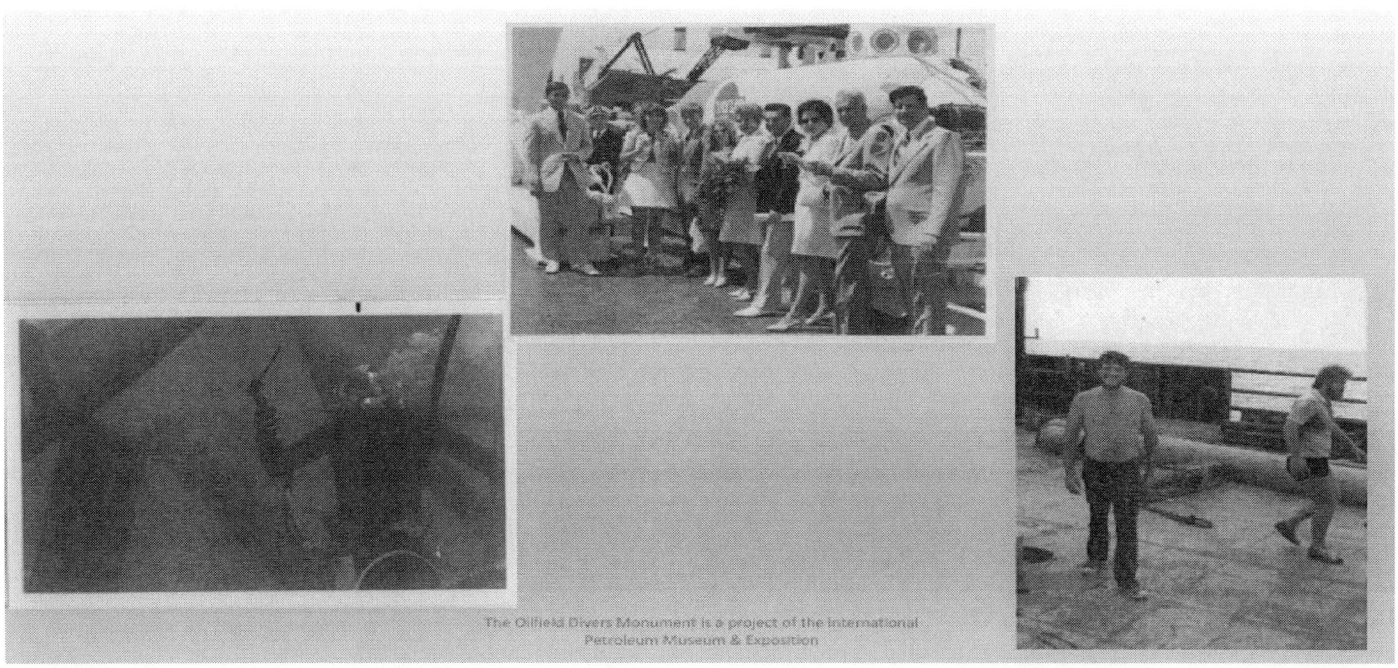
The Oilfield Divers Monument is a project of the International
Petroleum Museum & Exposition

The Oilfield Divers Monument is a project of the International Petroleum Museum & Exposition

The Oilfield Divers Monument is a project of the International Petroleum Museum & Exposition

The Oilfield Divers Monument is a project of the International Petroleum Museum & Exposition

Build The Best Team In The World!

Virgil Allen, Micha Allen, Jack Vilas III, Bryce Merrill, Rusty Wright

The Oilfield Divers Monument is a project of the International Petroleum Museum & Exposition

Fund Raising

BY DIVERS FOR DIVERS

The Oilfield Divers Monument is a project of the International Petroleum Museum & Exposition

The Oilfield Divers Monument is a project of the International Petroleum Museum & Exposition

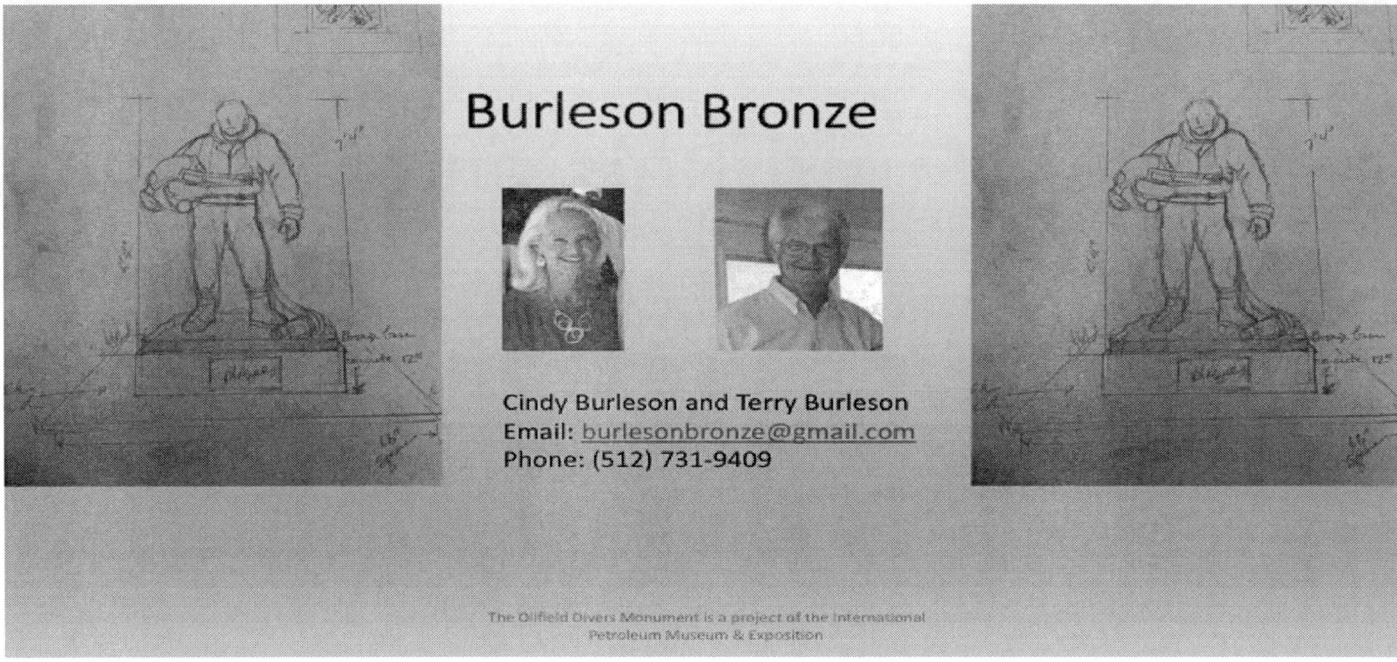
Burleson Bronze
Cindy Burleson and Terry Burleson
Email: burlesonbronze@gmail.com
Phone: (512) 731-9409
The Oilfield Divers Monument is a project of the International Petroleum Museum & Exposition

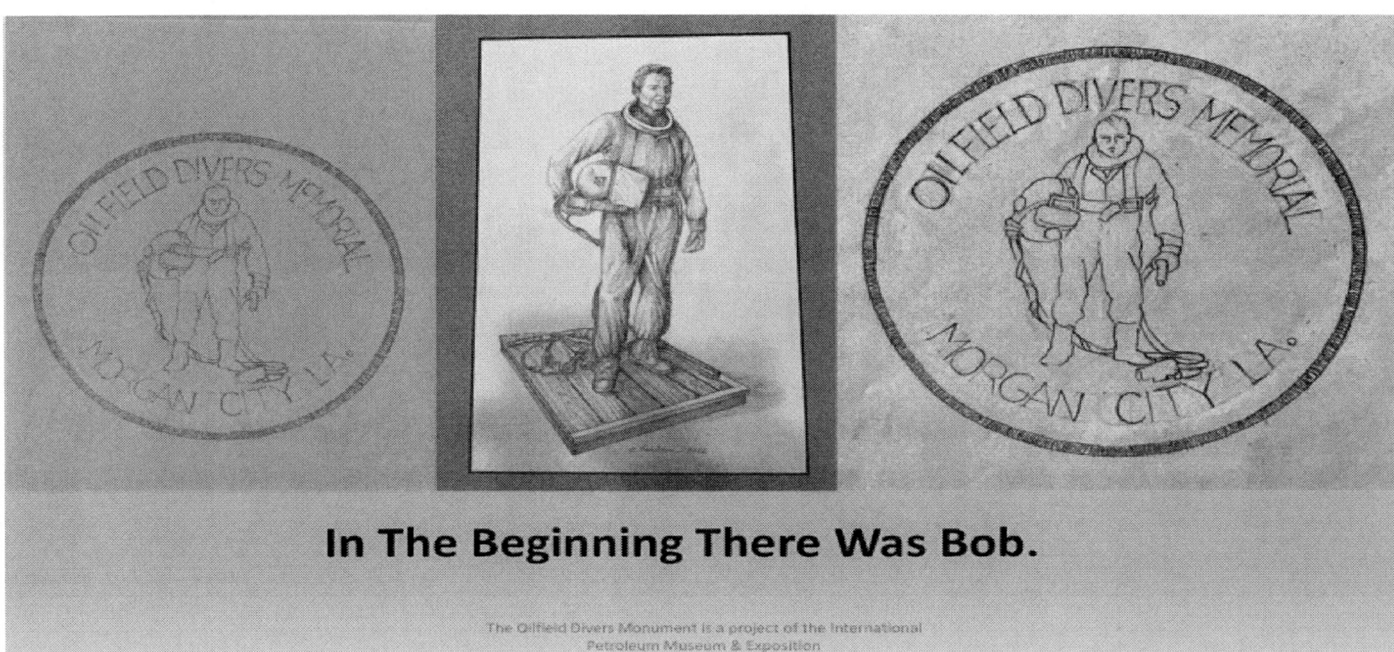
OILFIELD DIVERS MEMORIAL
MORGAN CITY LA.
OILFIELD DIVERS MEMORIAL
MORGAN CITY LA.
In The Beginning There Was Bob.
The Oilfield Divers Monument is a project of the International Petroleum Museum & Exposition

OILFIELD DIVERS MONUMENT
MORGAN CITY La.
The Oilfield Divers Monument is a project of the International Petroleum Museum & Exposition

The Oilfield Divers Monument is a project of the International Petroleum Museum & Exposition

Bryce Merrill
Rusty Wright
In the memory of Jerry Ard
Mike Fink
Walt Easley
Anonymous
Rusty Wright in Memory of Margaret Merrill
Bryce Merrill
Jim McClaugherty
Joe Walker
Mike Chiconsky
Dean Kinkel
Anonymous
Allen Dyson
Frank Hazen
Scott Miller
Randy Cudd in memory of Brad Declerk
Anonymous
Les Joiner in Memory of Jack Reedy
Anonymous
Elvia Chauvin
Sanford family in Memory of Joe Sanford, Sr.

Anonymous
Bruce Gilman
Allen Dyson
Les Joiner in Memory of Fred Achley
Neal Hainlin-Mitten
Chris Christensen in Memory of Ted Hess
Gary Smith Giving thanks to all who went before making diving safer.
TK BILLINGSLEY
Michael Barron
Mike Hughes
Robert Shannon
Drew Trent In Memory of David Dixon
Roger W. Smith in Memory of Jim Sugrue and Jude Herpin
Jack Vilas III
Timothy Gallagher
Sonny Freeman
Steve Struble
Anonymous
Frank Hazen
Anonymous
Ralph Tucker In Honor of Continental Divers

The Oilfield Divers Monument is a project of the International Petroleum Museum & Exposition

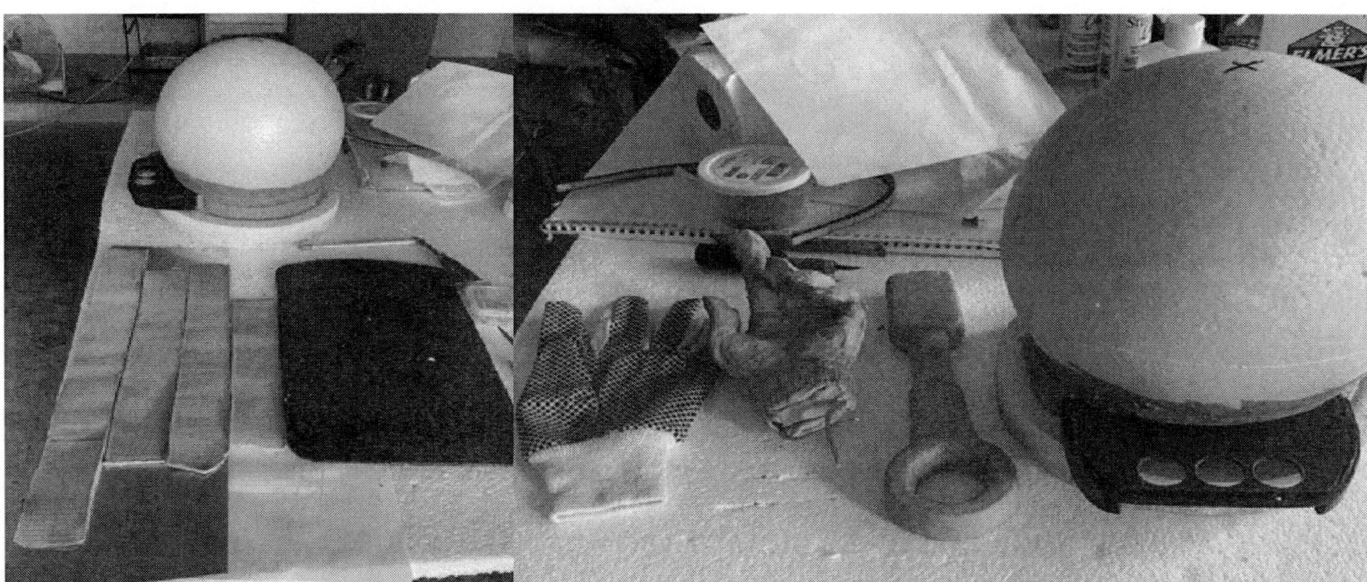

The Oilfield Divers Monument is a project of the International Petroleum Museum & Exposition

Anonymous
Jim McClaugherty
Anonymous
Anonymous
Anonymous
Anonymous
Mike Towler
Corey Sexton and I'm very proud to donate to such a great cause. Diving will always be in my blood and I will always be proud to be a diver. Thank you!
Raymond Wicklund
George Yax
Joe Farah
Anonymous
Anonymous
David Holt
John Hocutt In Memory of Tony Gage & Jude Herpin
Anonymous
George Hunt
Frank Connors
Christie Britt in Loving Memory of Sandy Wright
Jerry and Sharon Bailey in Loving Memory of Sandy Wright
Bryce Merrill in Loving Memory of Sandy Wright
Mark Diebold
Doug and Michelle Graupmann, giving thanks to all and in memory of Sandy Wright
Jason Mitchell
Randy Cudd in Honor of Gene Miller
Bruce Gilman
Anonymous
Karl Allen
Margaret Metcalf
Anonymous
Anonymous
Bruce Gallinger
Tom Sedar
Randy Cudd in Honor of Dale Fackler
David Krause
In Loving Memory of Norman Knudsen
Tom Halligan
Robert Cannon on behalf of Rusty Wright
In Loving Memory of Sandy Wright
Randy Cudd

The Oilfield Divers Monument is a project of the International Petroleum Museum & Exposition

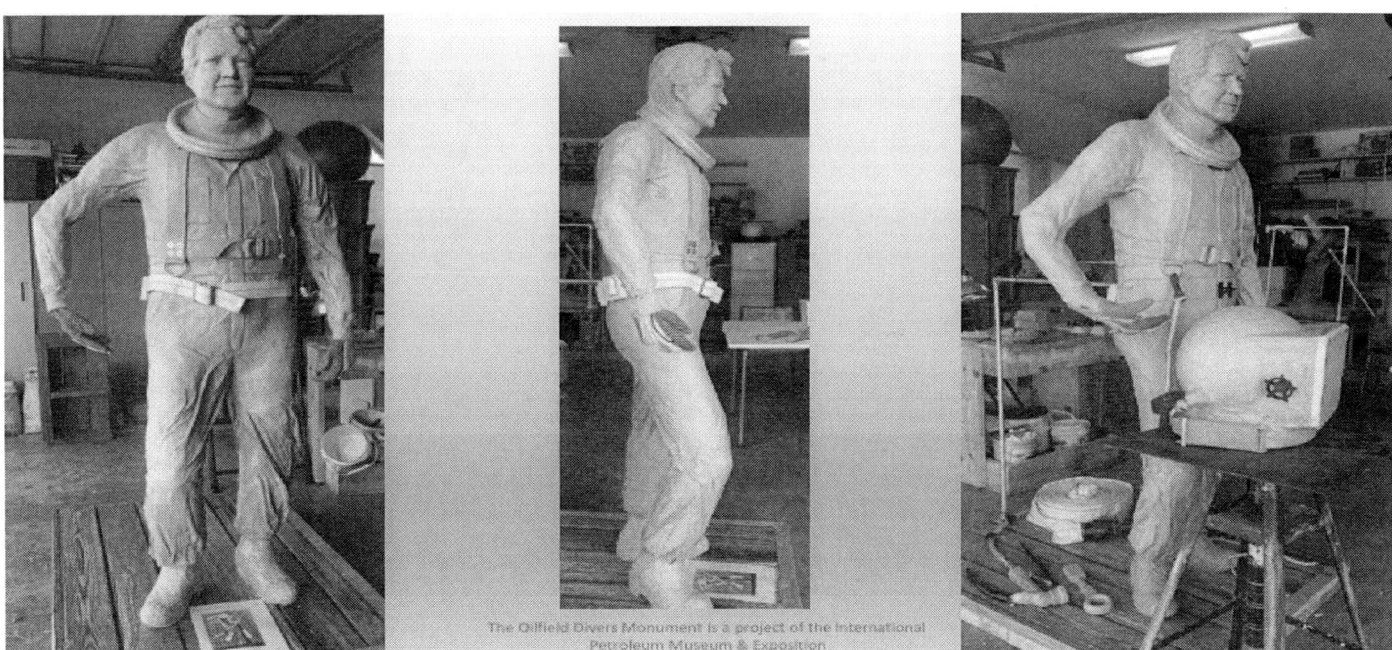

Wendy Smith In Memory of Sandy Wright and in honor of Rusty Wright! Beautiful people! And to the hard working divers - we thank you!!!
Anonymous
Charlie Wells
In Loving Memory of Sandy Wright
James & Jill Mount in Loving Memory of Sandy Wright
Alan Sanders
Thielst Family Trust
Anonymous
Tiffany & Bryan Flemming in Loving Memory of Sandy Wright
Brad Boerner
Bob Suggs
Anonymous
Gary Smith
Thomas Armstrong
Anonymous
Anonymous
Anonymous
Dawn Staten
Richard Smith
Terrence Rogers
John Thielst
David B Smith
Robert Woosley
Paul & Ruth Ehresman - In the Memory of Sandy Wright
Anonymous In Loving Memory of Sandy Wright
Anonymous
In The Memory of Randy Thompson
In The Memory of Riley Smith
Rick Lee
Anonymous
Diane Austin in recognition of oilfield divers everywhere.
Anonymous
Anonymous
In Memory of Thomas E. McGinnis
John Bidmead
Travis Detke
Sue Rubey
Anonymous
Anonymous
Rusty Wright in Loving Memory of Sandy Wright
Anonymous

The Oilfield Divers Monument is a project of the International Petroleum Museum & Exposition

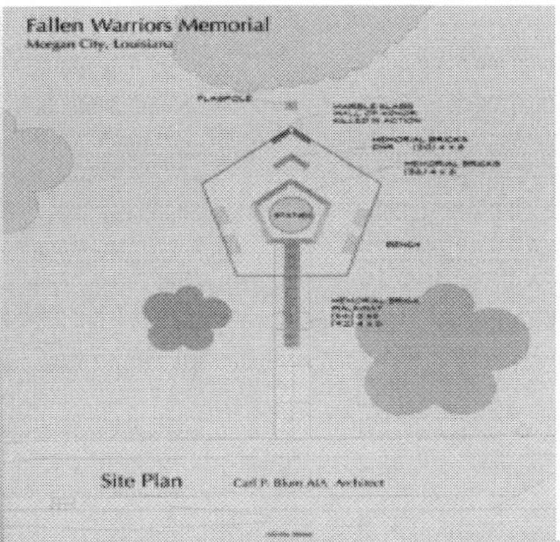

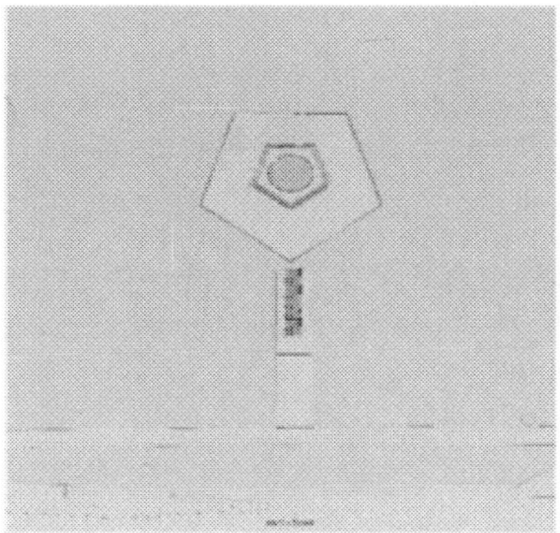

The Oilfield Divers Monument is a project of the International Petroleum Museum & Exposition

Ray Schwartz In Memory of Sea-Con Diver Duane "Chico" Canale
Anonymous
Jean Thomas Clark Thanks to the pioneers of the industry.
J.C. Spangler in Memory of Jay Shapcotte
Anonymous
Mike Fink
Anonymous
Josef Tillmanns
Mark Wieneke
Anonymous
Carrie Carnright, In Loving Memory of Dennis Carnright
Jim Mermis
Earl Liew In Memory Of Art Noble and Dennis Strickland
Rick & Jamie Rowen in loving Memory of Mary's Hideaway Lounge
Steven Brinkley
Cheryl Bourgeois Johnson, in loving Memory of Roland Leo Bourgeois, Jr
Anonymous
Denny & Sylvia Swartz
Tom Karnuta

Chris Prunty To all who paved the way!
Roger Tillman
Anonymous
Anonymous
Gary Kane
Anonymous
Anonymous
Jay Gauthier
Anonymous
Mac Kennedy
Gonzo-Rama
Armando Gonzalez
Anonymous
Kristin, Josh & Karen in Loving Memory of Jeff Duay
Anonymous
Paula & Reginald Drye in memory of Marc Begnaud
Anonymous
Anonymous
John D Smith
Dana Beebe in Memory of "Butch" Basinski
Anonymous
Joe Farah
Anonymous

The Oilfield Divers Monument is a project of the International Petroleum Museum & Exposition

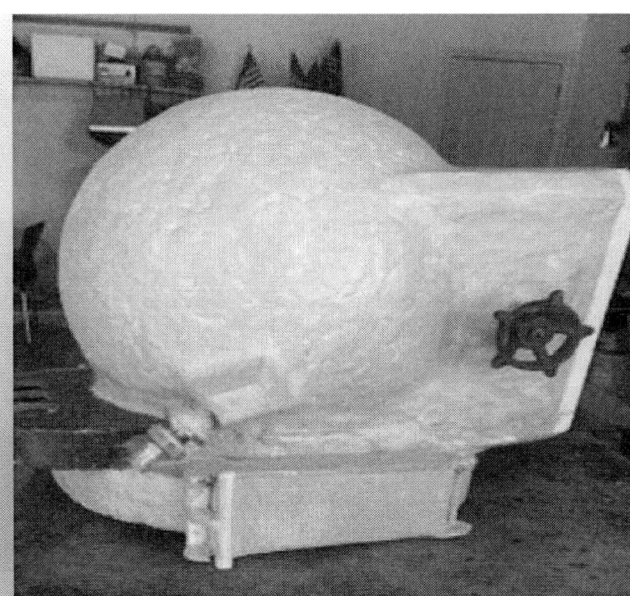

The Oilfield Divers Monument is a project of the International Petroleum Museum & Exposition

David Terry
Charlie & Jackie Rowen
Paul Carmody
Anonymous
Pat Bowman
Steven Brinkley
Anonymous
Anonymous
Tom Wright
Ed Thomas
Anonymous
Anonymous
James & Jill Mount in Loving Memory of Sherri Wright
Max Pearcy
In memory of Kevin Sass
Don Barthelmess in Memory of Andre Galerne
Donna Parr in Memory of Jeff Duay
Paul Guiver - Wishing you all the very best in a successful completion. Looking forward to seeing the monument in 2023. Well done to all involved - BZ! Best wishes from The Brits @ Monument Projects Ltd

Bing Strassburg - To all that came before and those that will be there later
Anonymous
Neal Hainlin-Mitten
Anonymous
Jim McClaugherty
Ken & Marilyn Theriot
Craig Ring
Eli Geminder
Tim Rompel
Wade Abdie to All the Brave Men and Women that improved the Safety of The Offshore Diving Industry.
Steve Struble
Anonymous
Anonymous
Mark Van Emmerik
Anonymous
Anonymous
Karl Witherow
Anonymous
JoAnn Eubank In Memory of Chip Eubank
Karen Smith in Loving Memory of Jeff Duay

The Oilfield Divers Monument is a project of the International Petroleum Museum & Exposition

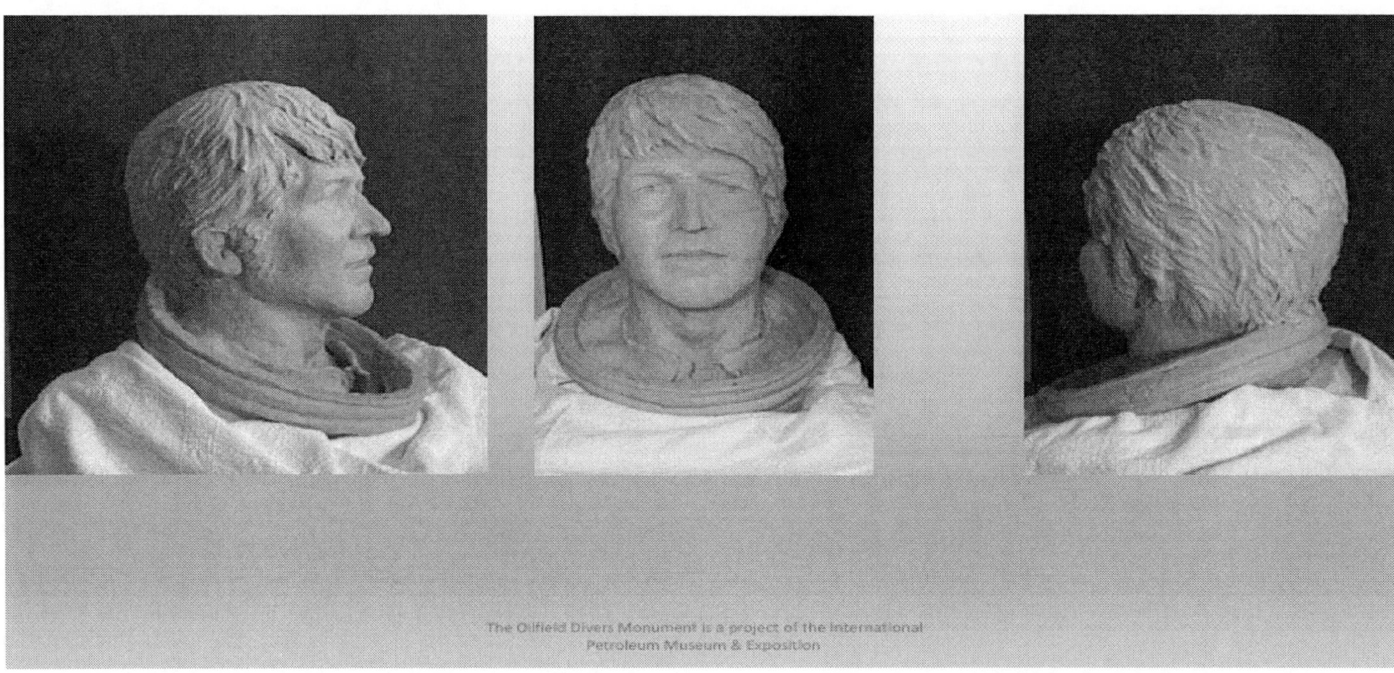

The Oilfield Divers Monument is a project of the International Petroleum Museum & Exposition

Anonymous
Gary Gibbs
Charlie Huff
Larry Bragg
Anonymous
Nancy Summers
Brooks LeBlanc in memory of Ben Wells
Tanner Bourg in memory of Ben Wells
Randy Cudd in Memory of Russell Guidroz
Randy Cudd in Memory of Steve Quinn
Randy Hebert
Robert Gruy
Dominique Vidrine
Barrie Miller
John Wood in memory of Don "Peanut" Holland
Matt Booth
David Barrow
Jerald & Sharon Lowrimore
Robert O'Dea
Howie Thompson
Randy Cudd in the memory of Joe Farah
Sonny Freeman in the memory of Joe Farah

Charles Stricker
Bill Farah
Creighton Burnett
Earl Liew In Remembrance of Bob Small, Always A True Friend and The Finest Tender God Ever Gave a Diver,
Anonymous for Blue Savoie
Anonymous for Bob Dunbar
Anonymous for Bob Mock
Anonymous for Bob Schmidt
Anonymous for Bobby Lee Clark
Anonymous for Buddy Ayers
Anonymous for Buddy Rogers
Anonymous for Carl Bock
Anonymous for Chuck Gates
Anonymous for Dick Evans, Professional Divers
Anonymous for Frederick Lawrence "Larry" Tomlinson, Professional Divers of N.O.
Anonymous for George "Smash" Saunders
Anonymous for George Condiff
Anonymous for Gerald Vincent
Anonymous for Jack Smith, S & H Subwater Salvage

The Oilfield Divers Monument is a project of the International Petroleum Museum & Exposition

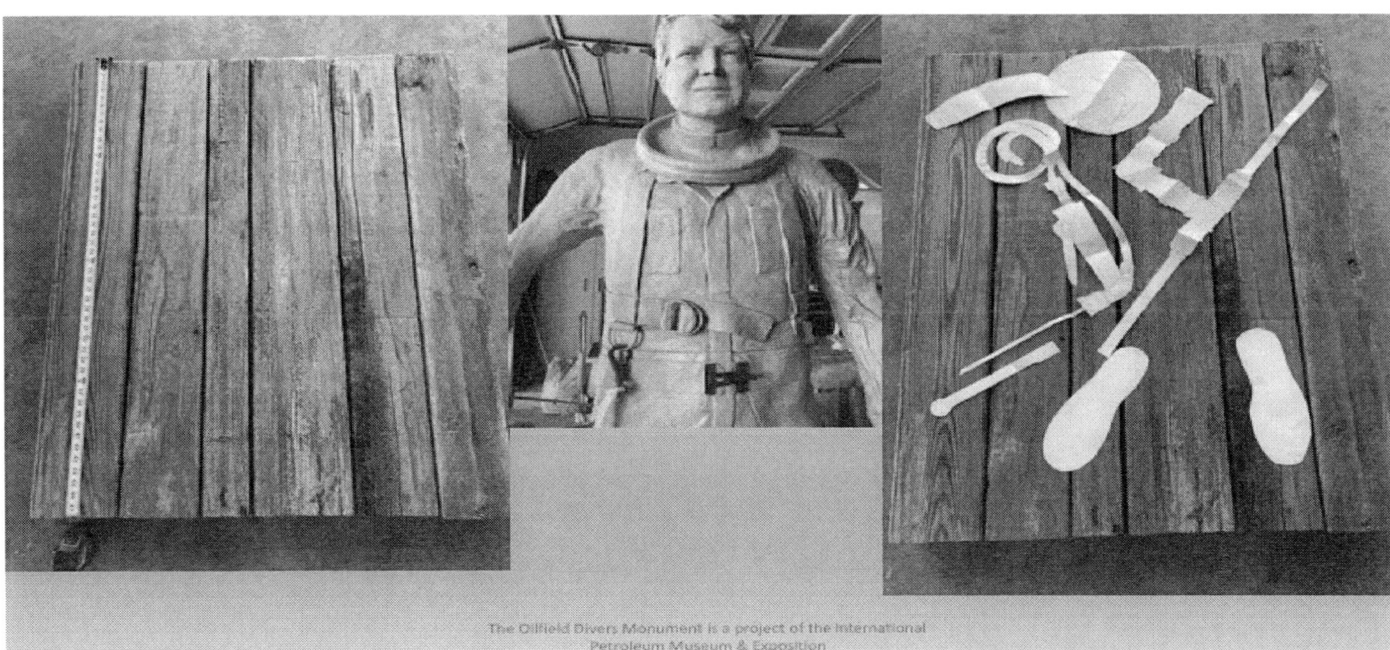

Anonymous for Jerry Neff
Anonymous for Jim Burgess
Anonymous for Joe "Seldom Seen" Vidrine
Anonymous for Joe Savoie
Anonymous for John Gray
Anonymous for John Haynie
Anonymous for Johnny Murphy
Anonymous for Kerry Gilbreath
Anonymous for Larry Byrd
Anonymous for Lawrence Troy Tomlinson
Anonymous for Lee Ingersoll
Anonymous for Lil's Elbow Room
Anonymous for Lyle Stockstill
Anonymous for Mark "Bear" Wallace
Anonymous for Neil Evans
Anonymous for P. G. Stockstill
Anonymous for Pat Savoie
Anonymous for Robert Neff
Anonymous for Stan Diket
Anonymous for Tonie Savoie Granger
Anonymous for Tony Crawford

Anonymous for Woody Cowart
Anonymous In Memory of Butch Mabes
Anonymous In Memory of Mike Lehman
Benny York
Anonymous
Jerry Cobb
Joe Cain
Hayley Feindel
Mark Trimmer
Kevin Erickson
Bob Leger
Charlie & Brenda Youngblood
Steve White
Jim Kober
Jeff Wilson
Tim Czapranski in memory of Duaine "Chico" Canale
Curt & Mary Cyr
Tim Czapranski for Jim O'Connell
Tim Czapranski in memory of Ben Wells
David Rosenberg
In honor of Mike Garrett
In the memory of Tim O'Connor

The Oilfield Divers Monument is a project of the International Petroleum Museum & Exposition

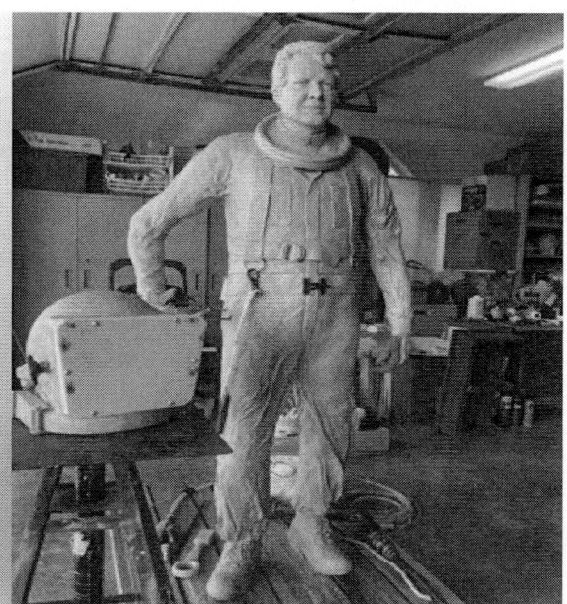

The Oilfield Divers Monument is a project of the International Petroleum Museum & Exposition

Howard Wall
Todd Baldridge
David Bull
In honor of George Williams
Wild Tom Biernatzki
Michael & Christine Farah Family Foundation In the memory of Joseph L. Farah, Jr.
Michael Ambrose
Patricc Prentiss Godspeed through God's Love. Blessed is a Commercial Diver. Thank you Virgil for supporting this!
Thomas Halligan In Memory of: Gerald (Wayne) Mears
Thomas Halligan In Memory of: Paul (Shorty) Vaughn
Thomas Halligan In Memory of: Terry Overland
Aaron Ott
Jennifer Hinkle in memory of Ted Hess
Robert Newman
Tim Gallagher in Memory of Lanny Falgout
Larry Bourg
Whitney Wells In memory of Ben Wells
In loving memoriam of Steve Guinn - Sat Diver
Mona Guidroz-Guinn In loving memory - Russell Guidroz (Continental Diving / Shop & Crew Supervisor)

David Archer
In memory of Joe Stubbs
In memory of Martel "Crip" Stevens
In memory of Michelle LeClair
In memory of Paul Tedder
Dennis Floyd to commemorate those divers, past and present, who make-up "The Brotherhood of the Oilfield Diver".
In the memory of Donald "Peanut" Holland, and to commemorate those divers, past and present, who made-up "The Brotherhood of the Oilfield Diver".
Sonny Freeman In memory of Dave Brower
John Jenco
Gene Lo Conte
Greg Rohr
Ray Spillers
Doug Yates
Naked Bob
Jose Vindeola III
Dwayne Roy and Dylan Roy, In Memory of Horace Pierre LeBoeuf Jr.
Nicholas Babbitt

The Oilfield Divers Monument is a project of the International Petroleum Museum & Exposition

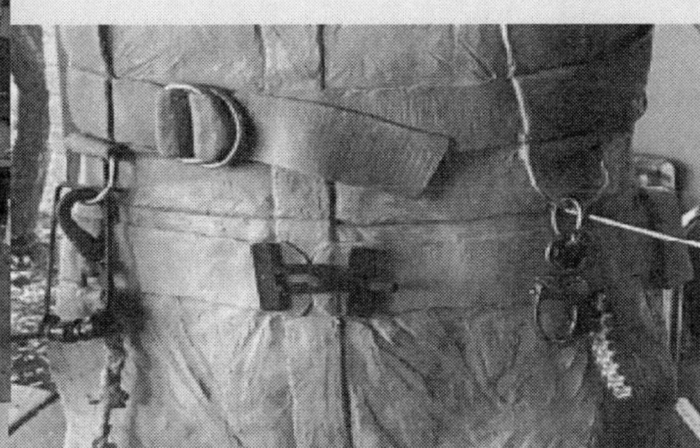

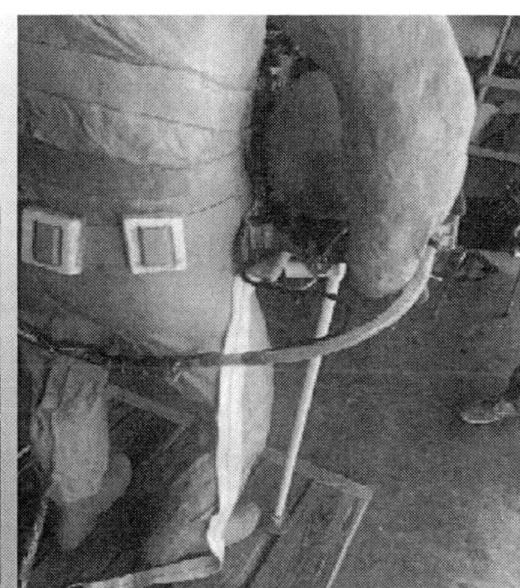

The Oilfield Divers Monument is a project of the International Petroleum Museum & Exposition

Tom Holmes in memory of Bob Holmes
Karl & Barry Witherow in the memory of David Witherow
Mike Hughes in Memory of William (Bill) Rowe Cook, Jr.
In Memory of Jay Brown
In Memory of J.C. "Skinney" Brown
Don Whitecotton
Kim Ware
Russ Altman
In Memory of Al Mikalow
In Memory of Don Byrd
In memory of Ricardo Chamberlin
Ryan Shawhan
Norb Gorman In Memory of Jay Schapcotte
James FintchIn Memory of Terrence Winston
Danny Felterman
Leslie Leaney
Christine Guidroz In Loving Memory of Russell & Connie Guidroz
Jerry Gaddis
Patricia Branch
Tom Bailey
Joe Adydan In memory of Jerry Gauthier and Roy Jewett

Rick Henley Jr.
Mike & Kathy Hobbs In memory of Rosco and Matt Bauer
Ron Duhame
Martha Gauthier
Eric Hofsommer
Elia Mears in memory of Wayne Mears
In Honor of Dr. Ann Bull: Dove 1992 Oil Platform Removal to document the fish population in support of the Rigs-to-Reefs Program.
Mike Middelton/Sandra Gaspard
In memory of Paul Tedder
Viola Hoetling
Anonymous

The Oilfield Divers Monument is a project of the International Petroleum Museum & Exposition

The Oilfield Divers Monument is a project of the International Petroleum Museum & Exposition

Chet Morrison Contractors, LLC
Aqueos Corporation
Oceaneering, Inc.
Bay-Tech Industries, Inc.
Triton Diving, LLC
Desco Corporation
Phoenix International, Inc.
ADCI
Your Name Could Be Here!

The Oilfield Divers Monument is a project of the International
Petroleum Museum & Exposition

The Oilfield Divers Monument is a project of the International
Petroleum Museum & Exposition

The Oilfield Divers Monument is a project of the International Petroleum Museum & Exposition

Your Logo Here!

The Oilfield Divers Monument is a project of the International Petroleum Museum & Exposition

Ron, Gene
BORN TO DIVE
FORCED TO WORK
The Oilfield Divers Monument is a project of the International
Petroleum Museum & Exposition

The Oilfield Divers Monument is a project of the International
Petroleum Museum & Exposition

POST DIVING JUICE
The Oilfield Divers Monument is a project of the International Petroleum Museum & Exposition

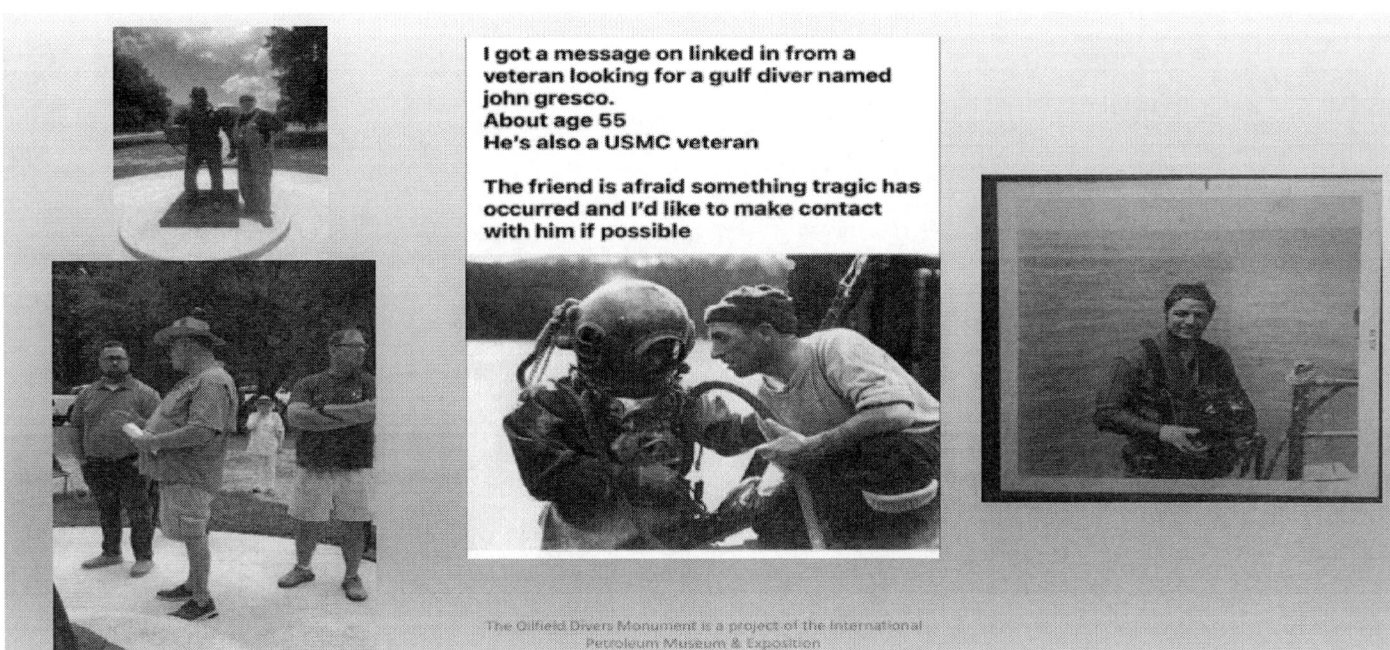
I got a message on linked in from a veteran looking for a gulf diver named john gresco.
About age 55
He's also a USMC veteran
The friend is afraid something tragic has occurred and I'd like to make contact with him if possible
The Oilfield Divers Monument is a project of the International Petroleum Museum & Exposition

The Oilfield Divers Monument is a project of the International
Petroleum Museum & Exposition

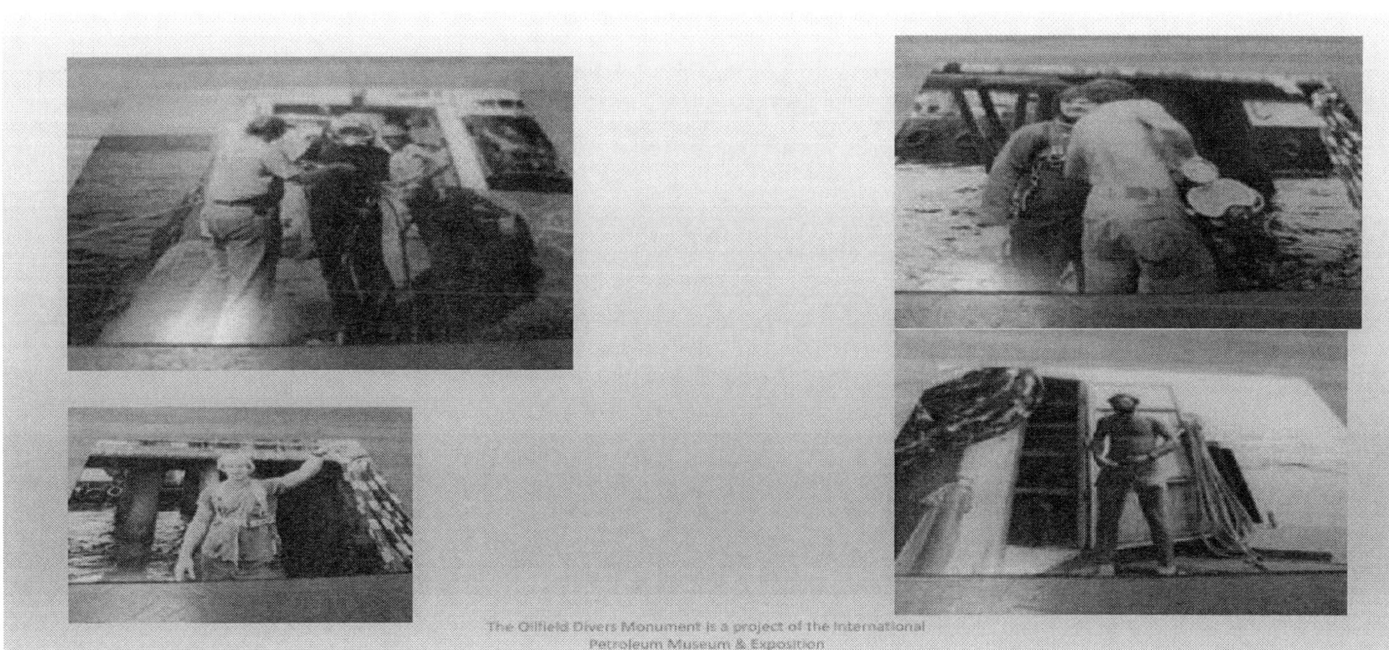
The Oilfield Divers Monument is a project of the International
Petroleum Museum & Exposition

The Oilfield Divers Monument is a project of the International
Petroleum Museum & Exposition

MORGAN CITY
MARTECH
The Oilfield Divers Monument is a project of the International
Petroleum Museum & Exposition

McDERMOTT DIVING DIV.
Jeff A. Duay
March 30, 1957 ~ May 22, 2006
The Oilfield Divers Monument is a project of the International Petroleum Museum & Exposition

The Oilfield Divers Monument is a project of the International Petroleum Museum & Exposition

Ray Schwartz In Memory of Sea-Con Diver Duane "Chico" Canale
Anonymous
Jean Thomas Clark Thanks to the pioneers of the industry.
J.C. Spangler in Memory of Jay Shapcotte
Anonymous
Mike Fink
Anonymous
Josef Tillmanns
Mark Wieneke
Anonymous
Carrie Carnright, In Loving Memory of Dennis Carnright
Jim Mermis
Earl Liew In Memory Of Art Noble and Dennis Strickland
Rick & Jamie Rowen in loving Memory of Mary's Hideaway Lounge
Steven Brinkley
Cheryl Bourgeois Johnson, in loving Memory of Roland Leo Bourgeois, Jr
Anonymous
Denny & Sylvia Swartz
Tom Karnuta
Chris Prunty To all who paved the way!
Roger Tillman
Anonymous
Anonymous
Gary Kane
Anonymous
Anonymous
Jay Gauthier
Anonymous
Mac Kennedy
Gonzo-Rama
Armando Gonzalez
Anonymous
Kristin, Josh & Karen in Loving Memory of Jeff Duay
Anonymous
Paula & Reginald Drye in memory of Marc Begnaud
Anonymous
Anonymous
John D Smith
Dana Beebe in Memory of "Butch" Basinski
Anonymous
Joe Farah
Anonymous

The Oilfield Divers Monument is a project of the International Petroleum Museum & Exposition

The Oilfield Divers Monument is a project of the International Petroleum Museum & Exposition

HydroTech Systems, Inc.
Oilfield Divers Monument
Morgan City La.
The Oilfield Divers Monument is a project of the International
Petroleum Museum & Exposition

10TH ANNIVERSARY
PHOENIX
INTERNATIONAL
UNDERWATER SOLUTIONS WORLDWIDE
The Oilfield Divers Monument is a project of the International
Petroleum Museum & Exposition

The Oilfield Divers Monument is a project of the International Petroleum Museum & Exposition

FARAH'S
WATER HOLE
seac
The Oilfield Divers Monument is a project of the International Petroleum Museum & Exposition

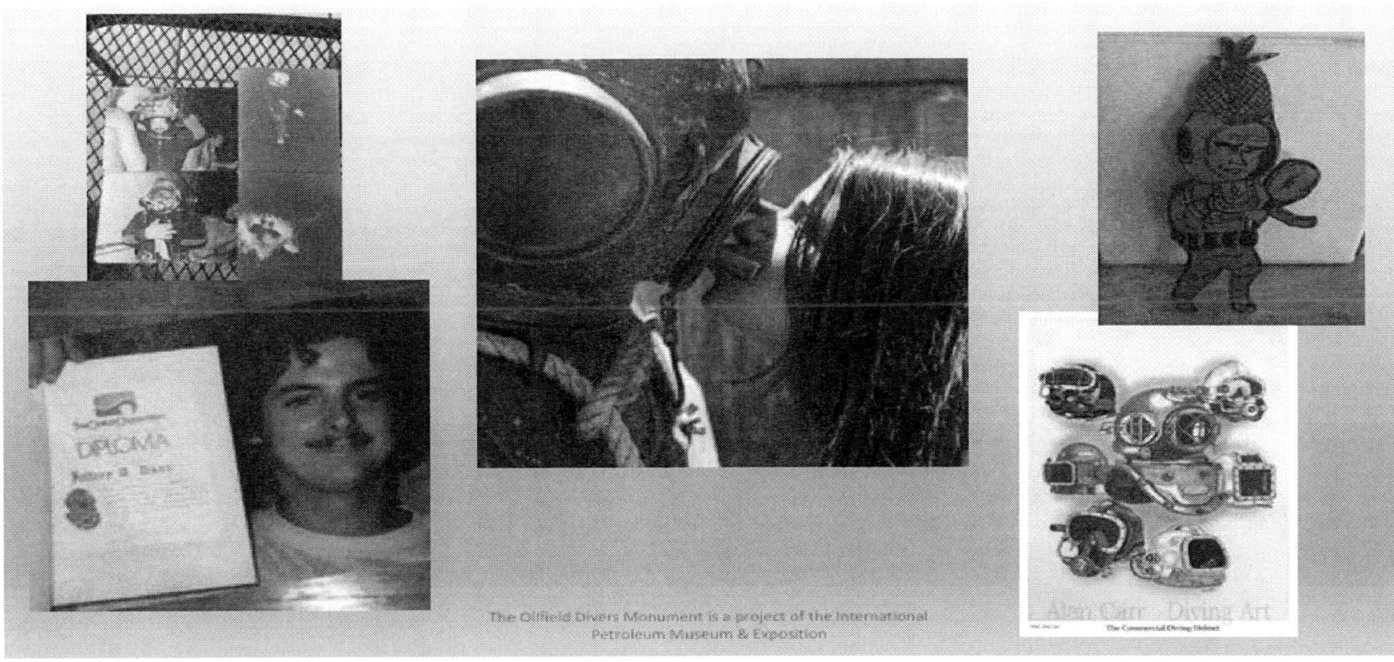
DIPLOMA
The Oilfield Divers Monument is a project of the International
Petroleum Museum & Exposition

The Oilfield Divers Monument is a project of the International
Petroleum Museum & Exposition

The Oilfield Divers Monument is a project of the International Petroleum Museum & Exposition

MORGAN CITY
MARTECH
OCEANEERING
The Oilfield Divers Monument is a project of the International Petroleum Museum & Exposition

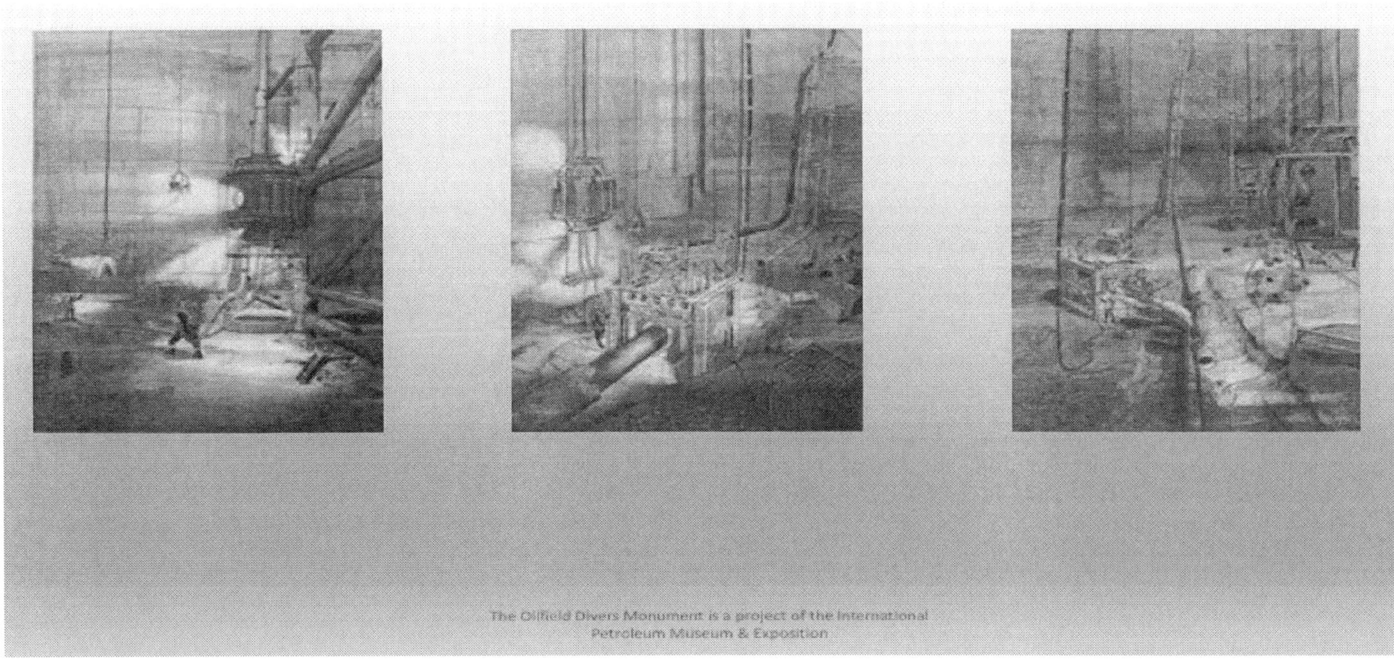
The Oilfield Divers Monument is a project of the International
Petroleum Museum & Exposition

The Oilfield Divers Monument is a project of the International
Petroleum Museum & Exposition

The Oilfield Divers Monument is a project of the International
Petroleum Museum & Exposition

The Oilfield Divers Monument is a project of the International
Petroleum Museum & Exposition

The Oilfield Divers Monument is a project of the International Petroleum Museum & Exposition

The Oilfield Divers Monument is a project of the International Petroleum Museum & Exposition

The Oilfield Divers Monument is a project of the International Petroleum Museum & Exposition

SEADIVE
INC.
SEADIVE
INC.
The Oilfield Divers Monument is a project of the International Petroleum Museum & Exposition

"i'll meet you for one drink but that's it"
22 drinks later:
The Oilfield Divers Monument is a project of the International Petroleum Museum & Exposition

COMMERCIAL DIVING CENTER
WILMINGTON, CALIFORNIA
The Oilfield Divers Monument is a project of the International Petroleum Museum & Exposition

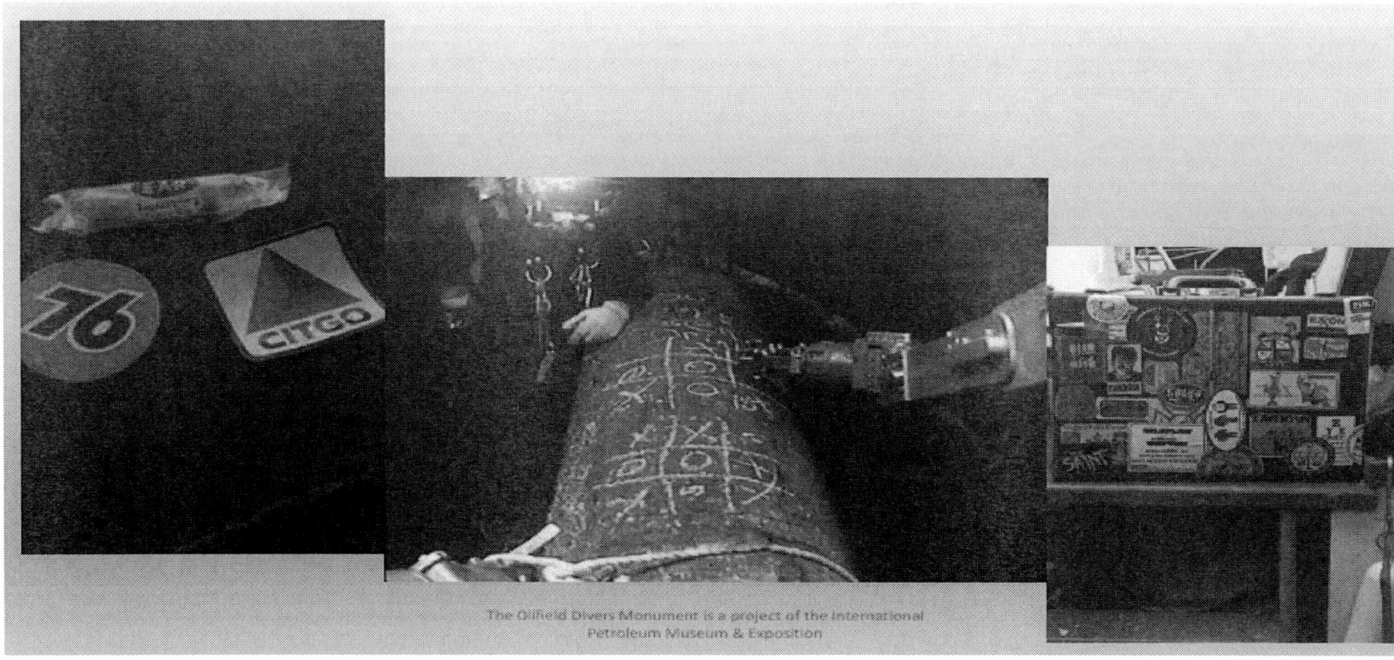
76
CITGO
The Oilfield Divers Monument is a project of the International Petroleum Museum & Exposition

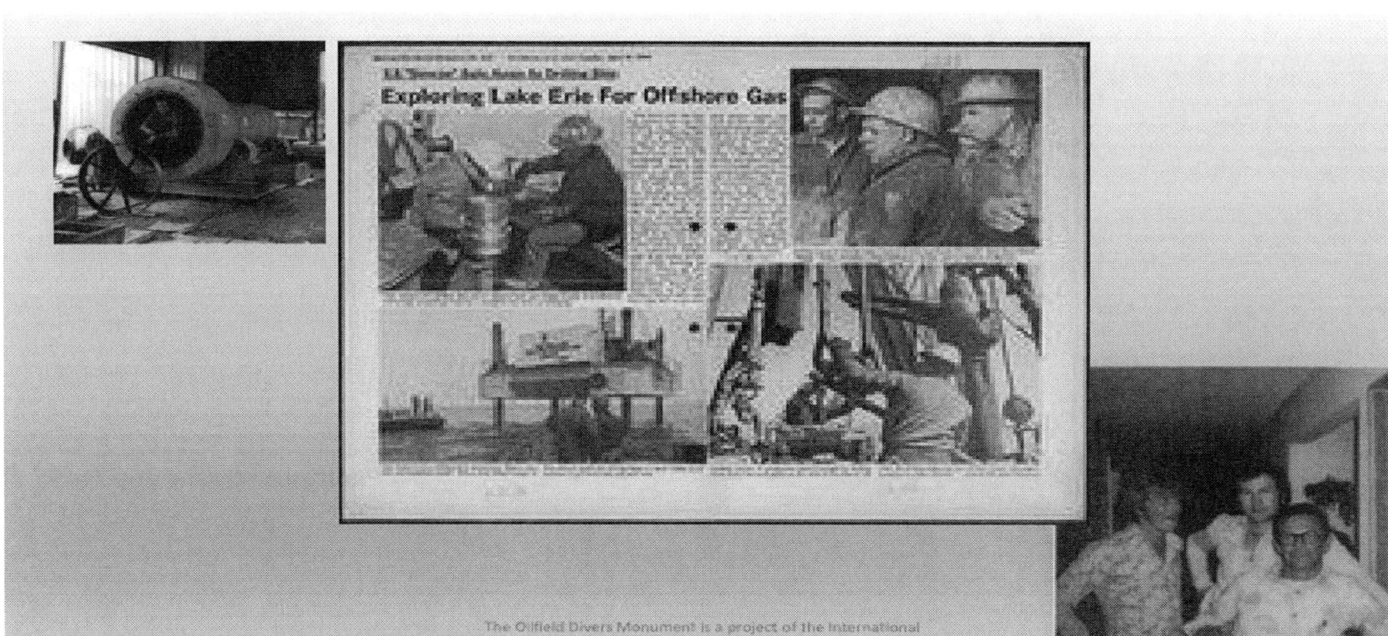
Exploring Lake Erie For Offshore Gas
The Oilfield Divers Monument is a project of the International Petroleum Museum & Exposition

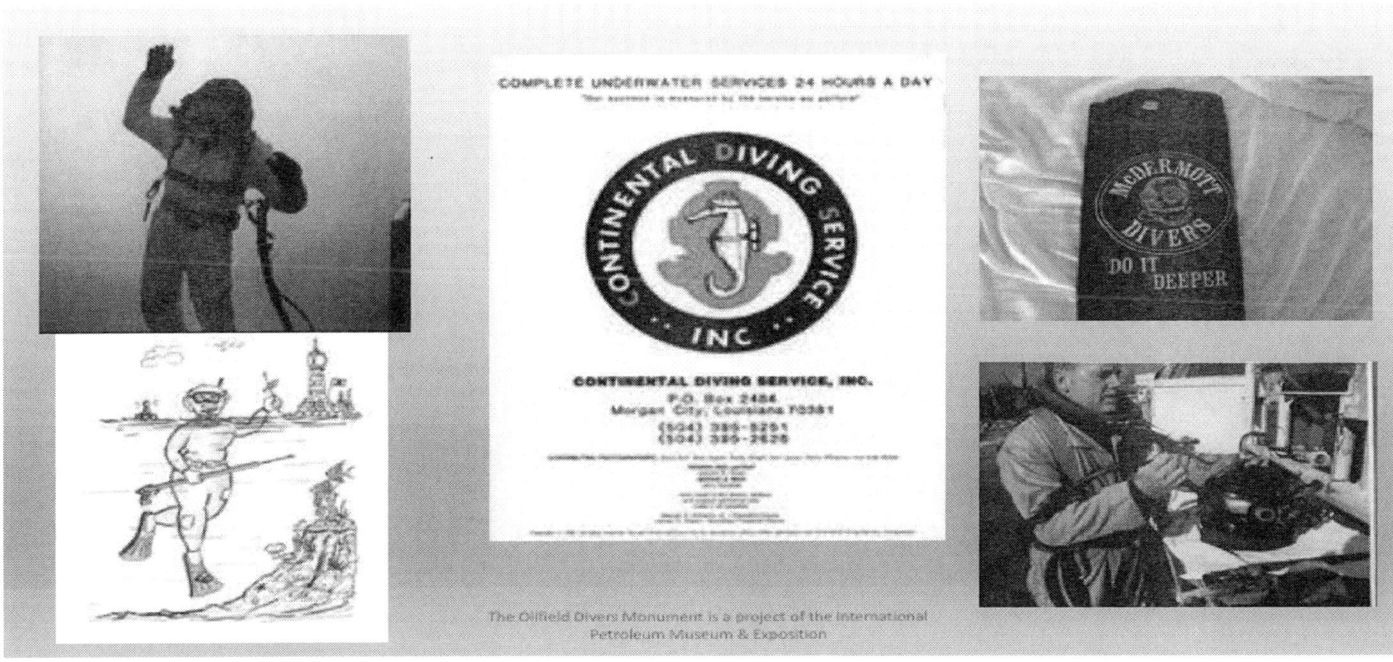
COMPLETE UNDERWATER SERVICES 24 HOURS A DAY
CONTINENTAL DIVING SERVICE INC
CONTINENTAL DIVING SERVICE, INC.
Morgan City, Louisiana
McDERMOTT DIVERS
DO IT DEEPER
The Oilfield Divers Monument is a project of the International Petroleum Museum & Exposition

AMERICAN OILFIELD DIVERS, INC.
The Oilfield Divers Monument is a project of the International Petroleum Museum & Exposition

MERRY CHRISTMAS
The Oilfield Divers Monument is a project of the International Petroleum Museum & Exposition

The Oilfield Divers Monument is a project of the International Petroleum Museum & Exposition

Grandpa
The Oilfield Divers Monument is a project of the International
Petroleum Museum & Exposition

The Oilfield Divers Monument is a project of the International
Petroleum Museum & Exposition

FRANK'S
LOUNGE
WORLD FAMOUS
BLOODY MARYS
The Oilfield Divers Monument is a project of the International
Petroleum Museum & Exposition

OCEANEERING INTERNATIONAL
SEA DOME
The Oilfield Divers Monument is a project of the International
Petroleum Museum & Exposition

The Oilfield Divers Monument is a project of the International
Petroleum Museum & Exposition

Bob R
Phil
The Oilfield Divers Monument is a project of the International
Petroleum Museum & Exposition

The Oilfield Divers Monument is a project of the International Petroleum Museum & Exposition

The Oilfield Divers Monument is a project of the International Petroleum Museum & Exposition

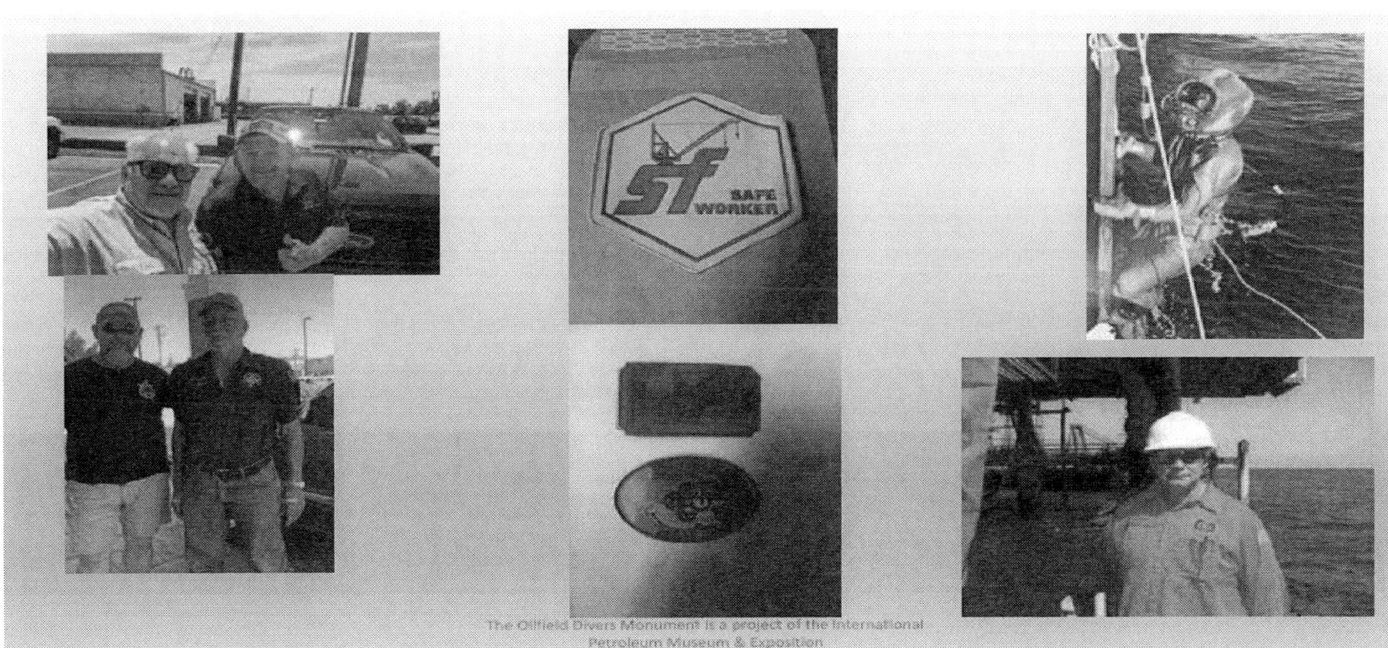
SF SAFE WORKER
The Oilfield Divers Monument is a project of the International Petroleum Museum & Exposition

"I'll meet you for one drink but that's it"
22 drinks later:
The Oilfield Divers Monument is a project of the International Petroleum Museum & Exposition

COMMERCIAL DIVING CENTER
WILMINGTON, CALIFORNIA
The Oilfield Divers Monument is a project of the International Petroleum Museum & Exposition

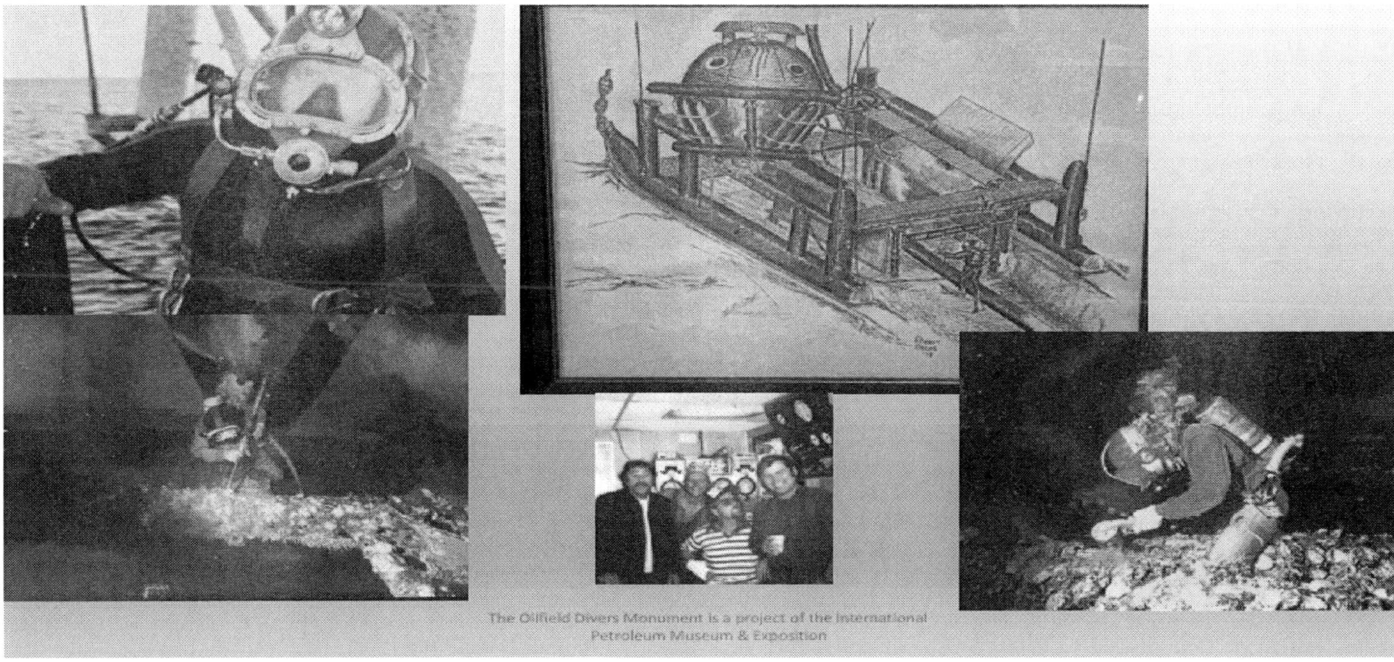
The Oilfield Divers Monument is a project of the International
Petroleum Museum & Exposition

MARTECH
Position DIVER
Name John Jenco
Signature

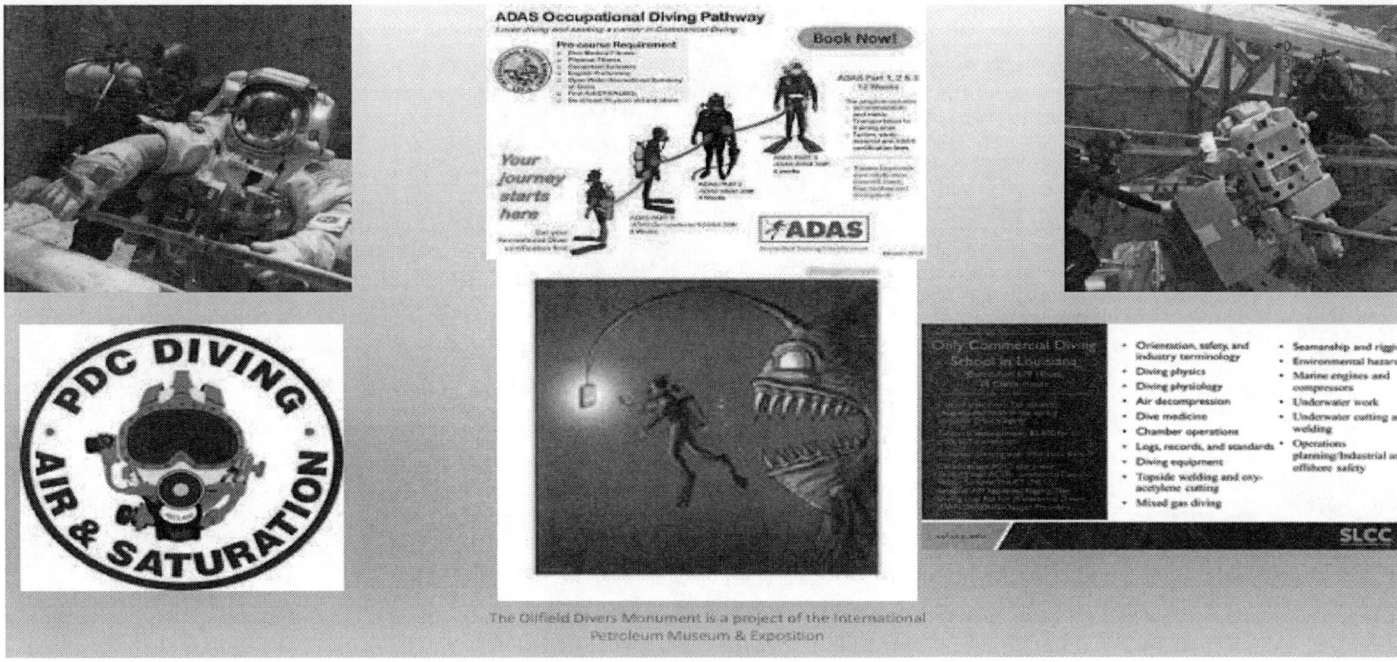

ADAS Occupational Diving Pathway
Book Now!
Your journey starts here
ADAS
PDC DIVING
AIR & SATURATION
Only Commercial Diving School in Louisiana
Orientation, safety, and industry terminology
Diving physics
Diving physiology
Air decompression
Dive medicine
Chamber operations
Logs, records, and standards
Diving equipment
Topside welding and oxy-acetylene cutting
Mixed gas diving
Seamanship and rigging
Environmental hazards
Marine engines and compressors
Underwater work
Underwater cutting and welding
Operations planning/Industrial and offshore safety
SLCC
The Oilfield Divers Monument is a project of the International Petroleum Museum & Exposition

s-media-cache-ak0.pinimg.com 2015
COMMERCIAL DIVER IS IN MY BLOOD
DIVERS ACADEMY INTERNATIONAL®
AIM HIGH, DIVE DEEP®
The Oilfield Divers Monument is a project of the International Petroleum Museum & Exposition

CAL DIVE
INTERNATIONAL
Oceanic
Stolt Offshore Inc.
TORCH
HORIZON
The Oilfield Divers Monument is a project of the International
Petroleum Museum & Exposition

DIVERS INSTITUTE OF TECHNOLOGY
SINCE 1968
SEATTLE, WA
CANADIAN WORKING DIVERS INSTITUTE
NEW ZEALAND SCHOOL OF
COMMERCIAL DIVER TRAINING
The Oilfield Divers Monument is a project of the International Petroleum Museum & Exposition

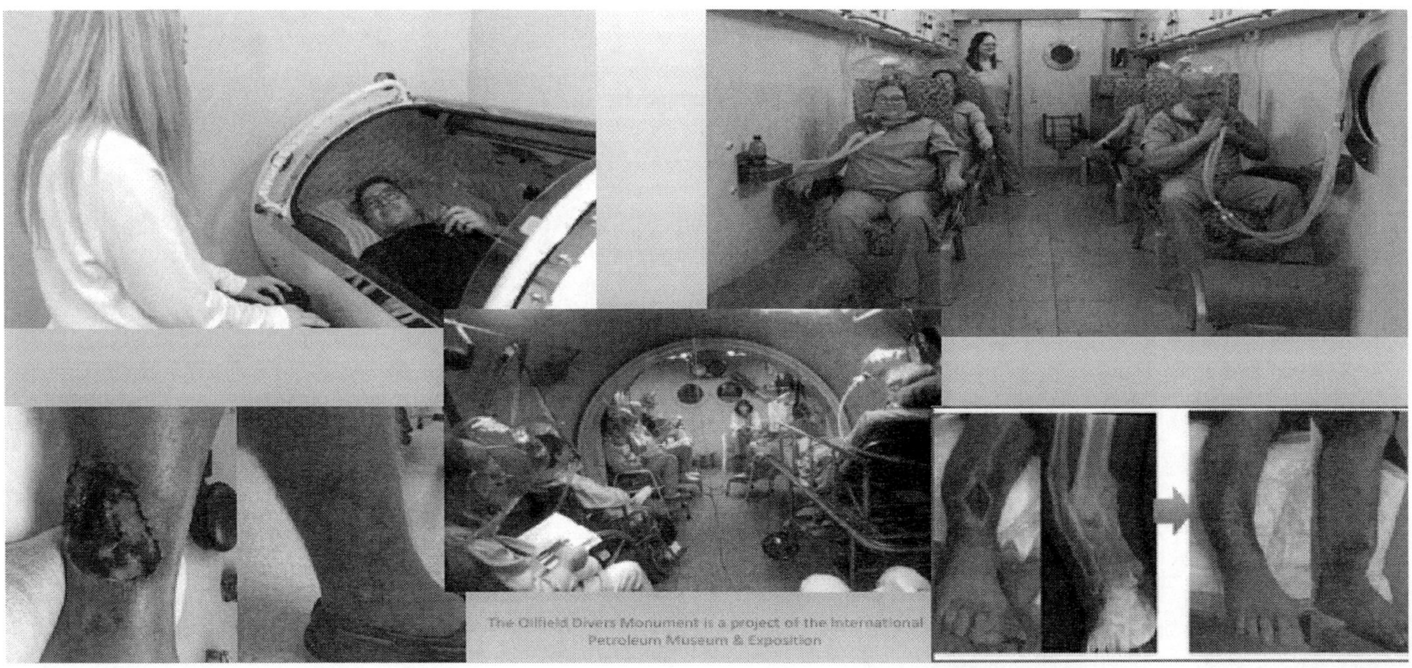
The Oilfield Divers Monument is a project of the International Petroleum Museum & Exposition

NAVAL DIVING AND SALVAGE
TRAINING CENTER
NAVY DIVER
COMMERCIAL DIVER
COASTAL SCHOOL OF DEEP SEA DIVING
OAKLAND - CALIF.
The Oilfield Divers Monument is a project of the International Petroleum Museum & Exposition

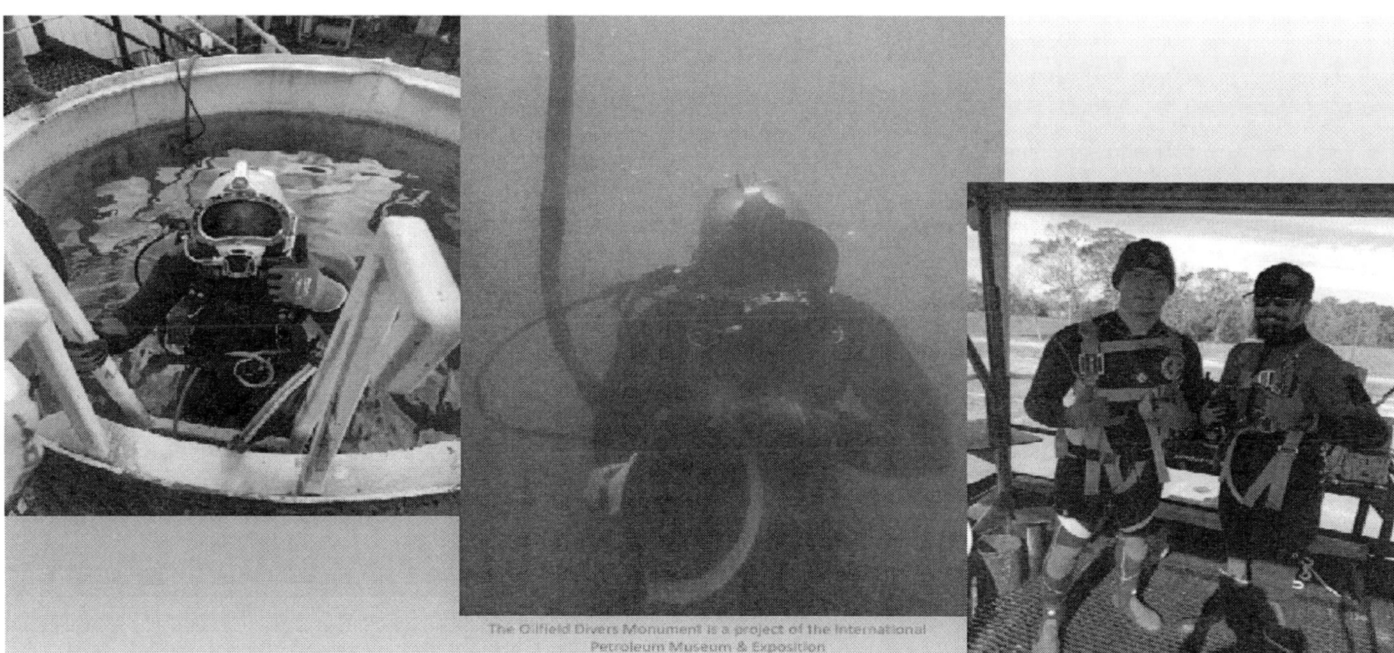
The Oilfield Divers Monument is a project of the International Petroleum Museum & Exposition

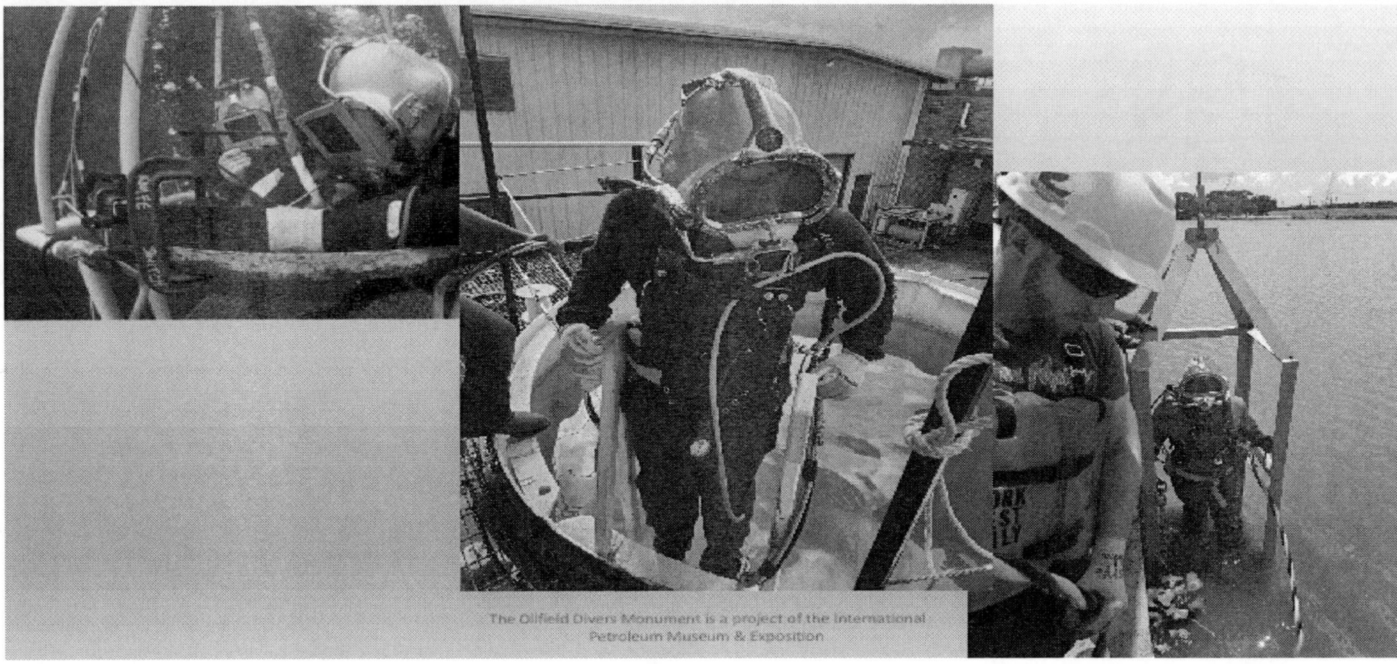

The Oilfield Divers Monument is a project of the International Petroleum Museum & Exposition

The Oilfield Divers Monument is a project of the International Petroleum Museum & Exposition

The Oilfield Divers Monument is a project of the International
Petroleum Museum & Exposition

The Oilfield Divers Monument is a project of the International
Petroleum Museum & Exposition

The Oilfield Divers Monument is a project of the International Petroleum Museum & Exposition

The Oilfield Divers Monument is a project of the International Petroleum Museum & Exposition

The Oilfield Divers Monument is a project of the International Petroleum Museum & Exposition

The Oilfield Divers Monument Project is taking a visionary step towards securing the future education of promising individuals by establishing an endowment for scholarships at Southern Louisiana Community College in Morgan City.

This approach ensures a lasting impact, providing assistance to scholars year after year. Beyond the altruistic benefits of contributing to education, **donors** also enjoy the advantage of tax deductions, making the act of endowing a scholarship both philanthropic and financially savvy.

The Oilfield Divers Monument is a project of the International Petroleum Museum & Exposition

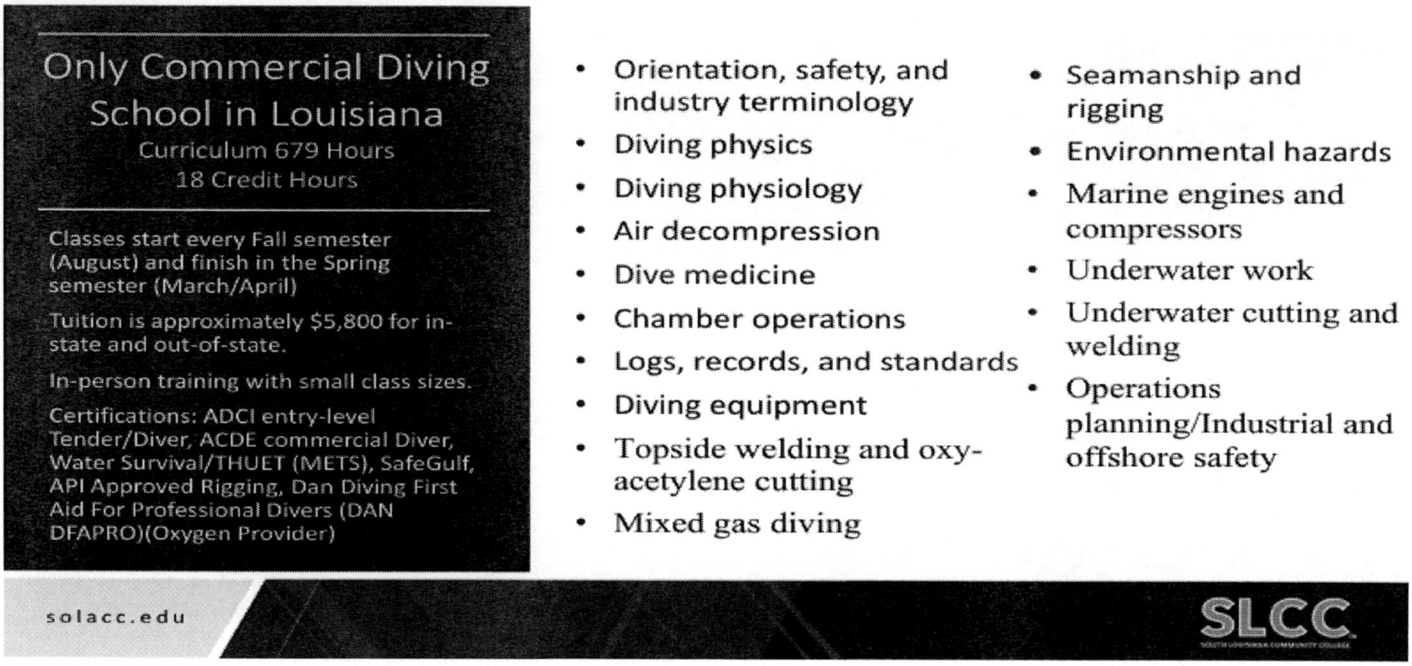
Only Commercial Diving School in Louisiana
Curriculum 679 Hours
18 Credit Hours
Classes start every Fall semester (August) and finish in the Spring semester (March/April)
Tuition is approximately $5,800 for in-state and out-of-state.
In-person training with small class sizes.
Certifications: ADCI entry-level Tender/Diver, ACDE commercial Diver, Water Survival/THUET (METS), SafeGulf, API Approved Rigging, Dan Diving First Aid For Professional Divers (DAN DFAPRO)(Oxygen Provider)
Orientation, safety, and industry terminology
Diving physics
Diving physiology
Air decompression
Dive medicine
Chamber operations
Logs, records, and standards
Diving equipment
Topside welding and oxy-acetylene cutting
Mixed gas diving
Seamanship and rigging
Environmental hazards
Marine engines and compressors
Underwater work
Underwater cutting and welding
Operations planning/Industrial and offshore safety
solacc.edu
SLCC

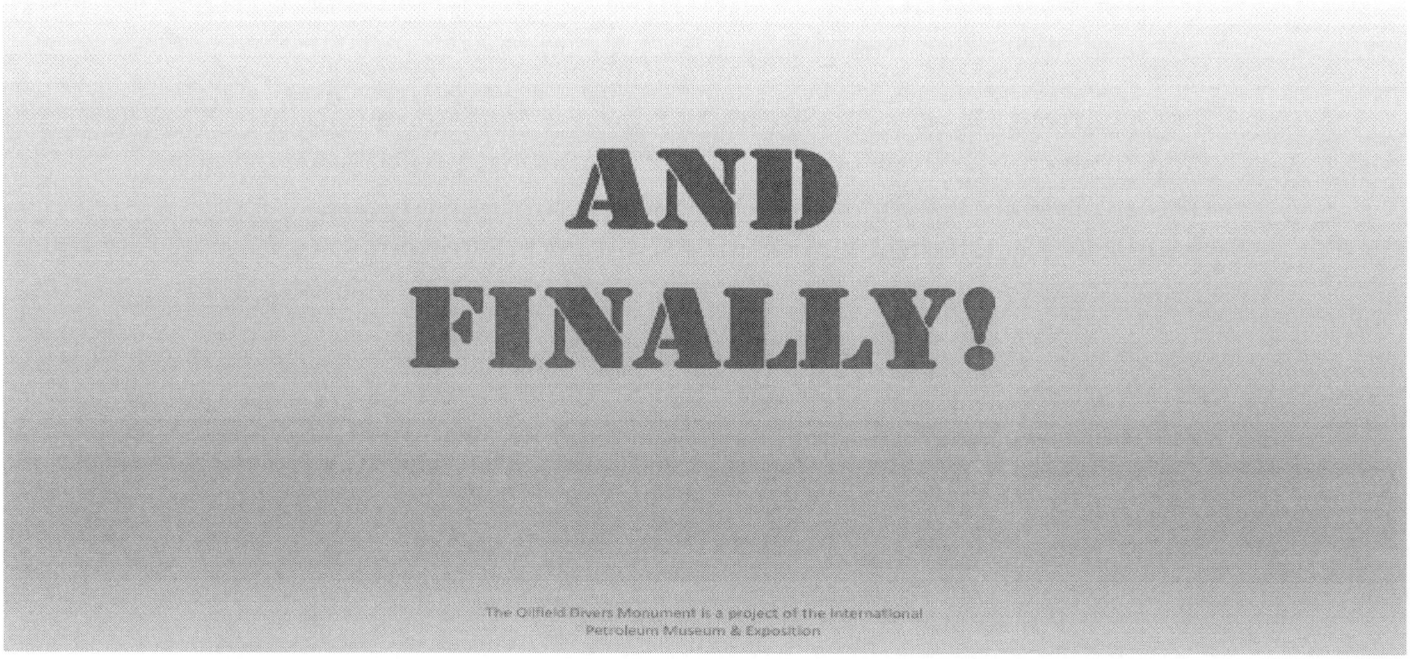
AND
FINALLY!
The Oilfield Divers Monument is a project of the International Petroleum Museum & Exposition

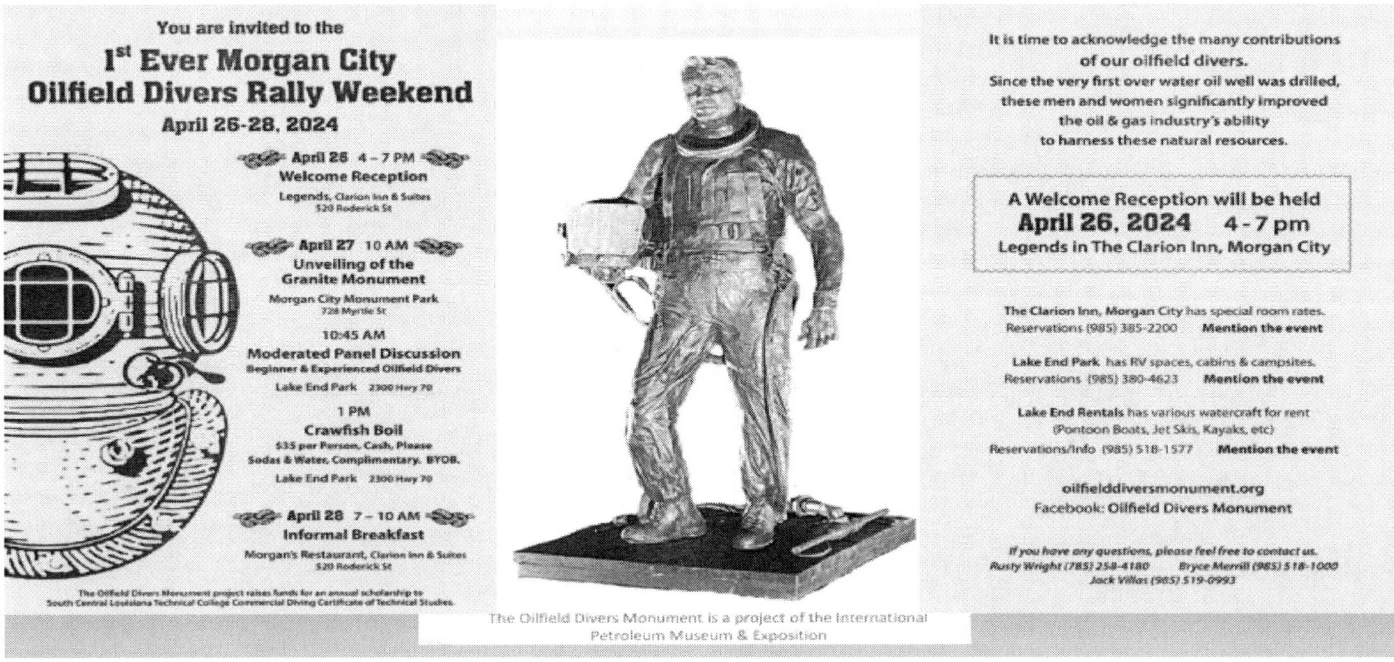

Virgil Allen, Micha Allen, Jack Vilas III, Bryce Merrill, Rusty Wright. Would like to thank all of our supporters.

BY DIVERS FOR DIVERS

The Oilfield Divers Monument is a project of the International Petroleum Museum & Exposition

MEETING THE DEMAND: THE POWER OF SCALING USV PRODUCTION

Brian Connon, VP Ocean Mapping

OUR MISSION

To collect the ocean data we need for the planet we want

SAILDRONE

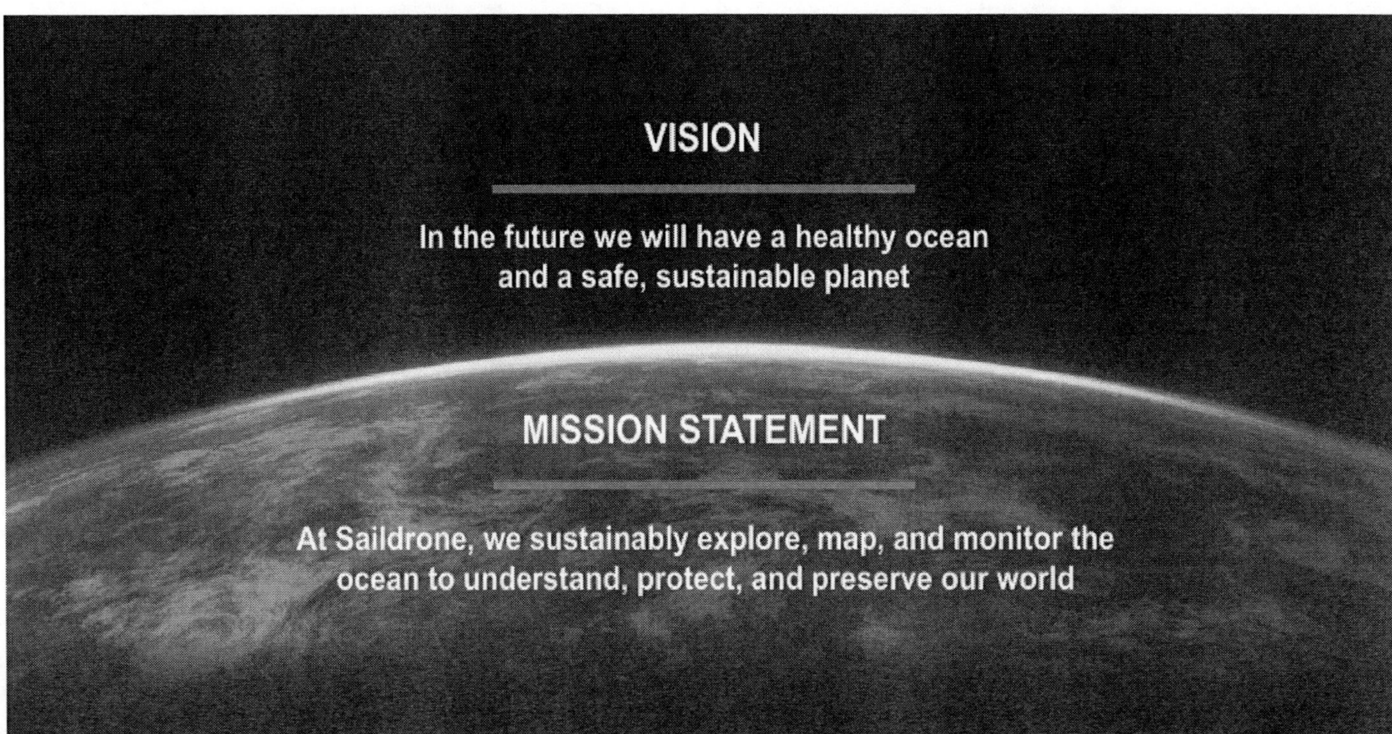

WHY UNCREWED SURFACE VEHICLES

Scalable. Capable. Safe.

SAILDRONE

Environmentally Friendly
Operations with low or no carbon footprint or environmental impact

Efficient
Capabilities delivered at a fraction of the cost of traditional approaches

Reduced Risk
No crewed support vessel required, reducing the operational cost, complexity and risk

Persistent
Long-distance/long-term duration without returning to port or support vehicle for maintenance or refueling

Safety
USVs can operate in harsh conditions where it would be hazardous to send crewed ships

ML Enabled
Many offerings are now supported by machine learning solutions

24/7 365

Real Time Data Access
24/7/365 real-time access to critical information

Proven Technology
USVs have been proven in the harshest of ocean conditions

SCALABILITY ENABLED

Why are USVs good candidates for scaling?

SAILDRONE

Design

- Size
- Modular
- Repeatable
- Easily upgraded and maintainable
- Low carbon footprint

Cost

- Affordable components
- Economical production
- Quick to market/mission

COPYRIGHT © SAILDRONE, INC. 2022. ALL RIGHTS RESERVED.
SAILDRONE, INC. PROPRIETARY – RELEASE BEYOND INTENDED RECIPIENTS IS PROHIBITED

SCALABILITY

What drives the decision?

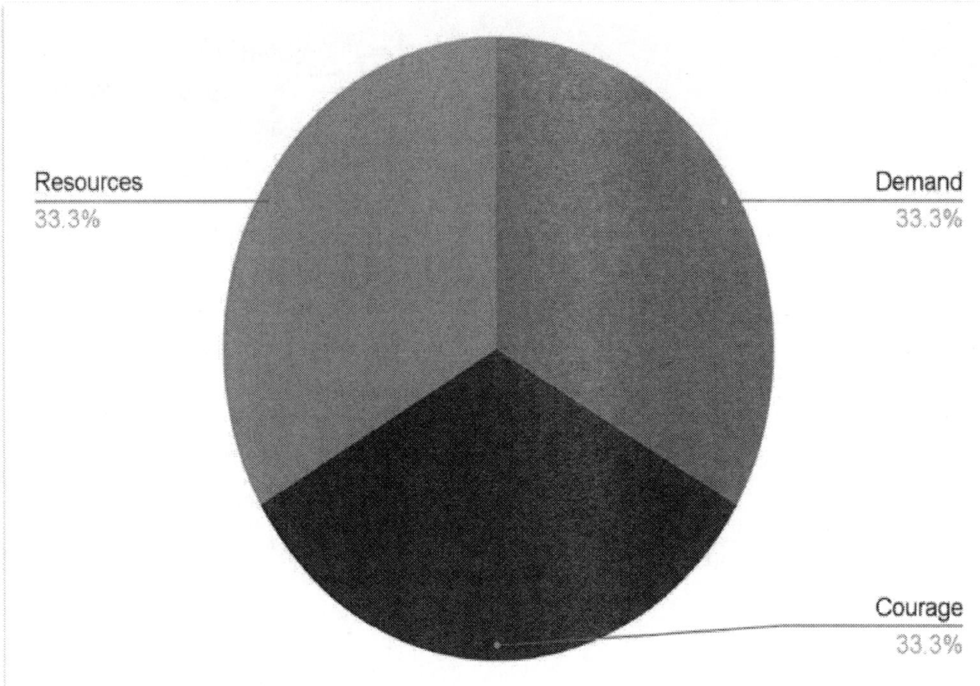

DEMAND

Is there a market?

SAILDRONE

Increasing Capability

- USVs can now tackle more than simple data collection
- Complex payloads
- Access to high bandwidth satcom

Expanding Missions

- Explosion of renewables
- IUU fishing and other crimes
- Climate observations

Capacity Gaps

- Traditional assets can't keep up

Focus on Safety and Environment

MISSIONS FOR UNCREWED SURFACE VEHICLES

What can they do?

SAILDRONE

Bathymetry
Multibeam operations with low or no carbon footprint or environmental impact

Fisheries
Survey capabilities delivered at a fraction of the cost of traditional approaches

Marine Mammal Tracking
No crewed support vessel required, reducing the operational cost, complexity and risk

Ice Operations
USVs can operate in harsh conditions where it would be hazardous to send crewed ships

METOC Observations
Augment or replace buoys to collect weather and lake data.

Water Quality Monitoring
Mobile platforms to identify and track critical water characteristics.

Maritime Surveillance
Expand eyes and ears on the water to report illegal actions, like IUU fishing.

RESOURCES

Can you do it?

SAILDRONE

Production

- People
- Facilities
- Supply Chain
- Outsource?
- Can you maintain quality?

Funding

- Big investment
- Economical production
- Quick to market/mission

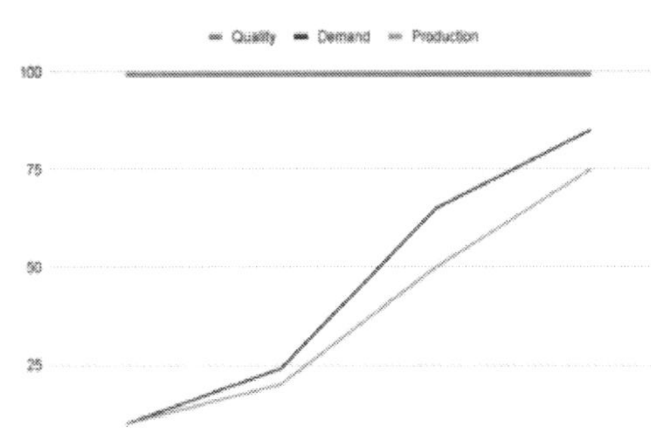

PARADIGM SHIFT IN THE COST OF OCEAN ACCESS

Reduced cost, risk, and carbon footprint

SAILDRONE

Ships	Saildrone
CAPEX: $100M – $1Bn	CAPEX: $0
• Fuel Burn: 1k – 5k gallons per day • Standard Deployment: 30 days	• Fuel Burn: Zero (Explorer) – 10 gallons per day (Surveyor) • Standard Deployment: 100 – 300 days
Significant risk for human operators Significant liability for employers	No risk to human life

97.3% REDUCTION IN CO2 EMISSIONS

2022 Aleutian Islands Mapping: San Francisco → Bering Sea → San Francisco

SAILDRONE

12,000 nm round trip
16,215 sq km of seabed mapped

Surveyor emitted just 2.7% of CO_2 compared to the ship alternative

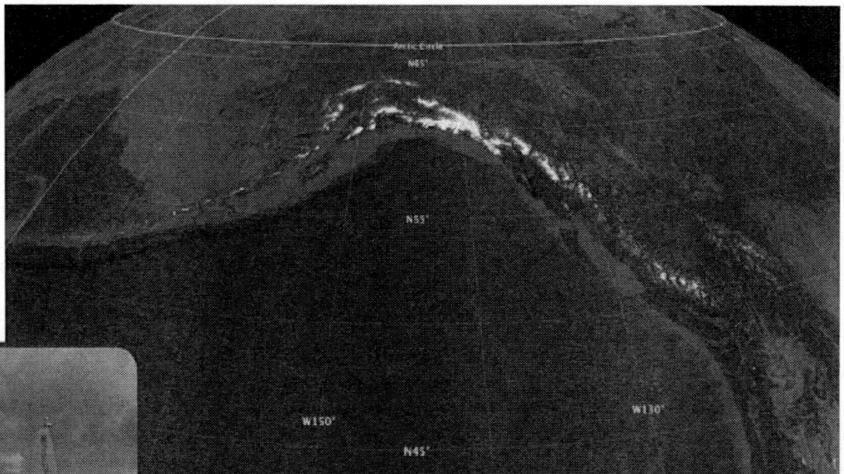

Total CO_2 emissions: 437.36 T

Total CO_2 emissions: 12.19 T

COURAGE
Is the risk tolerable?

SAILDRONE

No Guarantees
- Market could shift
- Competition
- Loss of funding

Timelines
- Can you hire quickly enough?
- How long can you wait for profitability?
- Can you scale quickly enough to catch the market?

Excess Capacity
- Can you afford USVs not deployed?
- More responsive to near term opportunities

10 YRS OF R&D ▶ PROVEN RELIABILITY
Significant barriers to entry, uncatchable technological lead

SAILDRONE

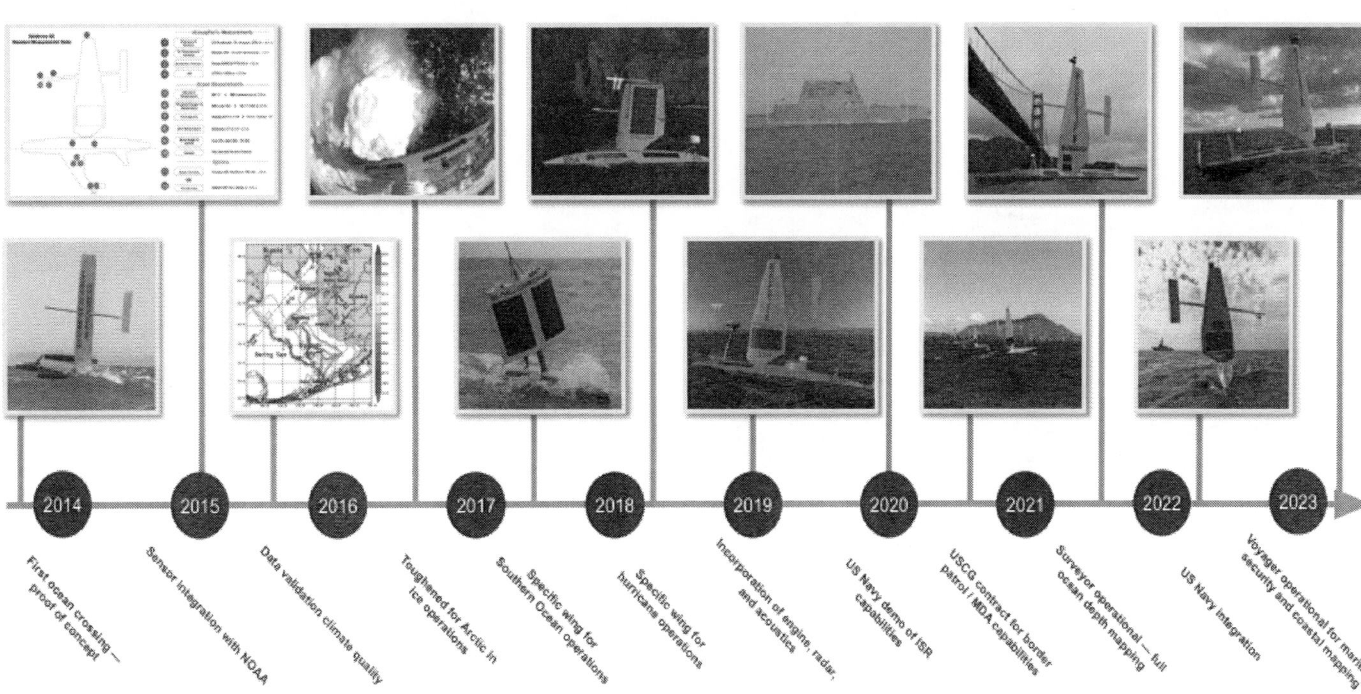

LAUNCH FROM ANY DOCK

Without the requirement of a crewed vessel, further reducing carbon footprint

SAILDRONE

EXCEPTIONAL DATA QUALITY

Rigorously tested by external collaborators

SAILDRONE

“Comparisons with shipboard measurements showed good agreement, inspiring confidence in these new instrument platforms.”

THE USE OF SAILDRONES TO EXAMINE SPRING CONDITIONS IN THE BERING SEA: INSTRUMENT COMPARISONS, SEA ICE MELTWATER AND YUKON RIVER PLUME STUDIES. OCEANS 2015

“The saildrones performed well in the harsh conditions of the Bering Sea and demonstrated the potential of this innovative platform to advance ecosystem research.”

ADVANCES IN ECOSYSTEM RESEARCH: SAILDRONE SURVEYS OF OCEANOGRPAHY, FISH, AND MARINE MAMMALS IN THE BERING SEA. OCEANOGRAPHY 30(2):113–115

“A platform that is ready for ocean research missions from the tropics to the Arctic.”

THE USE OF SAILDRONES TO EXAMINE SPRING CONDITIONS IN THE BERING SEA: VEHICLE SPECIFICATION AND MISSION PERFORMANCE. OCEANS 2015

A GLOBAL FLEET OF OCEAN DRONES

Wind and solar-powered, monitoring the planet in real time, above and below the surface.

SAILDRONE

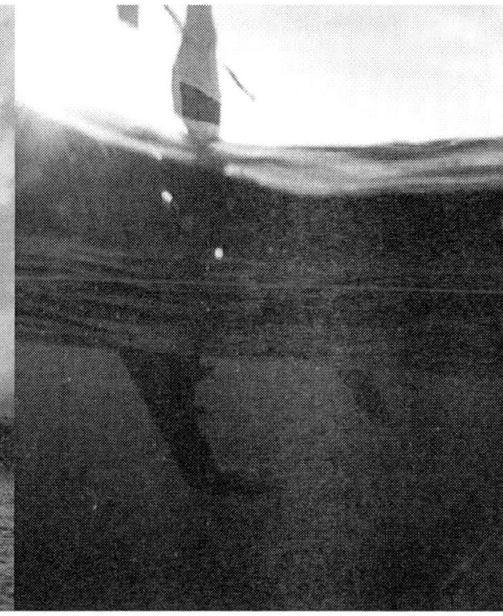

Maritime Domain Awareness

Intelligence, Surveillance & Reconnaissance (ISR)
Force Protection | Law Enforcement & Maritime Safety
Ecosystem Monitoring

Ocean Mapping

Single-beam and multibeam bathymetric data collection for navigation and charting, telecommunications, offshore energy, and physical oceanography to 23,000 feet (7,000 m) depth.

Ocean Data

Collecting essential ocean and climate variables.
Fisheries | Metocean Data Collection | Ecosystem monitoring | Satellite Calibration/Validation

MARITIME DOMAIN AWARENESS

Maximizing information and decision advantage

SAILDRONE

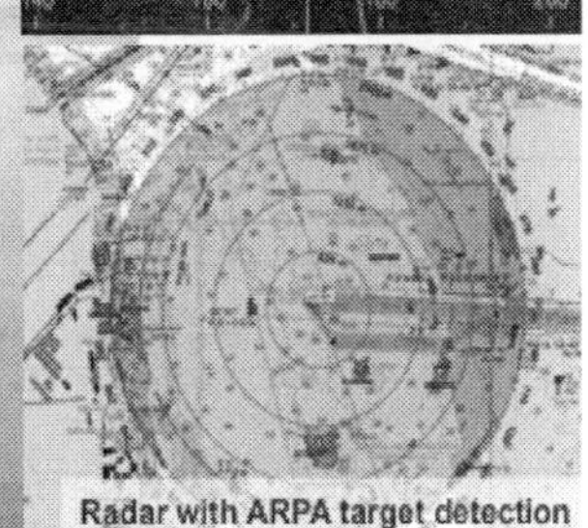

Missions

- IUU (Illegal, unreported and unregulated fishing)
- ISR (Intelligence Surveillance Reconnaissance)
- Law Enforcement and Maritime Safety
- Drug Interdiction
- Border & Harbor Security
- Ecosystem monitoring

Payload Options

- Radar array
- 360° optical array camera
- Onboard ML-enabled real-time detection and alerting
- Passive acoustics (under development)
- Class A – AIS transceiver

SAILDRONE BATHYMETRY PLATFORMS

USVs capable of global reach

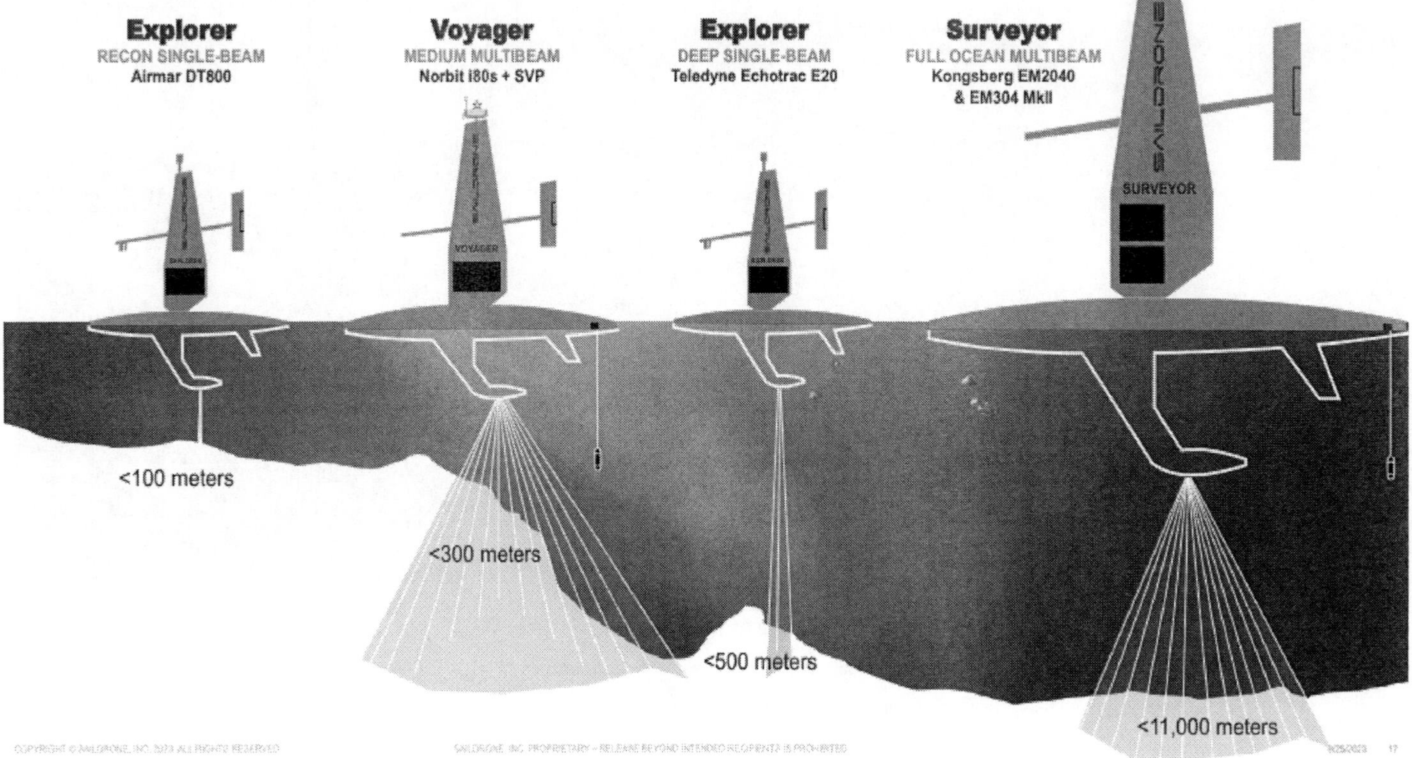

OCEAN DATA

Purpose-Built 24/7 Autonomous Ocean and Climate Data Solutions

SAILDRONE

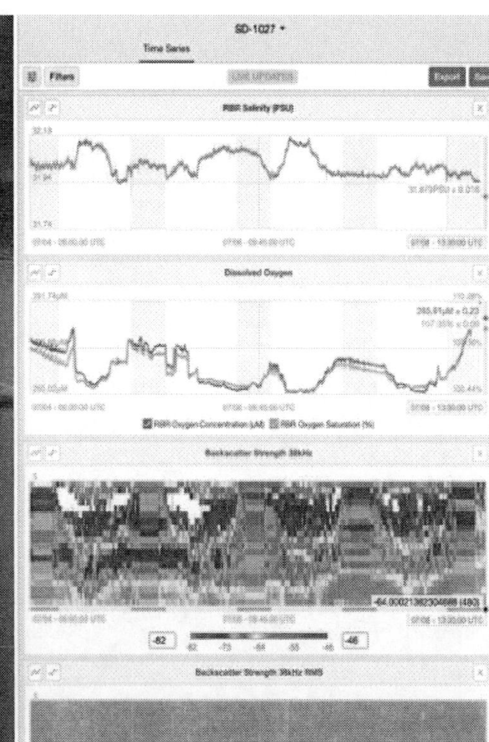

2021-22 ATLANTIC HURRICANE MISSION
Studying the rapid intensification of hurricanes

SAILDRONE

Real-time data, including air temperature and relative humidity, barometric pressure, wind speed and direction, water temperature and salinity, sea surface temperature, and wave height and period

- Total vehicles deployed: 12
- Cumulative mission days: 413 days
- Carbon emission: zero

RECOGNIZED!
SD 1045 in the Guinness World Records

SAILDRONE

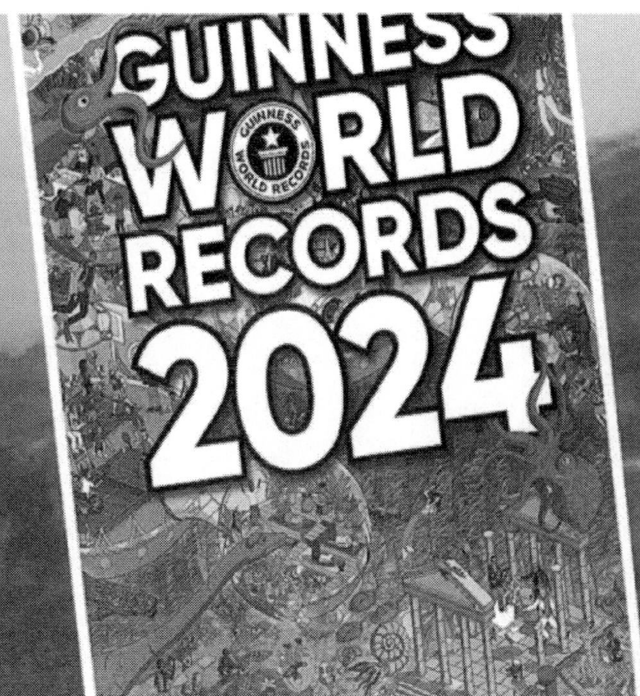

SD 1045 measured 109.83 knots (126.4 mph) in the eyewall of Category 4 Hurricane Sam in 2021.

ABS CLASSIFICATION

The new standard for USVs

Saildrone Voyager is the first autonomous asset built to American Bureau of Shipping (ABS) Class using the *ABS Requirements for Autonomous and Remote Control Functions.*

- Demonstrating Saildrone's commitment to safety, standardization, and reliability in its technology and operations
- Allows the Voyager to operate in the ports and waters of countries that require vessels to be classed by organizations such as ABS
- ABS classification helps companies to comply with regulatory requirements—Saildrone is working with ABS to define and build the regulatory guidance for USVs

> " Saildrone has spent three years maturing the Voyager design to be the industry leader in capability, reliability, and safety in the uncrewed vehicle sector. This classification from the American Bureau of Shipping defines the new gold standard for uncrewed systems and underscores the maturity of our technology. "
>
> *SAILDRONE FOUNDER & CEO RICHARD JENKINS*

SUMMARY

USVs at scale: a unique opportunity

Scalability can work!

- Requires patience and funding
- The market is ripe and already accepting new technology
- Shiptime is increasingly expensive and scarce
- USVs are an attractive alternative to traditional methods, such as ships and buoys

SAILDRONE

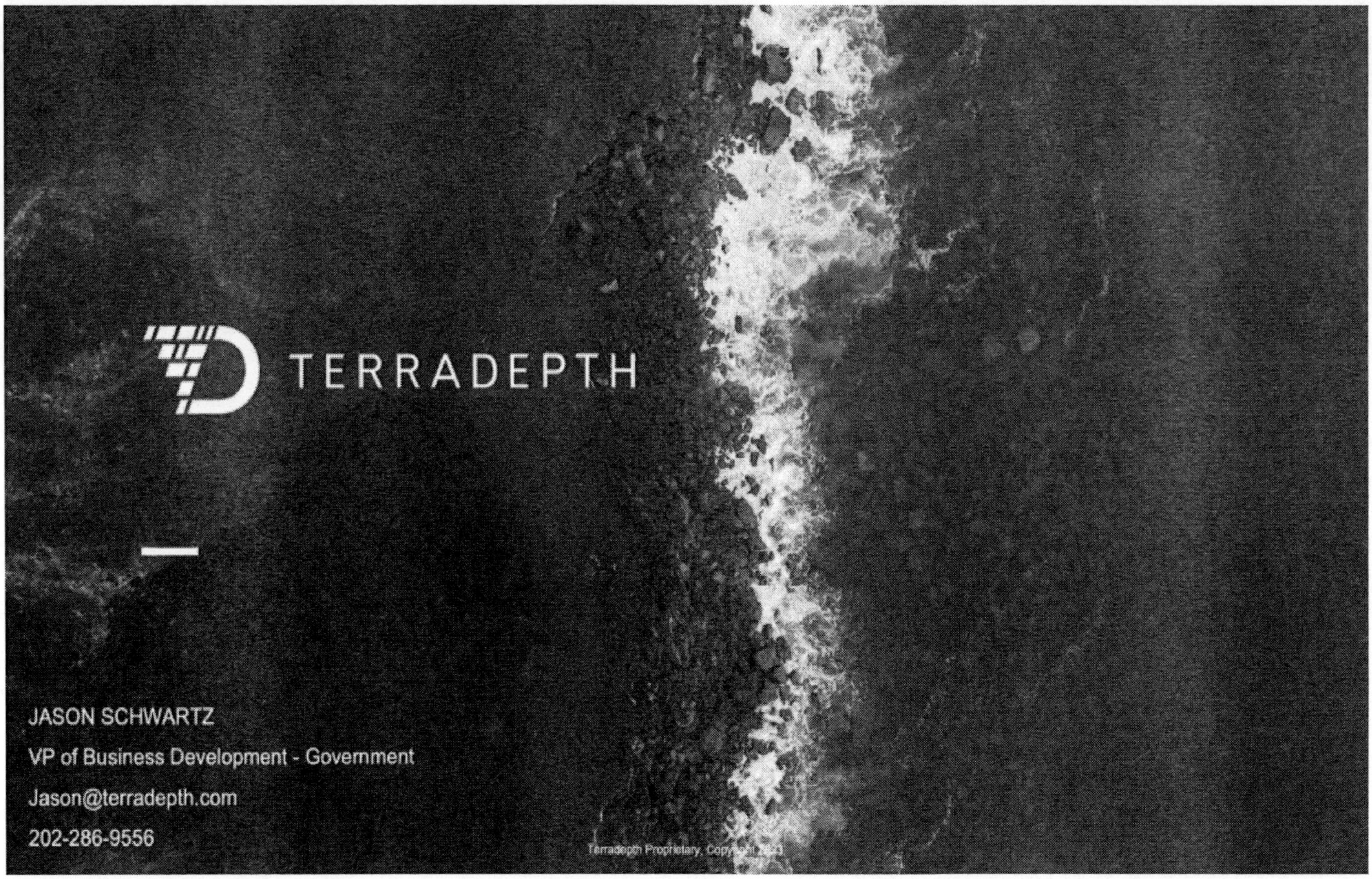
TERRADEPTH
JASON SCHWARTZ
VP of Business Development - Government
Jason@terradepth.com
202-286-9556
Terradepth Proprietary, Copyright 2023

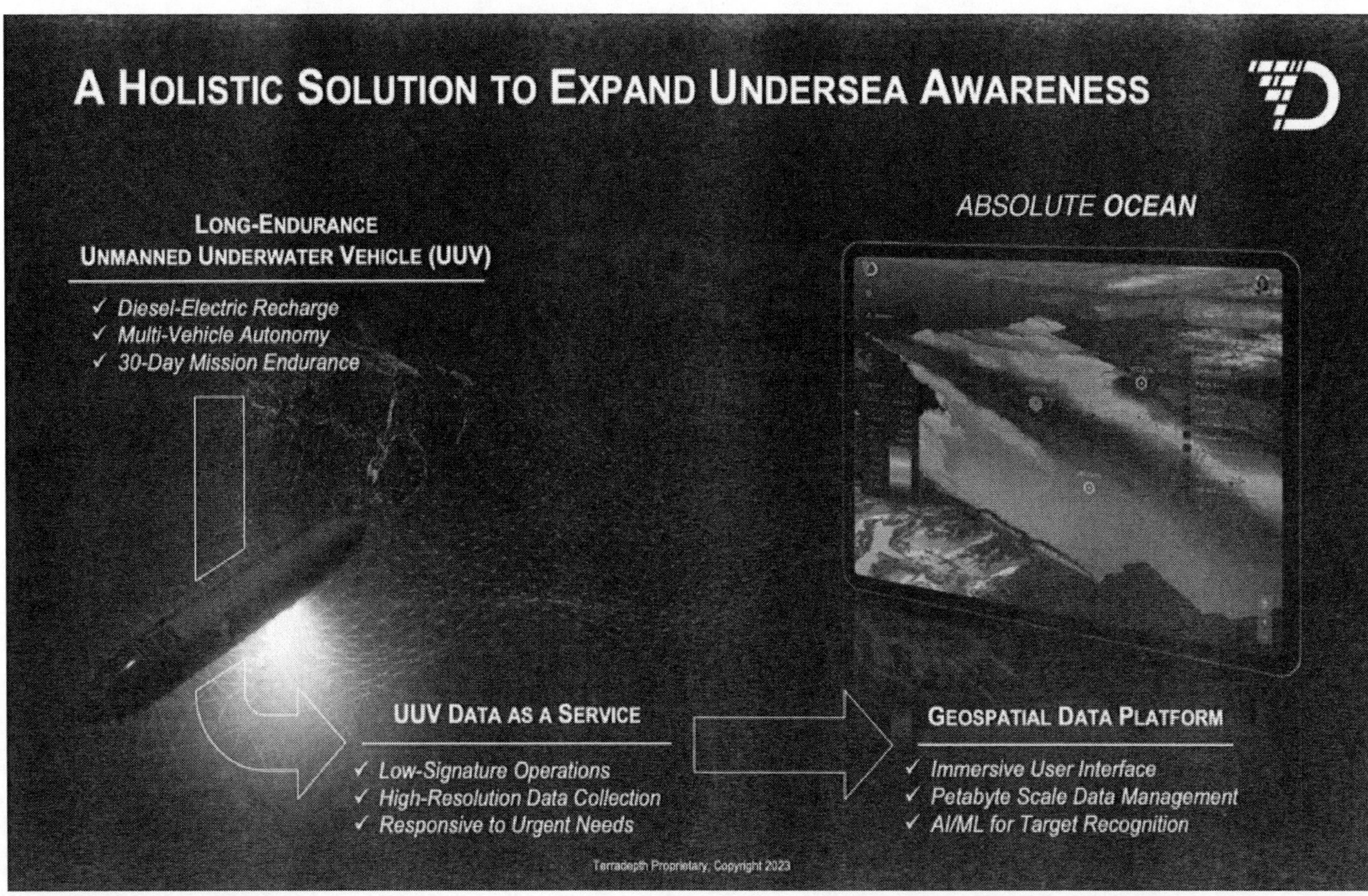
A HOLISTIC SOLUTION TO EXPAND UNDERSEA AWARENESS
LONG-ENDURANCE
UNMANNED UNDERWATER VEHICLE (UUV)
✓ Diesel-Electric Recharge
✓ Multi-Vehicle Autonomy
✓ 30-Day Mission Endurance
ABSOLUTE OCEAN
UUV DATA AS A SERVICE
✓ Low-Signature Operations
✓ High-Resolution Data Collection
✓ Responsive to Urgent Needs
GEOSPATIAL DATA PLATFORM
✓ Immersive User Interface
✓ Petabyte Scale Data Management
✓ AI/ML for Target Recognition
Terradepth Proprietary, Copyright 2023

UUV Platforms

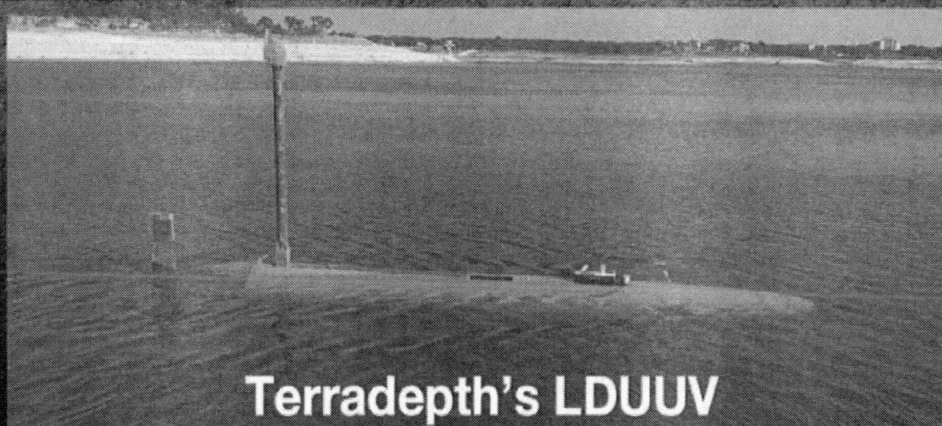

Long-Endurance LDUUV*

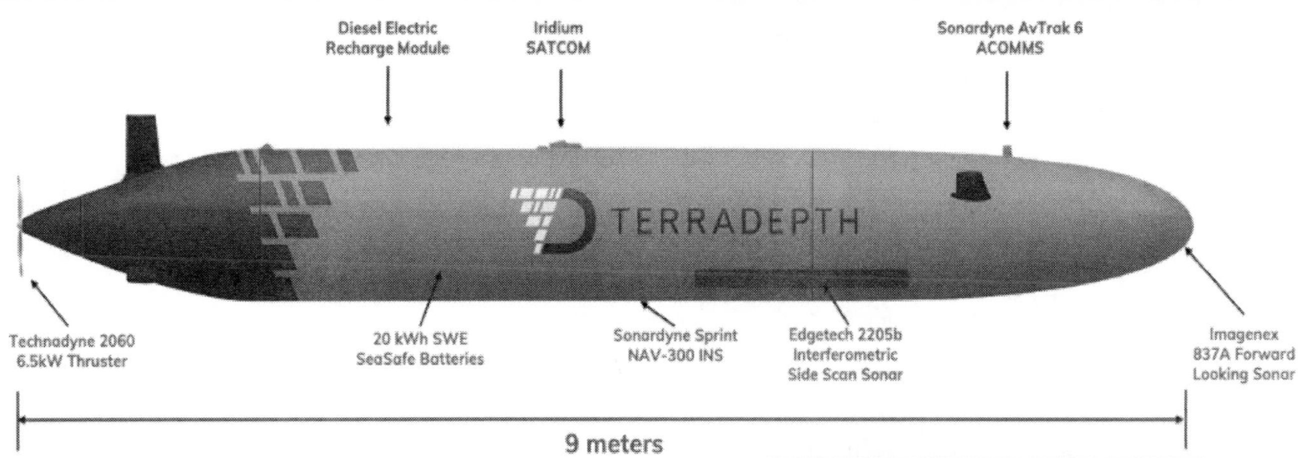

Specifications

System TRL: 6	**Size:** 9m x 1m	**Battery Capacity:** 20 kWh
Endurance: 7 Days	**Weight:** 8,000 lbs	**Fuel Capacity:** 13 gal Diesel
Depth Rating: 1,000m	**Speed:** 5 kts (max)	**Diesel Output:** 4.5 kW

* - Prototype undergoing sea trials

ABSOLUTE ***OCEAN***

FEATURES

- Secure Cloud Access
- Immersive User Interface
- Petabyte Scale Ocean Data Management
- Share Data Across the Enterprise
- AI/ML Automatic Target Recognition

Low-Logistics Survey

Low-Logistics Survey – USS Texas

Low-Logistics Survey – USS Texas

THANK YOU
JASON SCHWARTZ
VP of Business Development - Government
Jason@terradepth.com
202-286-9556
Terradepth Proprietary, Copyright 2022
9

UNDERWATER
INTERVENTION
Powered by WORKBOAT
OPENOCEAN
ROBOTICS
Navigating a Sustainable Future:
Empowering Offshore Wind Development
with Autonomous USVs
ANDY ZIEGWIED | DIRECTOR OF SALES
andy@openoceanrobotics.com

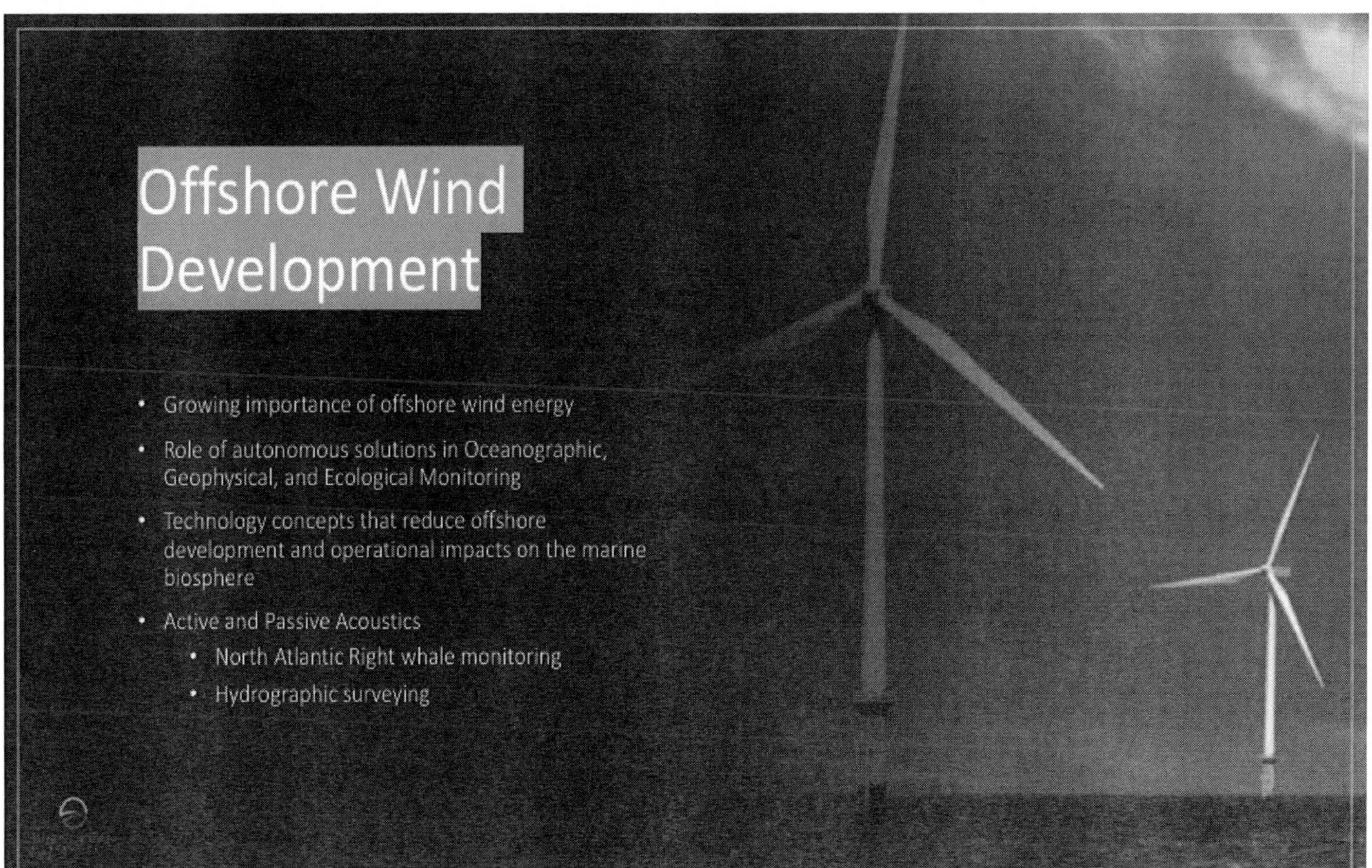
Offshore Wind
Development
Growing importance of offshore wind energy
Role of autonomous solutions in Oceanographic, Geophysical, and Ecological Monitoring
Technology concepts that reduce offshore development and operational impacts on the marine biosphere
Active and Passive Acoustics
North Atlantic Right whale monitoring
Hydrographic surveying

Growing Importance of Offshore Wind Energy
• Offshore wind is an abundant, efficient and clean alternative domestic energy resource with the potential to generate jobs and combat the effects of climate change.
• The challenges that face our nation and planet with climate change are forcing the rapid evolution of renewable energy, including extensive development for the harvesting offshore wind energy.
• Offshore wind capacity is expected to reach 234 GW by 2030, generating approximately 18% of the world's electricity.

Wind DataXplorer
The ship or shore launched Open Ocean Robotics DataXplorer delivers a broad range of remote sensing data acquisition, monitoring and analysis capabilities, with a quiet, persistent, zero operational carbon footprint.
Allowing developers to collect an array of environmental and wildlife data through sensors that enable the monitoring of bats, birds, fish and other marine mammals to determine the presence, frequency and distribution across the entire lease area - unlike fixed metocean buoys and towers.
OPENOCEAN

Autonomous Solutions for Offshore Wind Development

The data collected from day one will advance understanding of wind and wildlife patterns across the entire lease areas to inform the most environmentally responsible and efficient design, project layout and turbine siting.

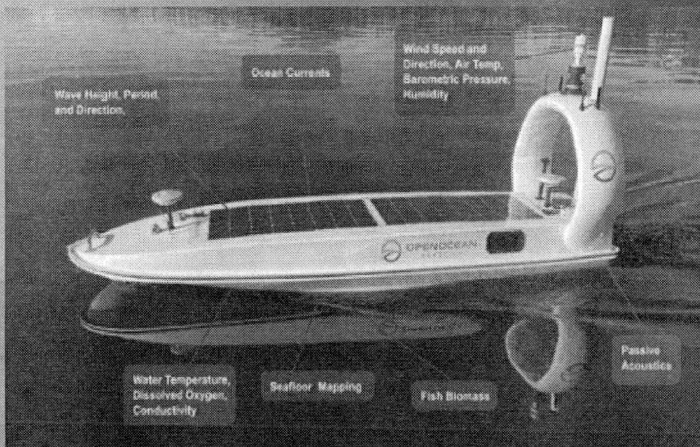

5 OPENOCEAN

Meeting Oceanographic, Geophysical, and Ecological Monitoring Needs

APPLICATION	DATA	SENSORS
Water Chemistry	• Hypoxia • Salinity • Turbidity • Chlorophyll • Bioinformatics	• Fluorometers • PH Sensors • DO Sensors • EDNA Samplers • pCO2 Sensors
Apex Predator Monitoring	• Wildlife Tracking	• Acoustic Tag Receivers
Marine Mammal Monitoring	• Wildlife Tracking • Ocean Noise	• Acoustic Tag Receivers • Cameras • Hydrophones
Cultural Heritage	• Photogrammetry • Photo / Video • Sonar Mosaic	• Multi-beam sonar • Side Scan Sonar • Gradiometer • Cameras
Fisheries Counts / Habitat Monitoring / IUU	• Bathymetry • Fish Biomass • Benthic Habitat	• Multi-beam sonar • Acoustic Modems • Hydrophone Arrays • Acoustic Tag Receiver
MetOcean	• Wind Speed and Direction • Air and Water Temperature • Sea State and Wave Energy • Atmospheric pressure • Humidity	• Anemometer • CTD • ADCP • Inertial magnetics
Seabed Mapping/UXO	• Georeferenced Imagery • Photo / Video	• Multi-beam Sonar • Side Scan Sonar • Gradiometer/ Magnetometer
Vessel Traffic Monitoring	• AIS Data • Imagery • Acoustic tracking data • Vessel positioning data • Commercial shipping traffic	• AIS receivers • Hydrophones • Radar • Cameras • FLIR

HARDWARE

DataXplorer ™

CAMERAS & SENSORS
360° cameras and sensors provide real-time insights

PATENTED SELF-RIGHTING SYSTEM
High sea state and surf zone capable

LONG ENDURANCE MISSIONS
Solar panels charge battery bank to repower at sea enabling extended duration deployments

CUSTOMIZABLE SENSOR PAYLOAD
Multiple modular sensor bays allow diverse sensor integration for broad market needs

SHALLOW WATER CAPABLE
Shallow draft and debris- shedding keel enable use in limited depth waters

Enabling safer, more affordable and sustainable ocean observations.

SOFTWARE

XplorerView ™

- Secure Multi-user System
- Real-time Data Visualization
- AI-driven Analytics
- Real-time Performance Monitoring
- Comprehensive Navigational Tools
- Precision Control

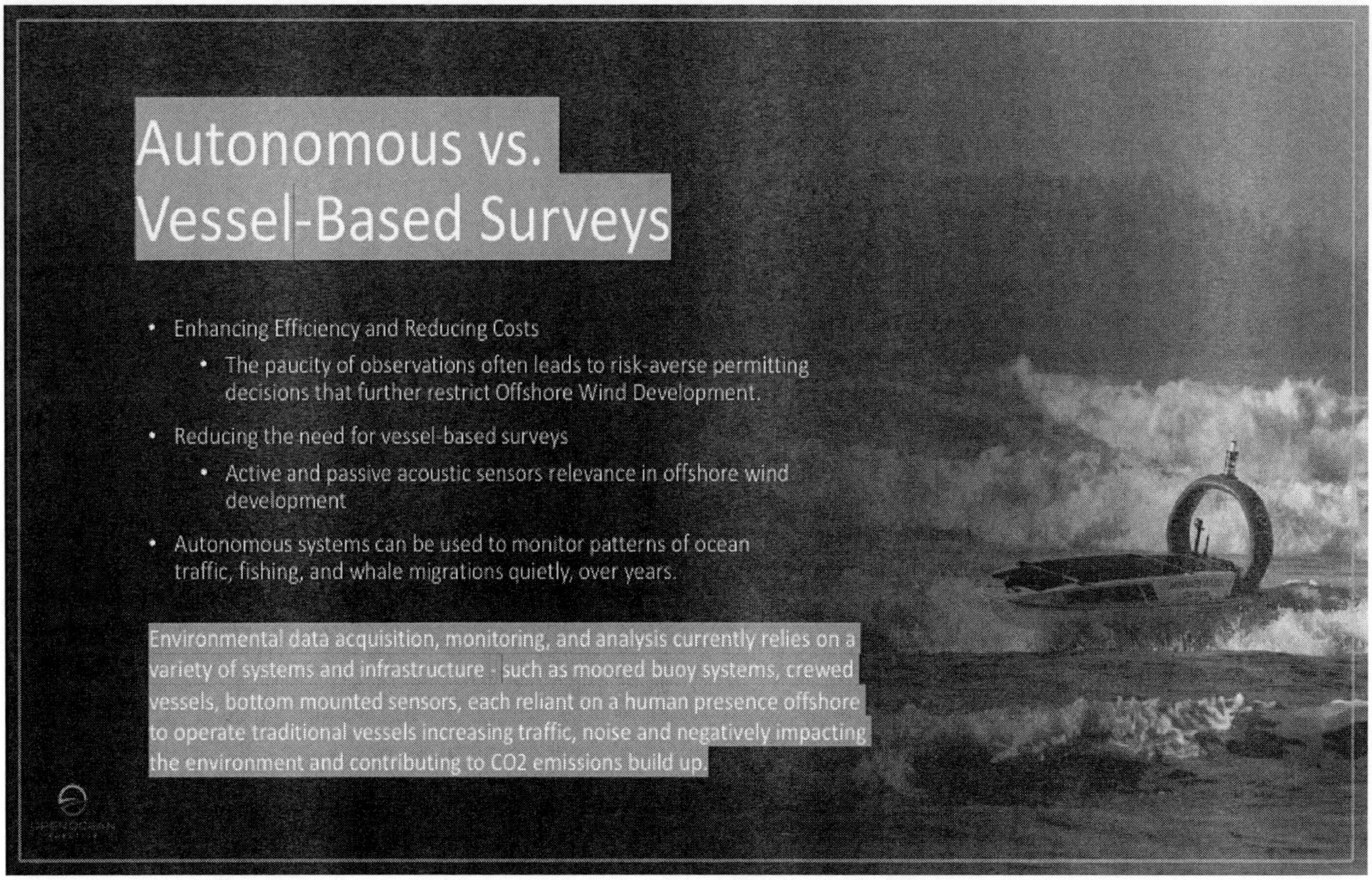
Autonomous vs. Vessel-Based Surveys
Enhancing Efficiency and Reducing Costs
The paucity of observations often leads to risk-averse permitting decisions that further restrict Offshore Wind Development.
Reducing the need for vessel-based surveys
Active and passive acoustic sensors relevance in offshore wind development
Autonomous systems can be used to monitor patterns of ocean traffic, fishing, and whale migrations quietly, over years.
Environmental data acquisition, monitoring, and analysis currently relies on a variety of systems and infrastructure - such as moored buoy systems, crewed vessels, bottom mounted sensors, each reliant on a human presence offshore to operate traditional vessels increasing traffic, noise and negatively impacting the environment and contributing to CO2 emissions build up.

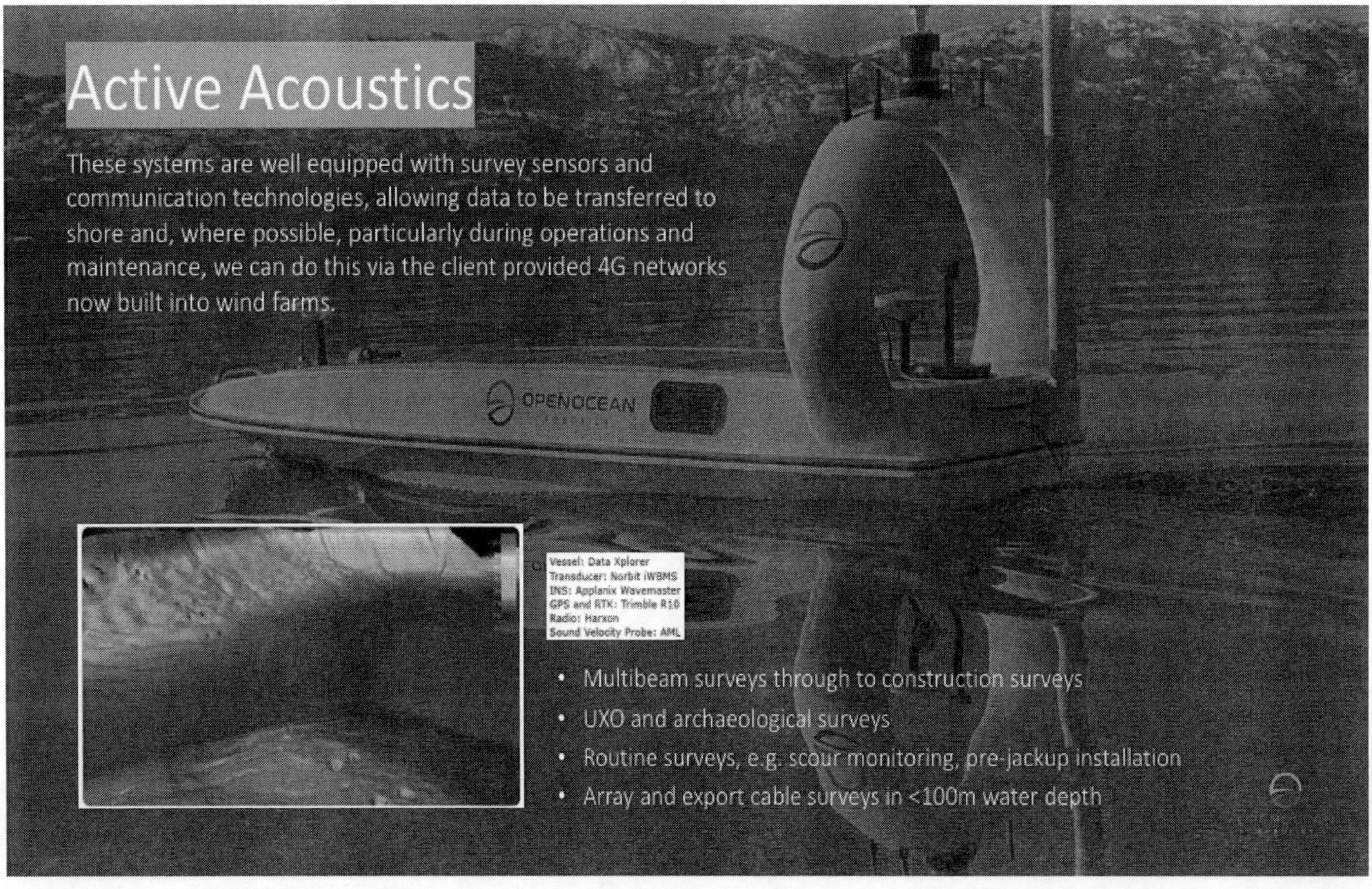
Active Acoustics
These systems are well equipped with survey sensors and communication technologies, allowing data to be transferred to shore and, where possible, particularly during operations and maintenance, we can do this via the client provided 4G networks now built into wind farms.
OPENOCEAN
Vessel: Data Xplorer
Transducer: Norbit iWBMS
INS: Applanix Wavemaster
GPS and RTK: Trimble R10
Radio: Harxon
Sound Velocity Probe: AML
Multibeam surveys through to construction surveys
UXO and archaeological surveys
Routine surveys, e.g. scour monitoring, pre-jackup installation
Array and export cable surveys in <100m water depth

Passive Acoustic Monitoring

- Passive acoustical monitoring (PAM) has been an important part of many NOAA & BOEM-funded studies to determine the occurrence, density, distribution, and migratory behavior of marine mammals and some species of fish.
- Studies of these animals and their habitats has always been challenged by weather, platform availability and their own presence (or absence).
- To distinguish between natural variability and changes caused by offshore energy development to observe change over many years and decades, it is a practical necessity to compare observations collected from many different data-gathering activities.
- New Strategy - Protecting North Atlantic Right Whales During Offshore Wind Energy Development

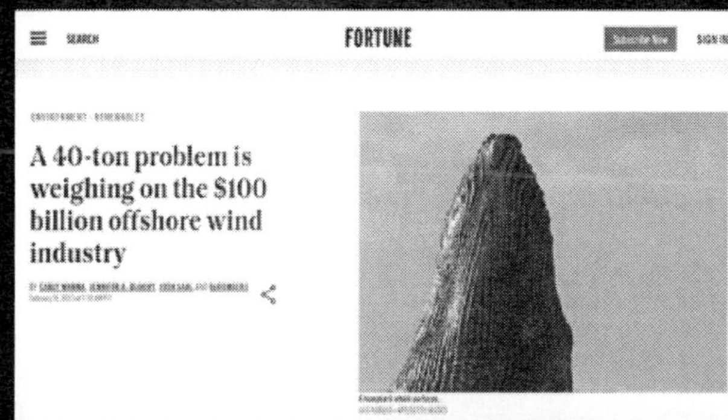
SEARCH FORTUNE SIGN IN

A 40-ton problem is weighing on the $100 billion offshore wind industry

PROJECT: PROTECTING ENDANGERED SOUTHERN RESIDENT KILLER WHALES

KEY OBJECTIVES:

- Real time detection of marine mammals
- Evaluate ability conduct acoustic mission while detecting other vessels and collecting ocean and atmospheric data
- Multiple data stream comparisons & analysis
- Evaluate and minimize USV self-noise

PROJECT SUPPORT:

- Dept. of Fisheries & Oceans (DFO) Canada Ecosystems and Oceans Science Contribution Framework
- Grant period: August 2021 - March 2023

PROJECT BACKGROUND – INTERIM SANCTUARY ZONES (ISZs)

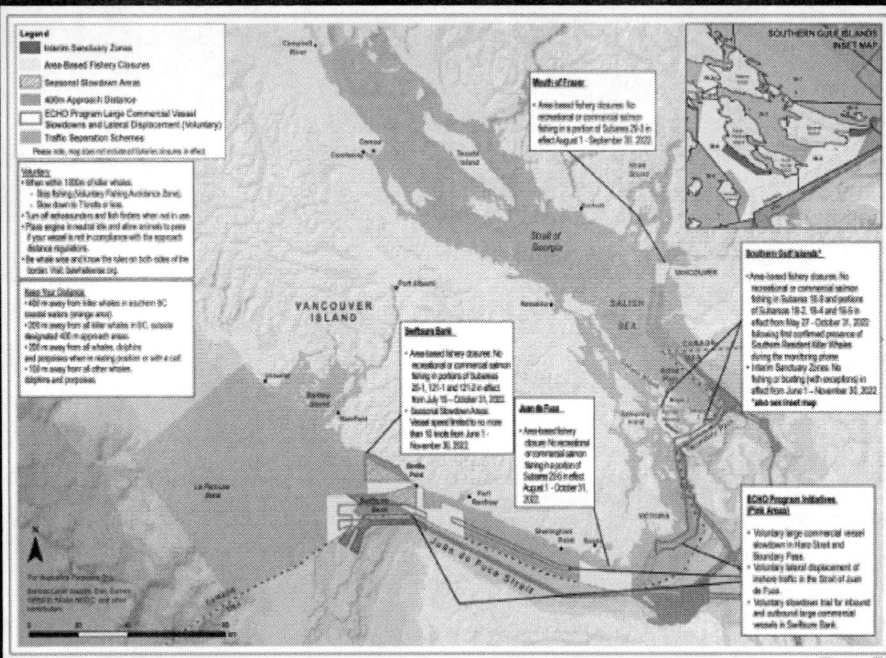

Overview of 2022 management measures to protect Southern Resident Killer Whales

THREATS TO SRKW:

- Increased noise exposure impacts hearing, communication, and foraging
- Overfishing in critical habitat

SOLUTION:

- Interim Sanctuary Zones (ISZs) created to reduce impacts
- ISZs are seasonally enforced (Jun-Nov) but hard to monitor

PROJECT AREA - PENDER ISLAND INTERIM SANCTUARY ZONE

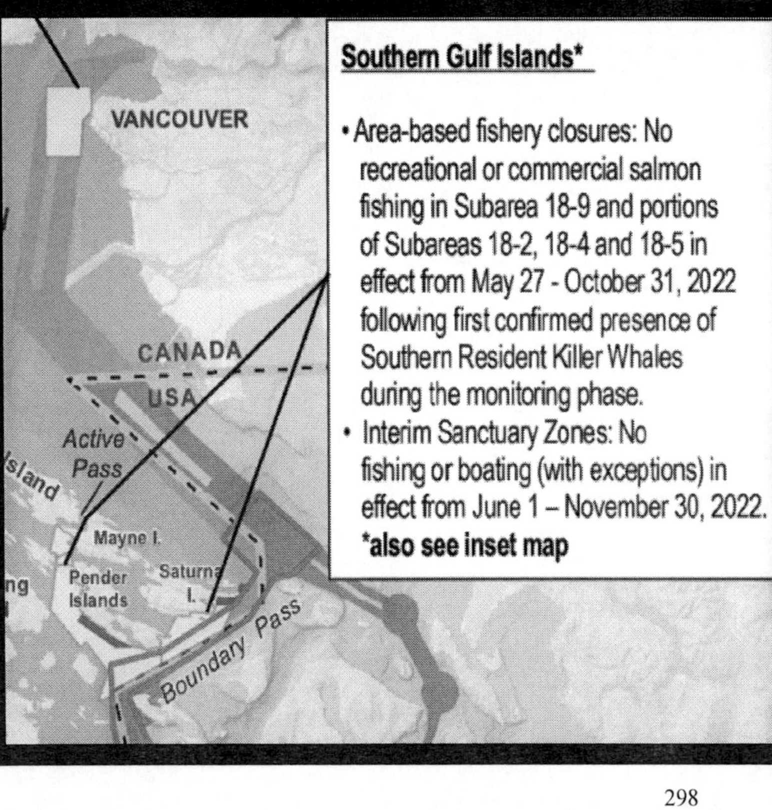

- Boat traffic prohibited in Pender Island ISZ during mission - October 2022
- USV operated outside ISZ with cameras, towed acoustic array, ocean and atmospheric sensors
- Detections included Grey whales and various vessels, including non-AIS.

USV MONITORING OF PENDER ISLAND ISZ – OCT. 2022

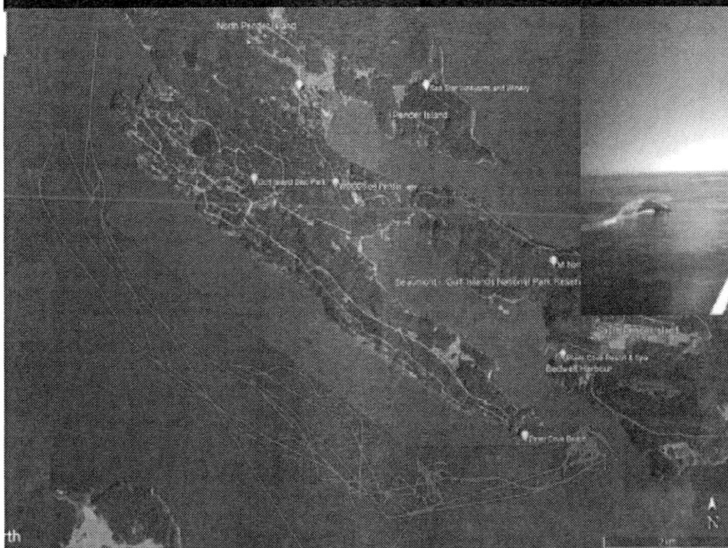

MISSION 4 STATS

DATES:	Oct. 3-7, 2022
MISSION LENGTH:	75 hours
DISTANCE TRAVELLED:	174 km (94 nm)
AVG. USV SPEED:	5.7 kph (3.1 kts)
AVG. WIND SPEED:	9.8 kph (5.3 kts)
MAX. WIND SPEED:	39 kph (21 kts)

SENSORS USED

Lucid Vision Labs,
Triton TRI054S

- Visible
- 4.7x angular resolution
- 4.7x detection radius
- 120dB HDR

Ocean Sonics,
icListen RB9

- Digital hydrophone
- 10 Hz-200 kHz bandwidth range
- 1 hydrophone element
- 50 m tow cable
- Power sources: USV, back-up battery

Teledyne FLIR,
Boson 640 LWIR

- Thermal
- 3.1x angular resolution
- 3.1x detection radius
- ITAR exempt

GeoSpectrum,
M512

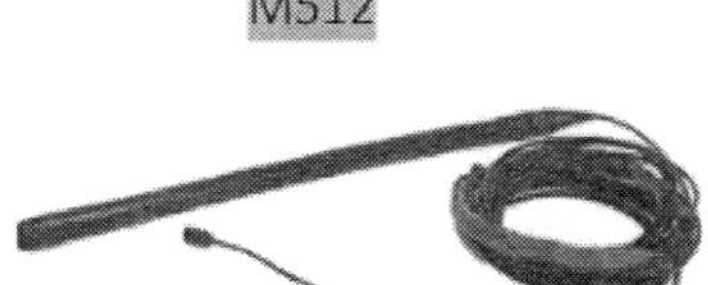

- Analog hydrophone
- Post-processing with DAQ
- 10 Hz-100 kHz bandwidth range
- 2 hydrophone elements
- 25m tow cable
- Power sources: USV

DETECTION OF MARINE MAMMALS WITH PASSIVE ACOUSTICS

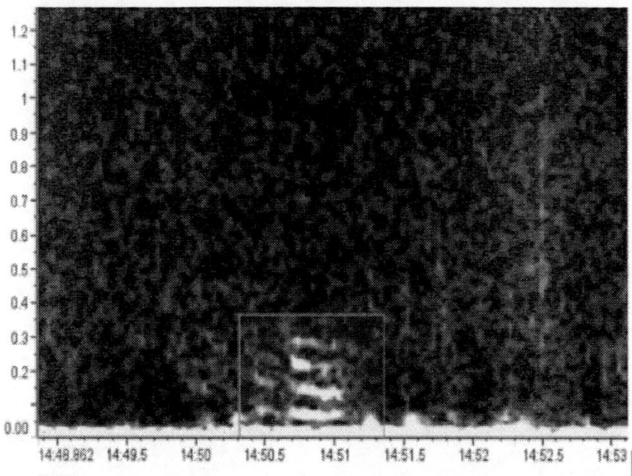

Spectrogram depicting possible Grey whale S1 calls (outlined by light blue box)

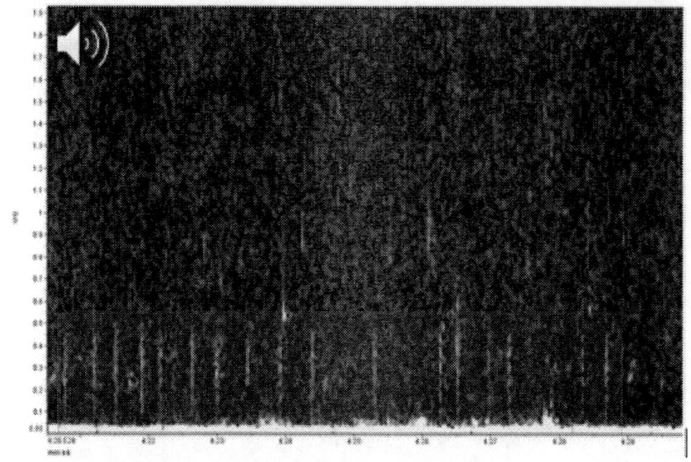

Spectrogram depicting a possible, unspecified Grey whale call (outlined by light blue box)

OPENOCEAN

MARINE MAMMAL RECORDINGS

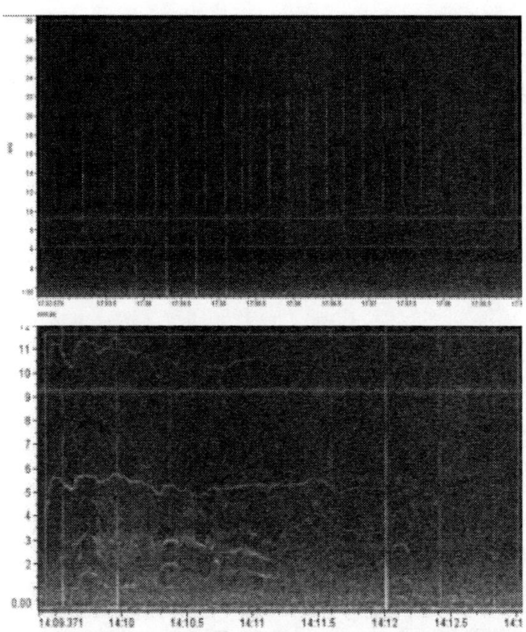

Spectrograms depicting killer whale echolocating clicks (ABOVE) and whistles (BELOW)

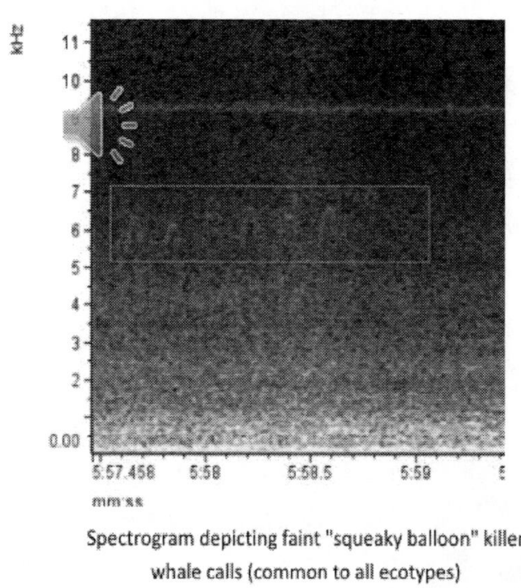

Spectrogram depicting faint "squeaky balloon" killer whale calls (common to all ecotypes)

OPENOCEAN

DETECTION OF NON-AIS VESSELS WITH CAMERAS AND PASSIVE ACOUSTICS

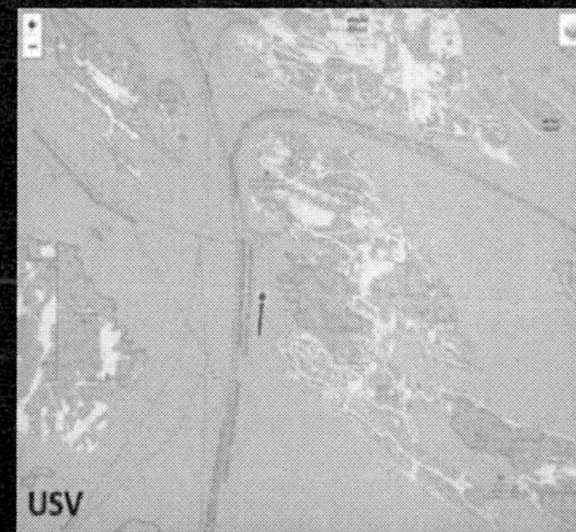

OOR watchkeepers identified 13 vessels not transmitting AIS signals during the visual data collection period.

Images are time-stamped and GPS tagged.

Acoustic data was analyzed 15 minutes before and after the estimated closest point of approach screenshot was taken.

MARINE MAMMAL SIGHTINGS

Sea lions sighted onshore at Race Rocks Ecological Reserve, February 2023. Red outline on the front wide angle camera (top) indicates the area seen in the narrow field visible (bottom left) and thermal (bottom right) cameras.

Title: "Navigating a Sustainable Future: Empowering Offshore Wind Development with Autonomous USVs

- **Abstract:** Embark on a voyage into the future of offshore wind development at the Workboat Show as we explore the game-changing potential of autonomous, solar-powered USVs. Unveil how these cutting-edge technologies are not only revolutionizing ocean monitoring and conservation but are also becoming instrumental in safeguarding marine life in synergy with offshore wind projects. Amid growing concerns over vessel speed limit regulations to protect North Atlantic right whales, discover how the vigilant eyes and ears of DataXplorer™ and other autonomous USVs are set to transform safe navigation, object detection, and avoidance. Gain insights into their pivotal role in enhancing the safety of crew transfer vessels (CTVs) and service operations vessels (SOVs), ensuring the successful construction, operation, and maintenance of offshore wind farms. Join us in shaping a sustainable and harmonious coexistence of renewable energy and marine ecosystems.

- **Learning Objectives:**

- Explore how solar-powered, autonomous USVs are revolutionizing offshore wind development by offering advanced capabilities in ocean monitoring, marine life protection, and vessel navigation.
- Understand the critical importance of object detection and avoidance in ensuring the safe navigation of uncrewed maritime systems and crew transfer vessels (CTVs) within the context of offshore wind projects.
- Examine the innovative use of automated detection, tracking, and counting of marine mammals, particularly whales, to enhance the safety and environmental responsibility of offshore wind operations.
- Engage with industry experts to learn about successful implementations and case studies showcasing the integration of autonomous USVs in offshore wind projects, emphasizing their role in achieving operational efficiency, safety, and compliance with marine conservation regulations.
- Gain insights into the collaborative efforts of Open Ocean Robotics and its partners in advancing maritime technology, contributing to sustainable offshore wind development while preserving the delicate balance of marine ecosystems.
-

The Acoustic Challenge or Distributed Signal Processing: find a reliable and persistent way to monitor whales.

- Exploring the art of possible from a distributed array..The primary driver is for Protected Species Observations for offshore wind and our Impact customers. (Secondary is MDA for MPA/IUU fishing).
- imagine a hydrophone on each (and every) node, and we have 3 (5 or 10) node in an wind Energy Area creating an (ad-hoc) array, would we be able to detect and calculate A: range and bearing on marine mammals. B: range and bearing on vessels. (images will be collected as well.)
- Assuming at least 3 nodes, and Iridium. So not line of sight.
- Mesh network for latency, bandwidth - more development and power.
- We will evaluate hydrophones and positions to determine detection range. We will focus on lower frequencies and whale species that are more challenging to detect from the surface.
- Take the Dominion area for test/demo area sizing, 112,800 acres.
- As for range, I understand there is a 1kM 3kM 5kM standard for the construction phase for the different mammals, 5kM for Right whales in particular I believe.
- Overall objective and frequencies detection should also include small fast boats (separate mission objective).

offshore wind, oceanography, and right whales

Table 4 – Vessel sighting summary from Data Xplorer™'s February 2023 mission.

Size	Number of Observations	Visual Range (NM)
Small (0-9 m, all non-AIS)	4	0.05-0.5 (estimate)
Medium (9-24 m)	3	0.14-1.46
Large (24+ m)	13	1.73-5.93

- offshore wind,
- oceanography, and
- right whales
- We are in the midst of planning research associated with collecting baseline data prior to the installation of wind farms in the MA/RI wind energy area, and at the same time, investigating oceanography associated with habitat use in that area by the highly endangered North Atlantic right whale.
- We are exploring options for sampling oceanography and potential right whale prey with echosounding, and I would be interested in speaking with someone about possibly employing a USV for these purposes.
- Automated detection, tracking, and counting of whales
- infrared cameras with image recognition software to automatically detect and count migrating gray whales

- While significant progress has been made in using autonomous vehicles to monitor the physical environment, developing these tools for the quantitative assessment of zooplankton, fish, and marine mammal populations has lagged. However, newly developed active and passive acoustic sensors for autonomous platforms make addressing these technological gaps now possible. Recent successes have demonstrated the ability for these systems to augment and completely replace vessel-based surveys. Through this proposed track we will assemble a variety of perspectives to discuss:

KONGSBERG

Addressing the Need to Grow the AUV Operator Workforce: Goal Based Mission Planning

Jacob Sobin
Business Development Manager
Kongsberg Discovery

The HUGIN Story

30 Years of collaboration, development and success

DEVELOPMENT

HUGIN was developed in partnership with FFI, Statoil and the Royal Norwegian Navy. Development began in the late 1980's.

DEPLOYMENT

HUGINs have been deployed all around the globe, covering more than 1 million line kilometres of survey.

COLLABORATION

We work with our customers to increase productivity, including sending personnel to assist with operations.

EVOLUTION

Our AUVs are continually changing with new payload sensors, data processing and in-mission autonomy.

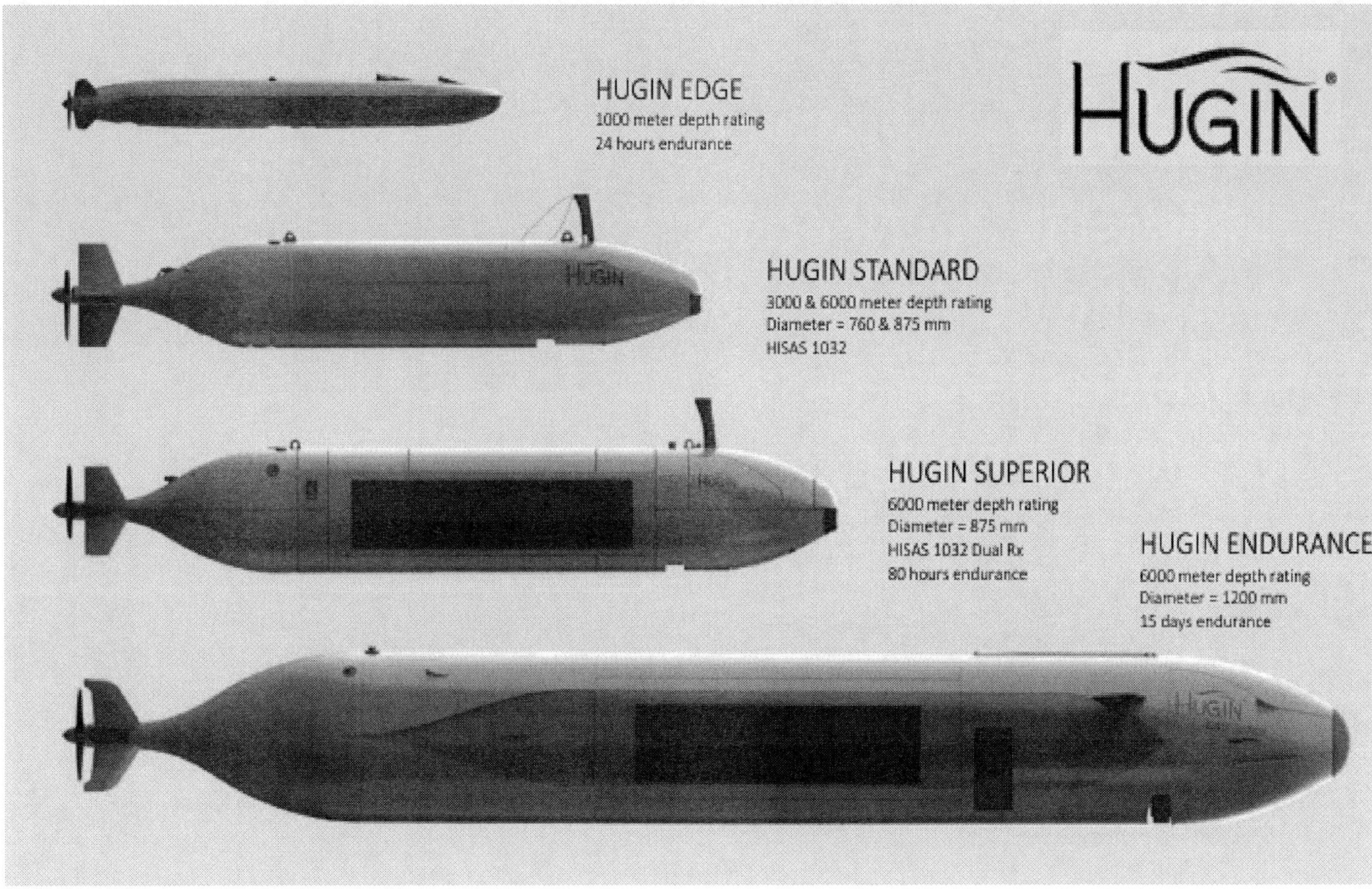

Addressing the Need to Grow AUV Operator Workforce

Paper made for OCEANS 23: Addressing the need to Grow AUV Operator Workforce.

Addressing the Need to Grow AUV Operator Workforce: Advanced Autonomy and Goal Based Mission Planning

KONGSBERG

4

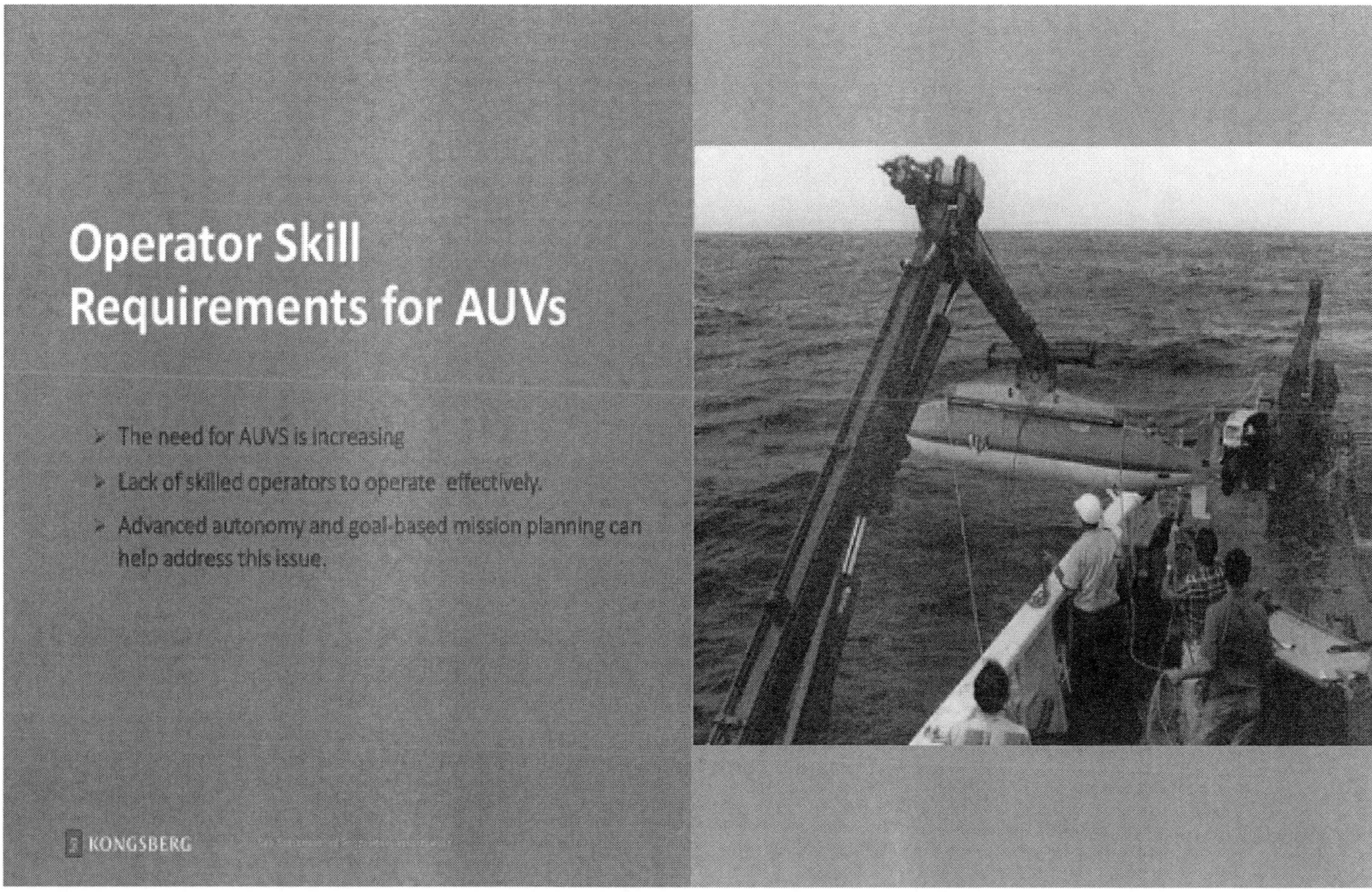
Operator Skill Requirements for AUVs
➢ The need for AUVS is increasing
➢ Lack of skilled operators to operate effectively.
➢ Advanced autonomy and goal-based mission planning can help address this issue.
KONGSBERG

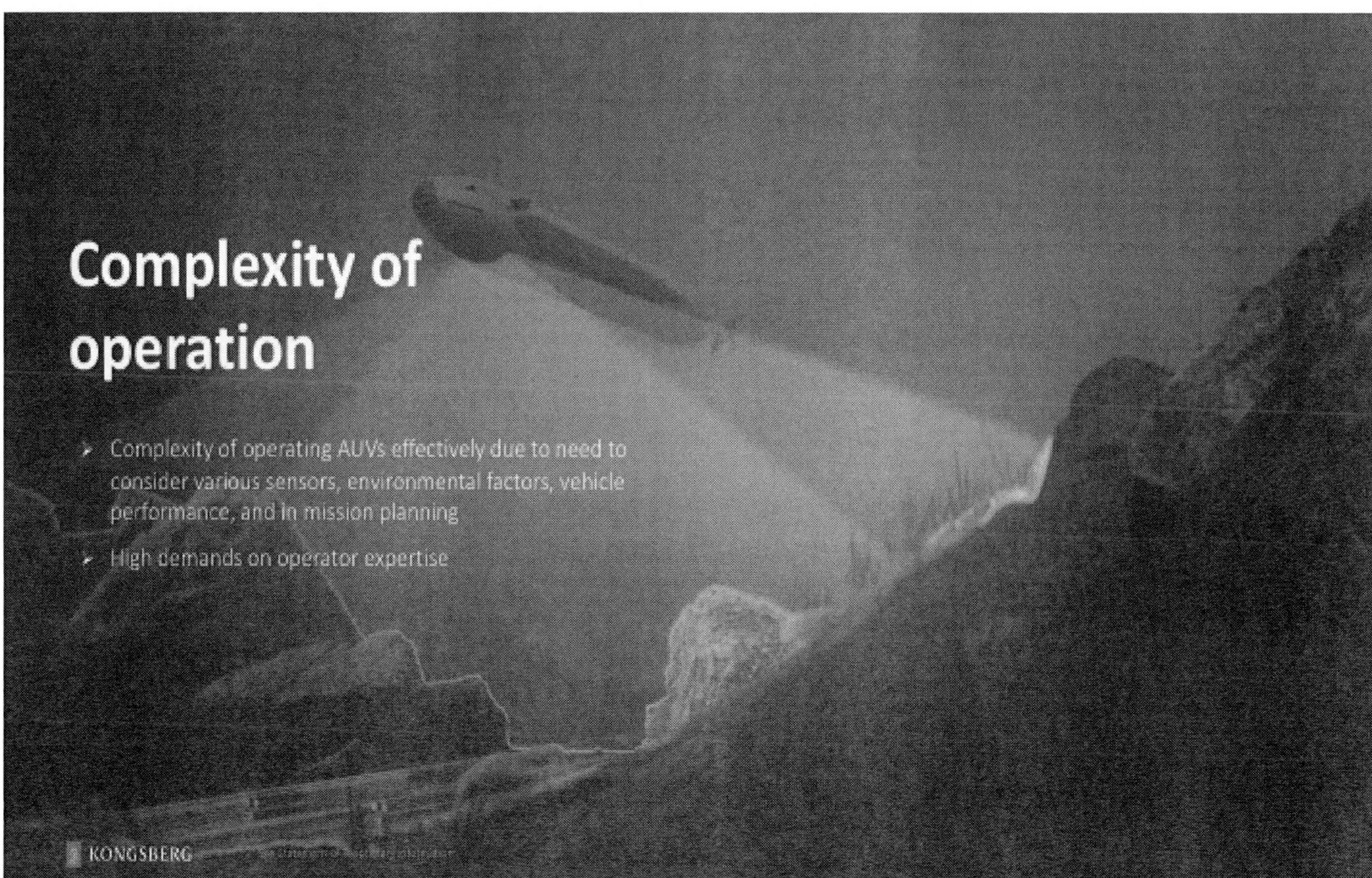
Complexity of operation
➢ Complexity of operating AUVs effectively due to need to consider various sensors, environmental factors, vehicle performance, and in mission planning
➢ High demands on operator expertise
KONGSBERG

Goal Based Autonomy

- Efficient and effective mission planning
- Enables AUVs to adapt
  - changing conditions
  - obstacles

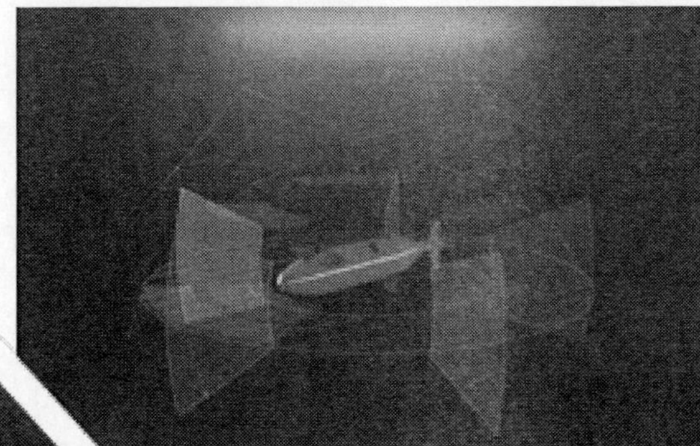

- Improves mission success rates and safety

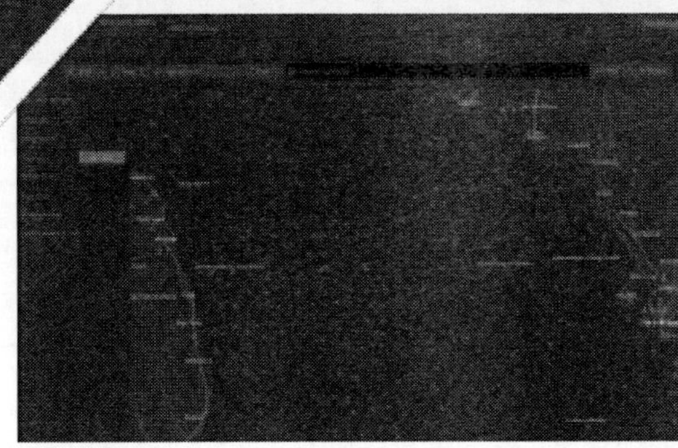

KONGSBERG

7

KONGSBERG

Importance of Advanced Autonomy

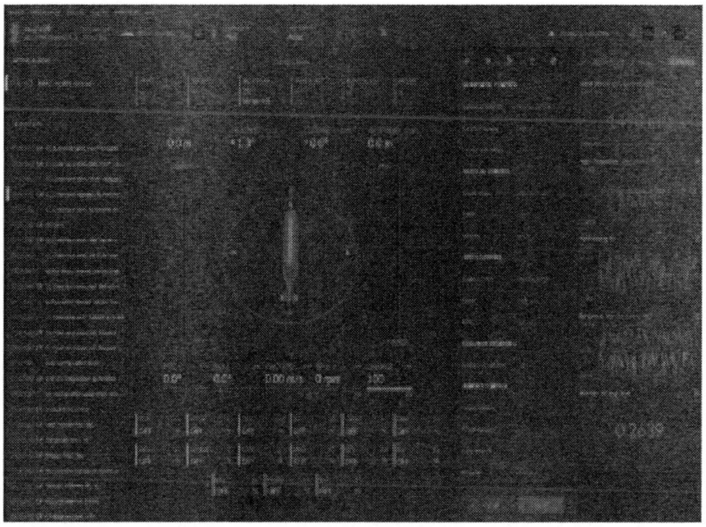

- Enables AUVs to operate more efficiently and effectively
- Reduces the need for specialized knowledge and skills to operate
- Increases mission success rates and safety

KONGSBERG 9

Mission Commander

This advanced autonomy aims to reduce operator expertise requirements and expand the potential workforce.

KONGSBERG See Statement of Proprietary Information

Interoperability

Interoperability between surface and underwater vehicles,

Automated launch/recovery systems.

KONGSBERG

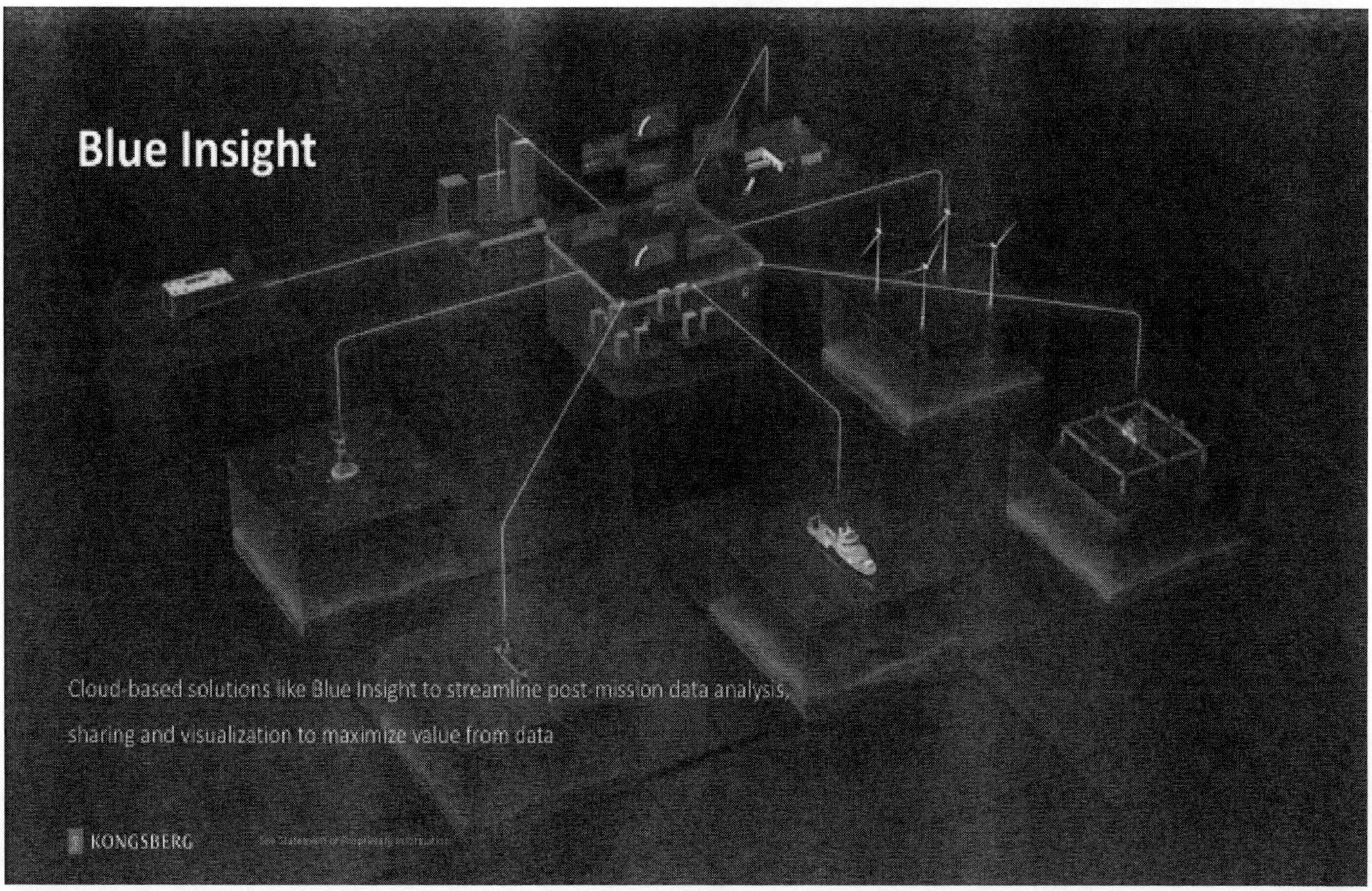

Role of Marine Robotics – Blue Economy

The past two decades has witnessed major developments from industry in the field of marine robotics.

Source: https://escaribbean.com/blue-economy-harnessing-blue-benefits/

KONGSBERG

Conclusion

- Advanced autonomy and goal-based mission planning are key to addressing the need for a skilled AUV operator workforce
- HUGIN AUV platform is an example of a system that utilizes these technologies
- These technologies have the potential to revolutionize the marine industry and enable new applications and business models.

KONGSBERG

Thank you!
Questions?

Jacob.sobin@kd.kongsberg.com

Supporting the New Blue Economy

Open Ocean Exploration and Science

By Philip L. Hoffman

December, 2023

What is The New Blue Economy

The New Blue Economy is a knowledge-based enterprise that builds on traditional ocean and coastal uses while promoting the gathering and use of new and enhanced ocean and coastal information to address societal challenges

NOAA Strategic Plan: 2022-2026

Goal #3: Accelerate Growth in an Information-Based Blue Economy

How is NOAA Investing in the New Blue Economy?

- Improving ocean-related data, predictions, and data access to drive sustainable, data-driven economic growth.
- Promoting the development of the Ocean Enterprise to develop innovative solutions for marine issues.
- Enhancing service delivery and decision support services through increased stakeholder engagement to co-develop ocean data products to meet regional and local needs.
- Building an expert and diverse workforce that is needed to advance the Blue Economy in a competitive and changing labor market.

OCEAN EXPLORATION

How is NOAA Set-up to meet this challenge?

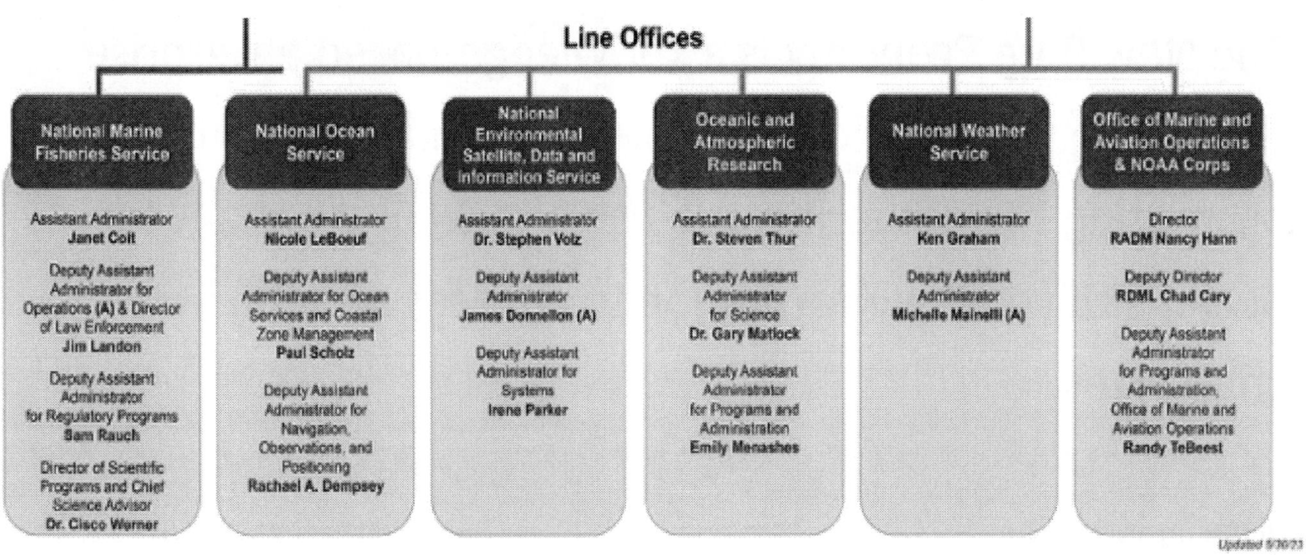

OCEAN EXPLORATION

How does NOAA Research contribute to the New Blue Economy?

Monitoring Modeling

Observations Assessment Forecasts & Products

OCEAN EXPLORATION

How Does Ocean Exploration Fit in the New Blue Economy?

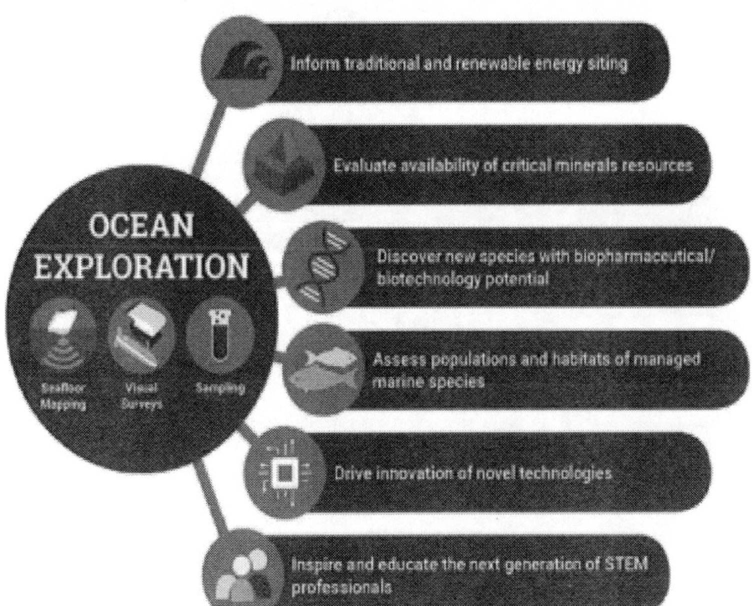

OCEAN EXPLORATION

We have a Lot of Work to Do:

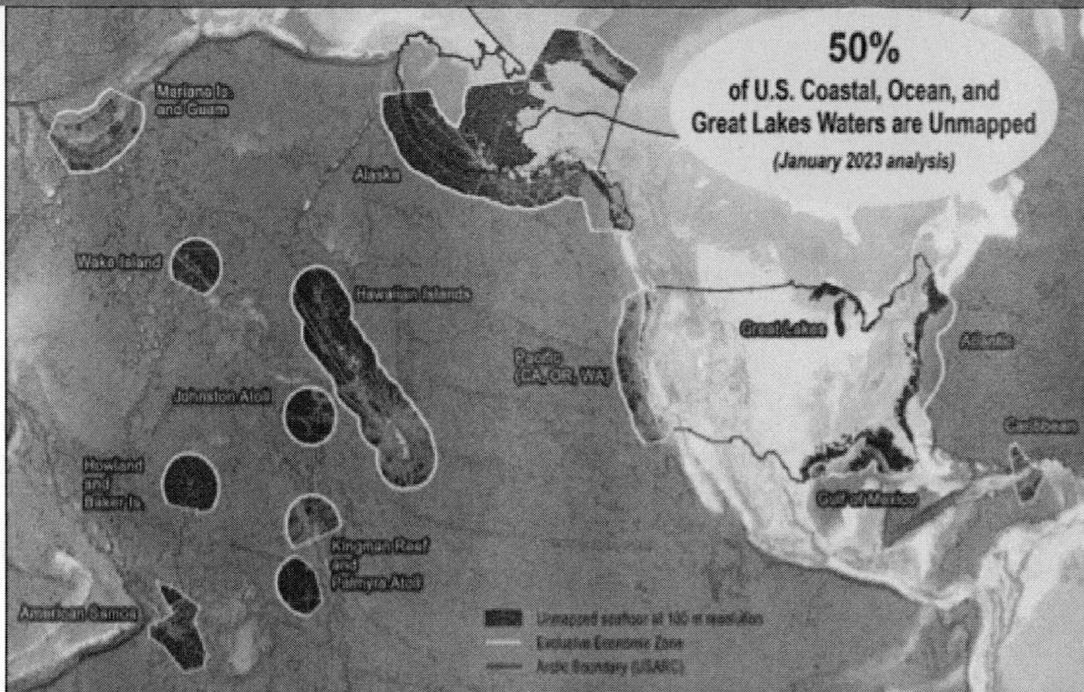

OCEAN EXPLORATION

What's Our Exploration Planning Process?

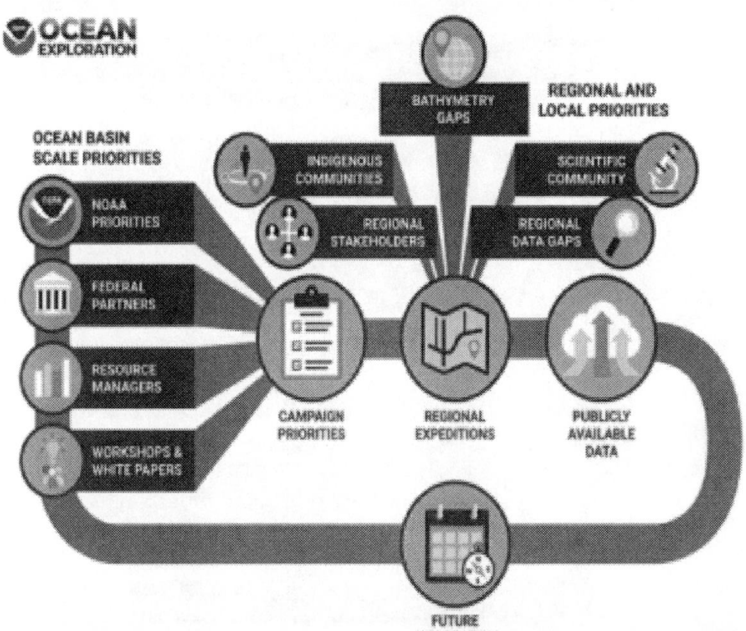

OCEAN EXPLORATION

NOAA Ship *Okeanos Explorer*

Reconditioned U.S. Navy ship Capable a T-AGOS class vessel initially retired in 2004; refitted and commissioned in 2008 for dedicated oceanographic research and exploration.

About NOAA Ship Okeanos Explorer

224 ft in length

18 crew

19 scientist berths

2 ROVs

Remotely Operated Vehicle **Deep Discoverer**: Technology: Submersibles

Remotely Operated Vehicle **Seirios**: Technology: Submersibles

OCEAN EXPLORATION

Our Current Mapping Technology

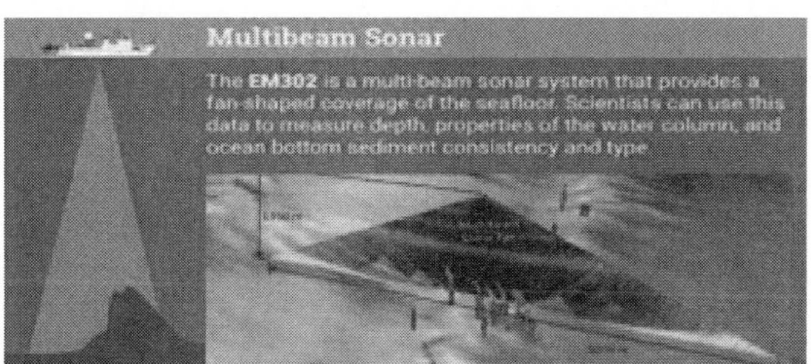

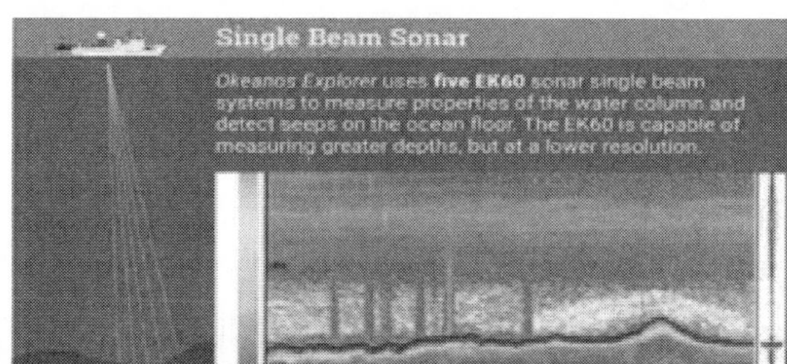

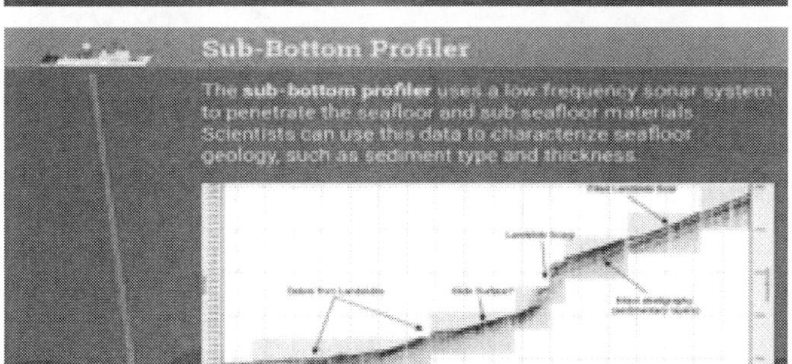

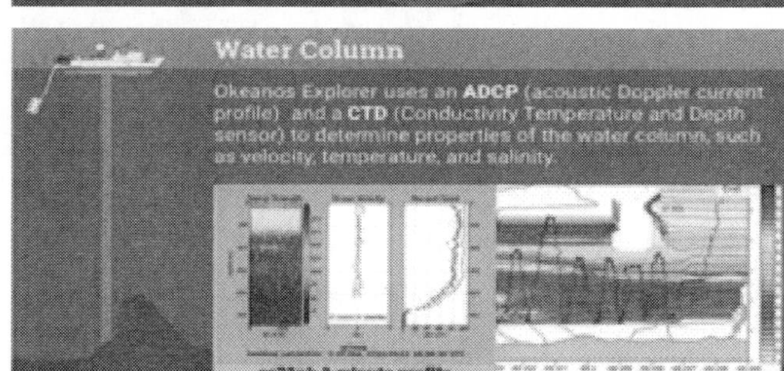

OCEAN EXPLORATION

Telepresence Opens Science to Everyone

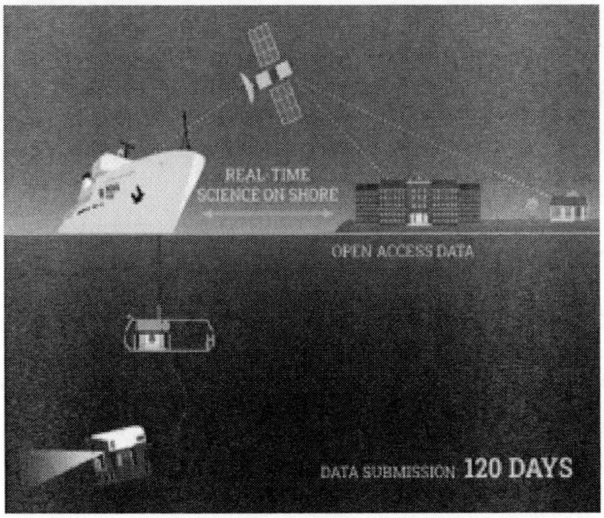

Toda

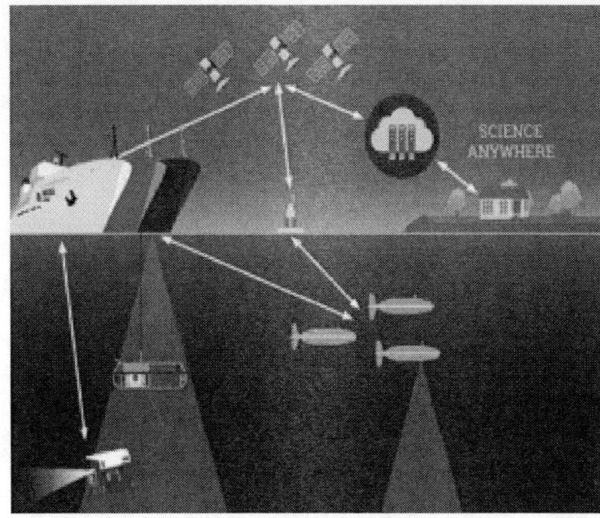

Tomorrow

OCEAN EXPLORATION

Committed to Partnerships

GLOBAL FOUNDATION FOR OCEAN EXPLORATION

CCOM JHC — Center for Coastal & Ocean Mapping Joint Hydrographic Center, University of New Hampshire

SCHMIDT OCEAN INSTITUTE

UCAR

OCEANX

USGS science for a changing world

NASA

ROGER F. WICKER CENTER for OCEAN ENTERPRISE

KRAKEN

BOEM

NAVSEA WARFARE CENTERS Panama City

NAVAL UNDERSEA WARFARE CENTER NUWC DIVISION NEWPORT, RI

DEFENSE POW/MIA ACCOUNTING AGENCY

FUGRO

Ocean Exploration Cooperative Institute 2022
259
DAYS AT SEA
17
EXPEDITIONS
193,405 SQKM
148,233 SQKM
120
DEPLOYMENTS
973
Samples with over 1000 sub samples
64
ROV
249
eDNA samples
56
UxS
Robotic exploration vehicles, from left to right: Nereid Under Ice, Mesobot and Drix. Far right, Molly Curran prepares a vehicle for deployment.
OCEAN EXPLORATION COOPERATIVE INSTITUTE
oeci
EXPLORING THE NATION'S BLUE FRONTIER
NAUTILUS LIVE
OCEAN EXPLORATION TRUST
THE UNIVERSITY OF SOUTHERN MISSISSIPPI
University of New Hampshire
WOODS HOLE OCEANOGRAPHIC INSTITUTION
THE UNIVERSITY OF RHODE ISLAND
OCEAN EXPLORATION
OCEAN EXPLORATION

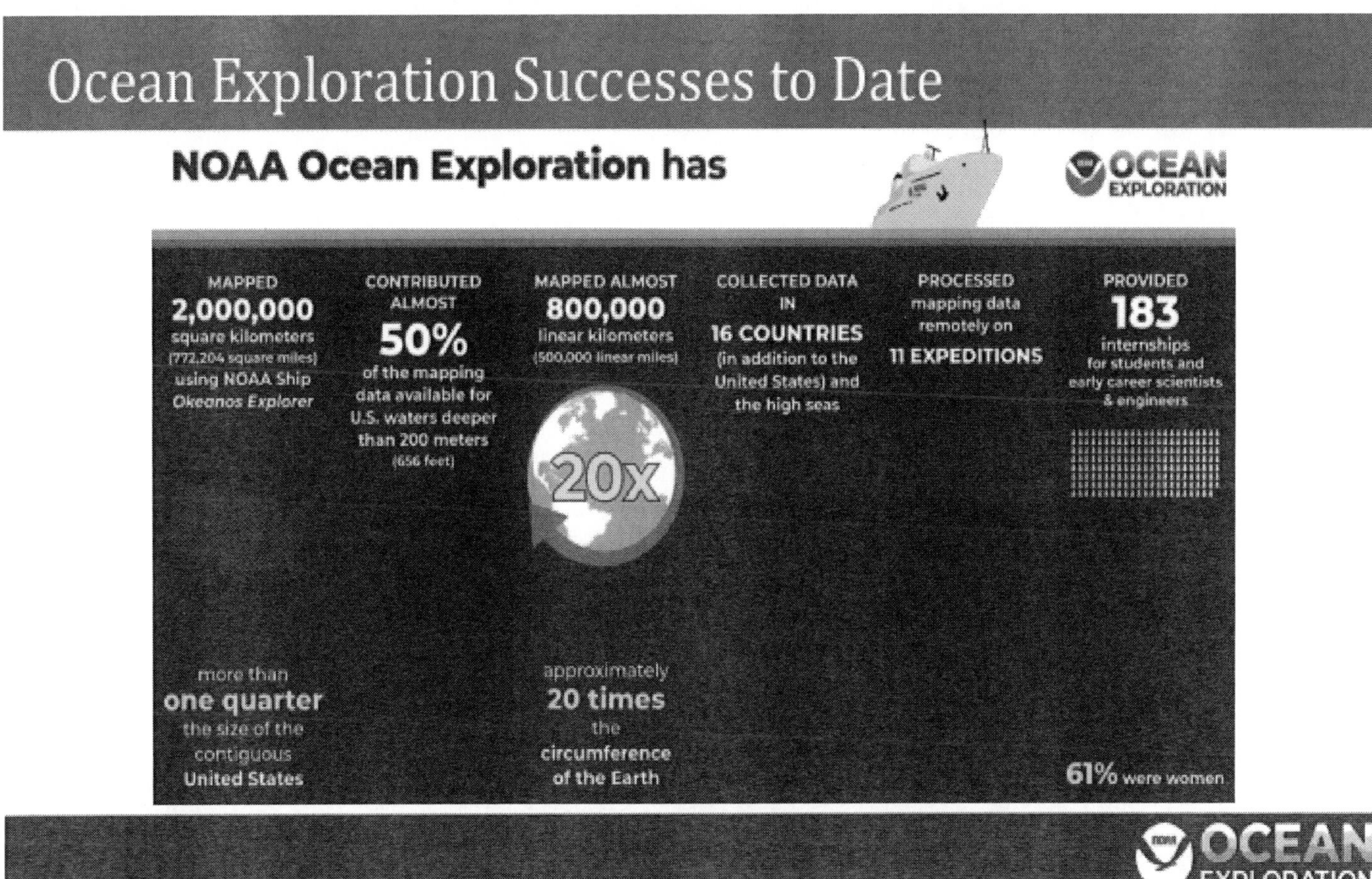

Ocean Exploration Successes to Date
NOAA Ocean Exploration has
OCEAN EXPLORATION
MAPPED
2,000,000
square kilometers
(772,204 square miles)
using NOAA Ship Okeanos Explorer
CONTRIBUTED ALMOST
50%
of the mapping data available for U.S. waters deeper than 200 meters
(656 feet)
MAPPED ALMOST
800,000
linear kilometers
(500,000 linear miles)
20x
COLLECTED DATA IN
16 COUNTRIES
(in addition to the United States) and the high seas
PROCESSED mapping data remotely on
11 EXPEDITIONS
PROVIDED
183
internships for students and early career scientists & engineers
more than
one quarter
the size of the contiguous United States
approximately
20 times
the circumference of the Earth
61% were women
OCEAN EXPLORATION

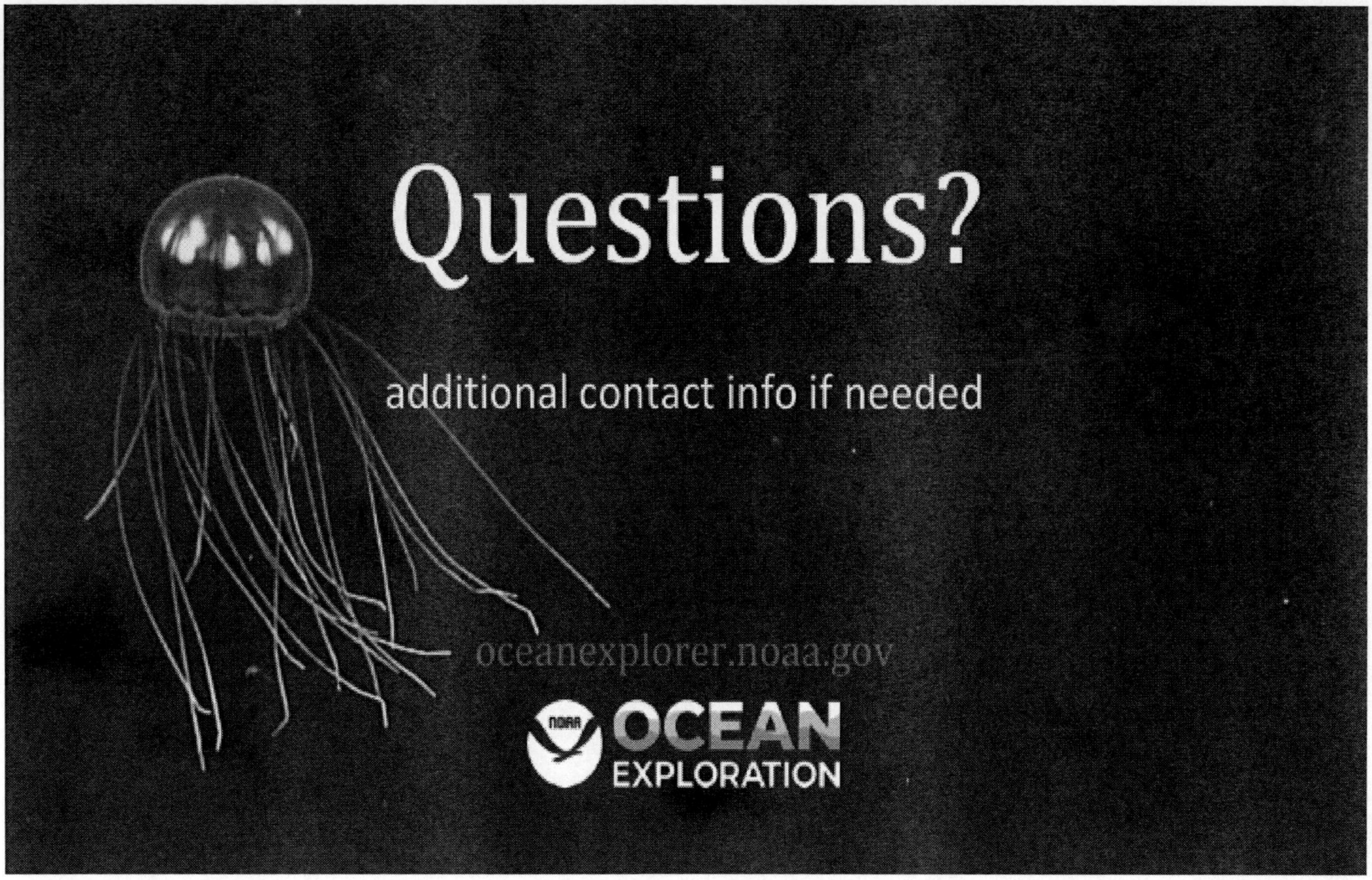
Questions?
additional contact info if needed
oceanexplorer.noaa.gov
NOAA
OCEAN
EXPLORATION

MATE FLOATS!
MARINE TECHNOLOGY CAMP

MATE Floats: Immersive Experience for Underrepresented Students in Marine Science and Technology

PRESENTED BY: Josie Adams, Rick Rupan,
Sasha Seroy, Ethan Campbell and Caleb Flaim

NCAT MTS marine technology society GO-BGC UNIVERSITY of WASHINGTON

Program overview and goals

- Weeklong summer program for community college students interested in marine science and technology
- Provide technical training **free-of-cost** in an oceanographic context using BGC Argo Floats as the primary case study
- Develop skills and understanding in how floats operate, the data they collect, and how they address modern scientific needs
- Provide career readiness preparation and mentorship

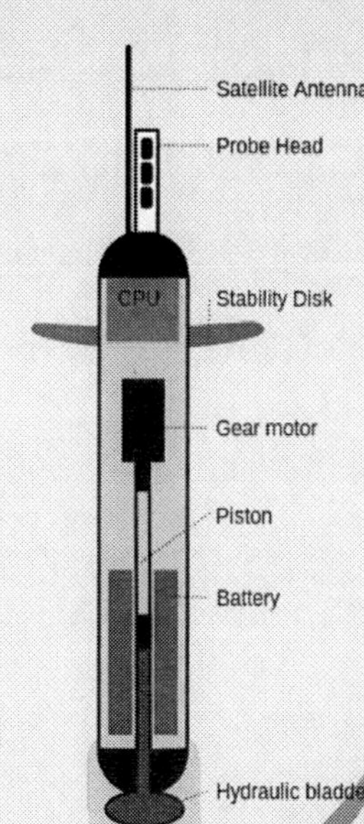

NCAT MTS marine technology society GO-BGC UNIVERSITY of WASHINGTON

Why do we need programs like MATE Floats?

- Geosciences are the least diverse of all STEM fields (Bernard & Cooperdock 2018)
- Estimated that at least 70% of all jobs are not posted publicly and are dependent on networking (Youngquist 2011, NPR)
- Early internships and mentoring are key to career advancement for young adults from underserved backgrounds (Ross et al. 2018)
- Fill needs of the rapidly-growing field of marine science and technology
- Train students with skills they need to enter the workforce

NCAT MTS marine technology society GO-BGC UNIVERSITY of WASHINGTON

How were students recruited?

- MATE ROV listservs
- Community and technical college networks
- GEODUC program for transfer students
- NCAT, MTS, and Go-BGC websites

NCAT MTS marine technology society GO-BGC UNIVERSITY of WASHINGTON

Student Demographics

n = 15

- 30% White
- 30% Hispanic/Latinx
- 30% Asian/Southeast Asian
- 8% Middle Eastern/North African

- 31% Female
- 69% Male
- 15% Genderqueer

- 15% Self-identified disabled

NCAT MTS marine technology society GO-BGC W UNIVERSITY of WASHINGTON

What skills do students need to get jobs in the geosciences?

Shafer et al. 2022, Geosphere

Common entry-level geosciences job requirements:

- **Field Skills (63%)**
- Written Communication (67%)
- Planning (53%)
- **Data Collection (48%)**

NCAT MTS marine technology society GO-BGC W UNIVERSITY of WASHINGTON

What skills do students need to get jobs in ocean technology?

Common entry-level requirements from a recent survey of 20 job postings in ocean technology (e.g. "technician"):

Experience working with instrumentation, sensors, calibration	70%
Computer skills (software)	70%
Troubleshooting	65%
Background in oceanography	65%
Electrical aptitude (basic wiring, multimeters, soldering)	60%
Communication (written and verbal)	45%
Experience at sea, or on small boats	45%

Data analysis, plotting, and statistics	45%
Mechanical aptitude (basic shop skills, power tools)	40%
Collaboration and working with a team	40%
Organizational skills	35%
Coding (Python, R, Matlab)	35%
Independence	25%
Attention to detail	20%

NCAT MTS marine technology society GO-BGC W UNIVERSITY of WASHINGTON

Program structure

Mon
- Basic ocean science
- Argo Float Introduction
- Puget Sound Model

Tues
- Basic Python
- Argo Float lab tour
- Seaglider demo
- Buoyancy lab

Wed
- Sensor building and deployment
- Data analysis in Python (time series)

Thurs
- Lab skills practice, bottle samples
- Data analysis in Python (depth profiles)

Fri
- Half-day cruise (R/V) Rachel Carson
- Career and resume workshop

NCAT MTS marine technology society GO-BGC W UNIVERSITY of WASHINGTON

Program: Teaching Team

- ➢ Mixed-tier teaching team
- ➢ Near-peer mentorship
- ➢ Diverse teaching team provided students with different perspectives and expertise

Rick Rupan
Research Engineer, UW

Sasha Seroy
Faculty, UW

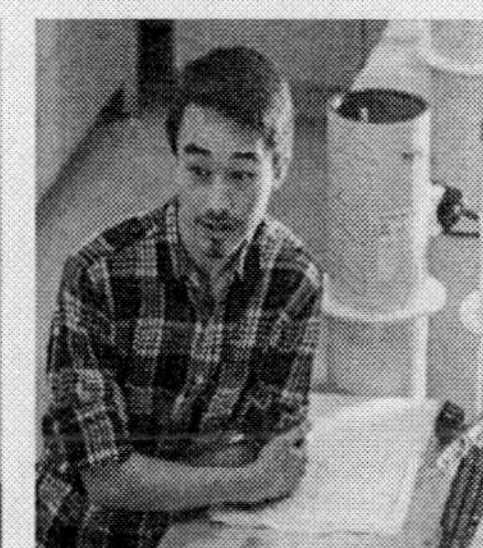

Ethan Campbell
Graduate Student, UW

Josie Adams
Undergraduate, UW

Caleb Flaim
Undergraduate, UW

NCAT MTS marine technology society GO-BGC W UNIVERSITY of WASHINGTON

Program: Day 1

Workforce skills
- ★ Background in oceanography
- ★ Collaboration

- Basic oceanography introduction provides scientific context
- Introduction to Argo Floats
- 3D Puget Sound Model enables students to apply oceanographic knowledge

NCAT MTS marine technology society GO-BGC W UNIVERSITY of WASHINGTON

Program: Day 2

Workforce skills
★ Experience with instrumentation
★ Coding
★ Collaboration

- Argo Float Lab Tour and Seaglider demo
- Basic Python skill building
- Buoyancy lab shows how Argo floats profile the water column

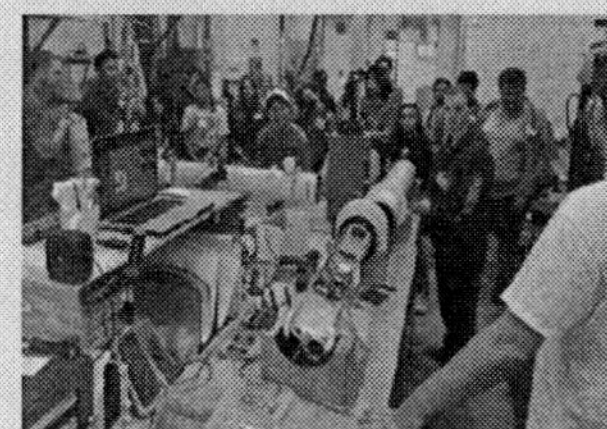

NCAT MTS marine technology society GO-BGC W UNIVERSITY of WASHINGTON

Program: Day 3

Workforce skills
★ Coding, data analysis, computer skills
★ Electrical aptitude
★ Experience with instrumentation, troubleshooting

- Hands-on sensor building, calibration and analysis of sensor performance
- Short-term deployment off UW dock
- Plotting sensor data in Python to test a hypothesis

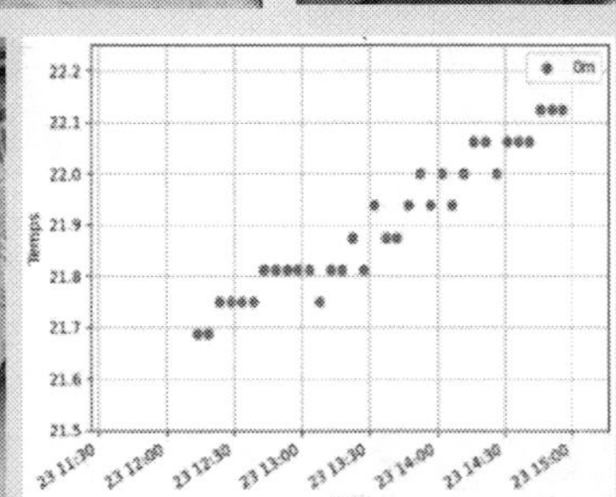

NCAT MTS marine technology society GO-BGC W UNIVERSITY of WASHINGTON

Program: Day 4

Workforce skills
★ Field skills and data collection
★ Data analysis and coding

- Lab techniques: oxygen titrations and chlorophyll analysis for ground-truthing sensors
- Plotting and comparing depth profiles of field site in Python from ships vs. Seagliders

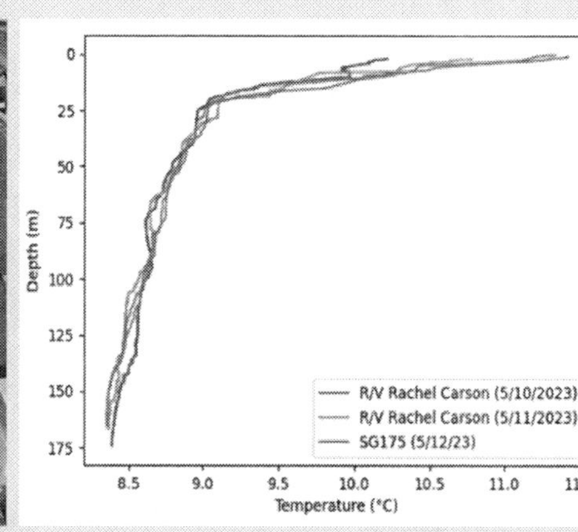

NCAT MTS marine technology society GO-BGC W UNIVERSITY of WASHINGTON

Program: Day 5

Workforce skills
★ Experience at sea, with instrumentation
★ Communication

➢ Research Vessel Rachel Carson research cruise to apply skills from the week

➢ Career workshop: networking with professionals, resume writing and job search preparation

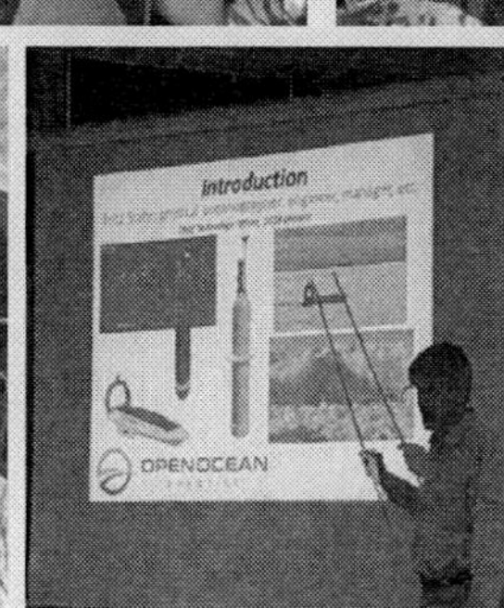

Industry partner Dr. Fritz Stahr, Open Ocean Robotics

NCAT MTS marine technology society GO-BGC W UNIVERSITY of WASHINGTON

Program outcomes

Students said:

- "I couldn't have asked for a better first look into the world of oceanography"
- "I am now sure of my decision to pursue a career in marine technology"
- "Working out in the field and applying the things we have learned really inspired me to become more involved in the marine tech industry"

NCAT MTS marine technology society GO-BGC UNIVERSITY of WASHINGTON

Program outcomes

I increased my knowledge and skills in..	Strongly Agree	Agree
How **Argo floats** monitor changing ocean conditions	14	
Designing technology to collect data to **test a hypothesis**	14	
Using **oceanographic instrumentation** to **collect data**	14	
Analyzing and interpreting **data using Python**	14	
Working effectively and professionally with an **interdisciplinary team**	13	1
Preparing for and **finding a career** in ocean technology	13	1
Basic physics of submerged bodies to govern ocean technology	13	1

n = 14, no responses that were "Neutral", "Disagree" or "Strongly Disagree

NCAT MTS marine technology society GO-BGC UNIVERSITY of WASHINGTON

Workforce skills developed during MATE Floats

Common entry-level requirements from a recent survey of 20 job postings in ocean technology (e.g. "technician"):

Experience working with instrumentation, sensors, calibration	70%
Computer skills (software)	70%
Troubleshooting	65%
Background in oceanography	65%
Electrical aptitude (basic wiring, multimeters, soldering)	60%
Communication (written and verbal)	45%
Experience at sea, or on small boats	45%

Data analysis, plotting, and statistics	45%
Mechanical aptitude (basic shop skills, power tools)	40%
Collaboration and working with a team	40%
Organizational skills	35%
Coding (Python, R, Matlab)	35%
Independence	25%
Attention to detail	20%

NCAT MTS marine technology society GO-BGC UNIVERSITY of WASHINGTON

MATE Floats! Future

- MATE Floats! 2024
- Expanding to different data types and sources, including the Ocean Observatories Initiative and Regional Cabled Array
- Build MATE Floats! Network
  - New students, create alumni listserv, build connections with industry partners
- Send students to conferences like UI

NCAT MTS marine technology society GO-BGC UNIVERSITY of WASHINGTON

Acknowledgements

- Jill Zande, Matt Gardner
- MATE FLOATS! Teaching team (Rick Rupan, Dr. Sasha Seroy, Ethan Campbell, Caleb Flaim)
- The student participants and their mentors
- R/V Carson crew
- Argo Lab members (Greg, Andrew, KyungJae, Aadu)
- Drs. Steve Riser and Alison Gray
- UW Ocean Technology Program for materials and space
- NCAT, MTS, Go-BGC, UW

Citations

- Bernard, R.E., Cooperdock, E.H.G. No progress on diversity in 40 years. Nature Geosci 11, 292–295 (2018). https://doi.org/10.1038/s41561-018-0116-6
- Ross, M., Anderson Moore, Murphy, K., Bateman, N., DeMand, A., and Sacks,V. Pathways to high-quality jobs for young adults. Brookings Metro (2018). https://www.brookings.edu/wp-content/uploads/2018/10/Brookings_Child-Trends_Pathways-for-High-Quality-Jobs-FINAL.pdf
- Shafer, G.W., Viskupic, K., and Egger, A.E. Critical workforce skills for bachelor-level geoscientists: An analysis of geoscience job advertisements: Geosphere, v. 19, no. 2, p. 628–644 (2023) https://doi.org/10.1130/GES02581.1.
- https://www.npr.org/2011/02/08/133474431/a-successful-job-search-its-all-about-networking

MATE Floats! Research Workshop

MATE FLOATS!
MARINE TECHNOLOGY CAMP

PRESENTED BY:

Riley Smith

Oregon State University

Marine Technology Focus

NCAT MTS marine technology society GO-BGC W UNIVERSITY of WASHINGTON

Introduction

Riley Smith

- Student At Linn-Benton Community College
- Undergraduate Student of Oregon State University
  - Bachelor of Science - Oceanography
  - Bachelor of Science - Electrical and Computer Engineering
  - Graduating - Spring 2026
- Prior to MATE Floats!
- Finding the Opportunity

Linn-Benton Community College

Oregon State University

NCAT MTS marine technology society GO-BGC W UNIVERSITY of WASHINGTON

MATE Floats! Workshop

- Lessons
  - Module 1: Intro lab tour / Basic physics and engineering (R. Rupan)
  - Module 2: Intro to shipboard oceanographic data collection tools and resources
  - (J. Adams)
  - Module 3: Intro to Python for data analysis (E. Campbell)
  - Module 4: Underwater vehicle designs and failures (R. Rupan)
  - Module 5: Sensor design and implementation (S. Seroy)
  - Module 6: Sensor design and implementation (S. Seroy)
  - Module 7: Python for data analysis (E. Campbell)
  - Module 8: In situ vs. physical sample data analysis (J. Adams) / Careers (R. Rupan)
  - Module 9: R/V Carson / Data analysis
  - Module 10: R/V Carson / Data analysis
- Research Vessel Experience

NCAT MTS marine technology society GO-BGC UNIVERSITY of WASHINGTON

What Its Like

NCAT MTS marine technology society GO-BGC UNIVERSITY of WASHINGTON

Data Analysis

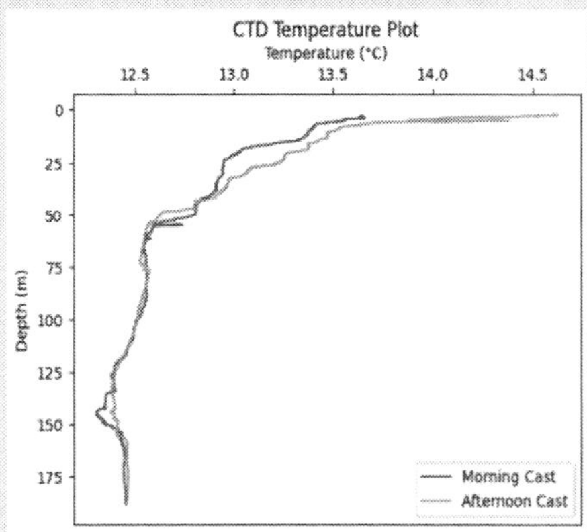

Figure 1

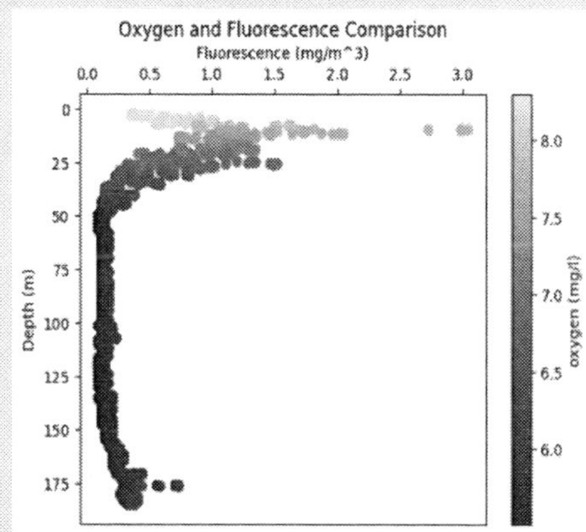

Figure 2

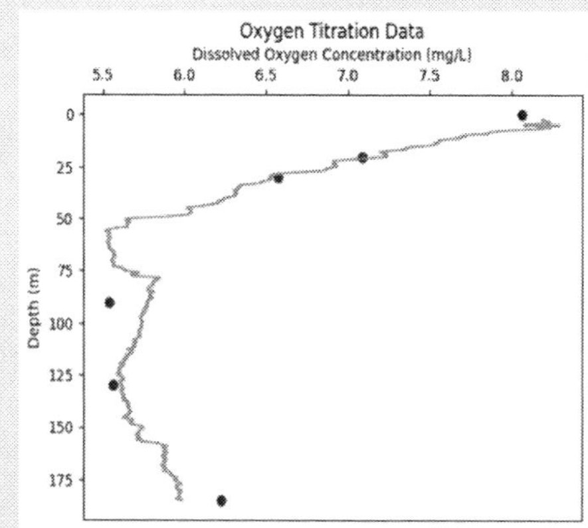

Figure 3

NCAT MTS marine technology society GO-BGC UNIVERSITY of WASHINGTON

Graphs created in Python – skills learned at MATE Floats!

Knowledge & Skills Gained

- Resume Building
- The Scientific Method
- Sensor Design and Implementation
- Software development for research
- Interdisciplinary Communication
- Research Vessel Experience
- Industry Networking and Connections

NCAT MTS marine technology society GO-BGC UNIVERSITY of WASHINGTON

Career Impact

Networking
- Exposure to industry leaders
- Career opportunities

Exposure
- Applicable skills and knowledge gained
- Skills can be applied in a variety of industries

Passion
- Inspirational industry exposure
- Fit my interests well

Education
- Changed my degree focus
- Was considering a minor in Oceanography
- Added B.S. in Oceanography

NCAT MTS marine technology society GO-BGC UNIVERSITY of WASHINGTON

Thank you!

NCAT MTS marine technology society GO-BGC UNIVERSITY of WASHINGTON

My *MATE Floats!* Workshop Experience

PRESENTED BY: M Elwood-Dieu

NCAT MTS marine technology society GO-BGC UNIVERSITY of WASHINGTON Oregon State University Linn-Benton Community College

Camp: Oceanography modules, experiential learning, and resume-building

- Modules on core oceanographic methods (sensor building, system modeling, data collection, and data analysis)
- Career skills: interviews of Rick Rupan and other lab employees/camp organizers

NCAT MTS marine technology society GO-BGC UNIVERSITY of WASHINGTON

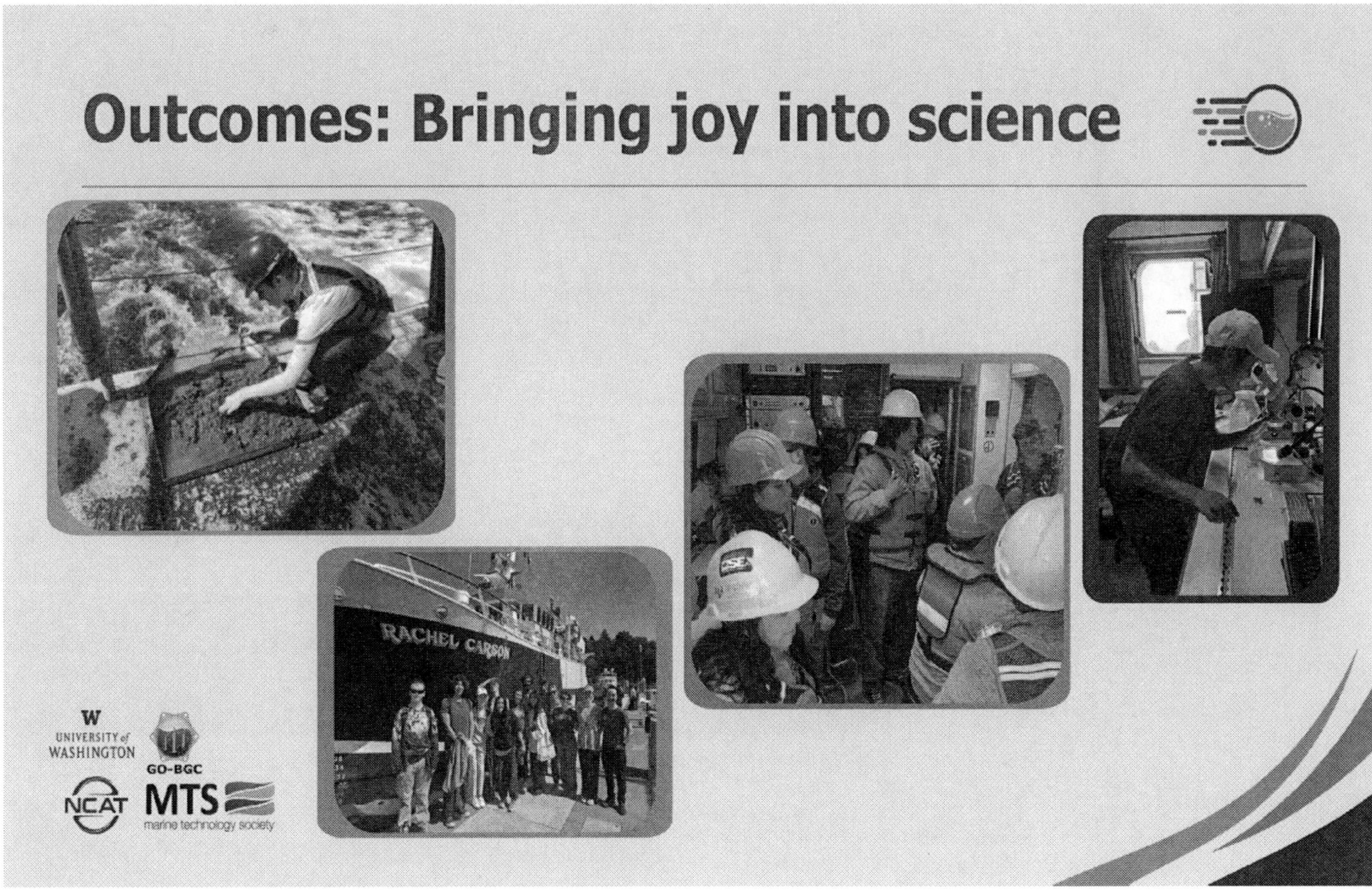

Outcomes: Bringing joy into science
RACHEL CARSON
W
UNIVERSITY of WASHINGTON
GO-BGC
NCAT
MTS
marine technology society

MATE FLOATS!
MARINE TECHNOLOGY CAMP

Cristo Chavira's MATE floats Experience

NCAT MTS marine technology society GO-BGC W UNIVERSITY of WASHINGTON

A little about me...

NCAT MTS marine technology society GO-BGC W UNIVERSITY of WASHINGTON

A little about me...

Electrical Engineering at University of California Santa Cruz.

Electrical lead for Slugbotic's MATE team.

"Exercises in Empathy" Student Participant.

NCAT MTS marine technology society GO-BGC W UNIVERSITY of WASHINGTON

Why MATE floats?

MATE FLOATS! MARINE TECHNOLOGY CAMP

NCAT MTS marine technology society GO-BGC W UNIVERSITY of WASHINGTON

MATE FLOATS! - Marine Technology Camp is taught by UW School of Oceanography Float Engineer Rick Rupan with assistance from UW undergraduate students. During this camp, students will have the opportunity to sail on board the UW R/V Rachel Carson. A detailed camp schedule with activities and learning objectives will be provided later this spring.

There will be minimal costs to the participants. Student travel, housing on the UW campus, and most meals are being covered by NCAT. Ship time on board the R/V Rachel Carson is being covered by the GO-BGC project.

MATE ROV COMPETITION

NCAT MTS marine technology society GO-BGC UNIVERSITY of WASHINGTON

What I experienced...

MATE FLOATS! MARINE TECHNOLOGY CAMP

NCAT MTS marine technology society GO-BGC UNIVERSITY of WASHINGTON

Hands-on Learning

- Oceanography(Science).
- Collecting data from ocean-water samples (Chemistry).
- Circuit Building (Engineering).
- Data interpretation (Programming)
- Creating simple buoyancy machines(Physics).
- Interacted with FLOAT Engineers

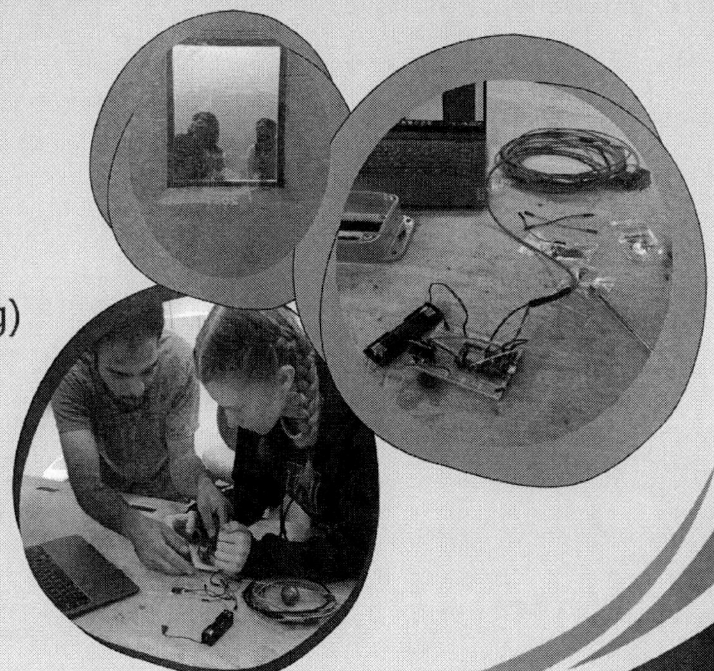

NCAT MTS marine technology society GO-BGC UNIVERSITY of WASHINGTON

Data Interpretation

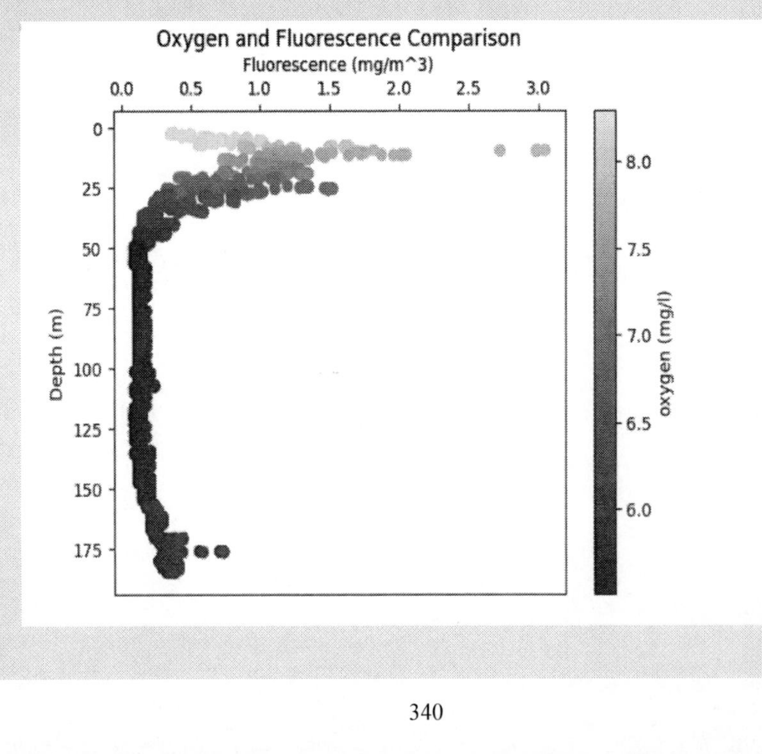

–Riley Smith

RV Rachel Carson

- **Field Work!**
  - CTD Sensor/Water sample collector.
  - Lab Work.
  - Shiptime.

NCAT MTS marine technology society GO-BGC W UNIVERSITY of WASHINGTON

What I learned…

MATE FLOATS!
MARINE TECHNOLOGY CAMP

NCAT MTS marine technology society GO-BGC W UNIVERSITY of WASHINGTON

What I learned...

- There are options.
- How to work in multidisciplinary environments.
- On-Site Skills.
- Community

NCAT MTS marine technology society GO-BGC W UNIVERSITY of WASHINGTON

MATE FLOATS! MARINE TECHNOLOGY CAMP

Thank You!

MTS marine technology society NCAT GO-BGC W UNIVERSITY of WASHINGTON

AUTHOR INDEX

Underwater Intervention
5206 Cypress Creek Parkway
Suite 202
Houston, Texas 77069
USA

ISBN 978-1-7138-8680-8

9 781713 886808

CURRAN ASSOCIATES INC.
proceedings.com

23rd International Conference on Condensed Matter Nuclear Science 2021

Published in the Journal of Condensed Matter Nuclear Science (JCMNS) Volume 36

Online
9 – 11 June 2021